The Spanish Military and Warfare from 1899 to the Civil War

José Vicente Herrero Pérez

The Spanish Military and Warfare from 1899 to the Civil War

The Uncertain Path to Victory

palgrave
macmillan

José Vicente Herrero Pérez
Zaragoza, Spain

ISBN 978-3-319-54746-6 ISBN 978-3-319-54747-3 (eBook)
DOI 10.1007/978-3-319-54747-3

Library of Congress Control Number: 2017944561

Cover illustration: © Hi-Story/Alamy Stock Photo
Cover design: Henry Petrides

Printed on acid-free paper

This Palgrave Macmillan imprint is published by Springer Nature
The registered company is Springer International Publishing AG
The registered company address is: Gewerbestrasse 11, 6330 Cham, Switzerland

To my parents

Preface

The Spanish Civil War (1936–1939) was one of the greatest military conflicts between the two world wars of the twentieth century. The ideological nature and the involvement of foreign powers in the conflict have been attractive subjects for historians. Nevertheless, Spain's civil conflict was essentially fought by Spanish armies (navies and air forces are beyond the scope of this monograph). Surprisingly, this fact seems to have gone unnoticed by most academic historians. This situation stems from the scarce interest in the military history of Spain after the Napoleonic Wars that they displayed until recently. It is not that the Spanish army is missing from their works, but they deal with the military as a political actor, a social group or a bureaucratic institution. Unfortunately, such scholarship has not fostered a similar interest in warfare-related issues.

It is an exaggeration to say that the Spanish armies of the Civil War are unknown, but there is more than a grain of truth in it, especially in relation to military professional matters such as tactics, logistics, training, small unit leadership and so on. On the other hand, the regular officers in both sides were those of the pre-war Spanish army under different caps. Therefore, the performance of the armies on the battlefield depended very much on the way the Spanish military prepared for war before 1936. This last issue has been neglected in academic scholarship and is precisely the main subject which this monograph is to address.

This book intends to explore the attitudes of the Spanish military towards the evolution of warfare during the early decades of the twentieth century and their influence on the armies of the Spanish Civil War. The campaigns fought in Morocco by the Spanish army during this period have received only limited attention, since their influence on Spanish battlefield doctrine was limited. A partial exception has been made with the campaign of Melilla (1909) because it was a testing ground for new infantry tactics and the machine gun. On the other hand, although the emphasis is on professional topics, the present author thought necessary to include a survey of the civil–military relations, which influenced—and were influenced by—the internal problems of the officer corps during the period under consideration.

A problem found during the research for this monograph was that the records available in the military archives for the period 1899–1936 were fragmentary and dispersed, whereas the present author lacked enough funding and time to delve into these archives exhaustively. This fact makes it very difficult to reach definitive conclusions about some of the topics discussed in this monograph. By contrast, there are lots of records of the Civil War waiting to be exploited, but the same financial and time constraints only allowed the present author to examine a limited number of files. Nevertheless, it is hoped that, together with the published primary sources, enough evidence has been found and presented as to broaden significantly the scholarship on the Spanish army. So far as the present author is aware, no other academic work dealing with this subject has been written before.

The present author is grateful to the staff of the following institutions, whose collections were the foundation of this book: Archivo General Militar de Ávila; Archivo General Militar de Madrid; Archivo General Militar de Segovia; Biblioteca de Aragón, Saragossa; Biblioteca Central Militar, Madrid; Biblioteca Nacional, Madrid; Centro de Documentación, Ministerio de Defensa, Madrid; Hemeroteca Municipal, Saragossa; and King's College London Library. Special mention must be made of the staff of the library of the Academia General Militar, Saragossa, whose collections were extremely helpful in the final phase of research for this book.

The present author is also grateful to the publisher and journals for allowing him to use contents of the works mentioned below. Chapters 2 and 6 include contents which were first published in 'Enseñanza superior militar y corporativismo de Estado Mayor en el Ejército español, 1893–1932', in *La enseñanza militar en España: 75 años de la Academia*

General Militar en Zaragoza (Madrid: Ministerio de Defensa, Secretaría General Técnica, 2003), pp. 365–371. Most of the content of Chap. 8 was first published in 'De la División Orgánica a la Brigada Mixta. La evolución de las Grandes Unidades en España, 1914–1939', *Revista de Historia Militar*, 93 (2003), pp. 109–134. Chaps. 9 and 10 include contents which were first published in 'The Spanish Military and the Tank, 1909–1939', *Journal of Military History*, 80 (2016), pp. 757–780.

The present author wishes to express his special gratitude to several persons who played key roles in turning an undeveloped idea into this book. Professor Philip Sabin, of the Department of War Studies at King's College London, accepted to supervise a doctoral dissertation on what was a rather arcane subject at the time and put up with the annoyance of the long-distance supervision of a very unpredictable research student. This book owes much to his advice and insightful criticism. The enthusiasm and encouragement of Professor Geoffrey Jensen, of the Virginia Military Institute, turned an informal exchange of views on common scholarly interests into a decision to write this monograph. His comments on several draft chapters and his assistance in looking for a publisher were invaluable as well. At Palgrave Macmillan, the present author would like to thank Emily Russell, who got the proposal accepted, and Carmel Kennedy, who put up patiently with delays, clarified doubts, and helped to sort out difficulties. Last but not least, the present author's deepest gratitude is to his parents, Vicente and María Pilar. This book would never have been possible without their never-ending understanding and generosity.

Zaragoza, Spain — José Vicente Herrero Pérez

Contents

Abbreviations

AGMA-CGG	Archivo General Militar de Ávila, Fondos del Cuartel General del Generalísimo (N)
AGMA-EN	Archivo General Militar de Ávila, Fondos del Ejército Nacional
AGMA-EP	Archivo General Militar de Ávila, Fondos del Ejército Popular
AGMM-CAD	Archivo General Militar de Madrid, Colección Adicional de Documentos
AGMS	Archivo General Militar de Segovia
AME	*Anuario Militar de España*
CEDA	*Confederación Española de Derechas Autónomas* (Spanish Confederation of Autonomous Rightist Groups)
CGG	*Cuartel General del Generalísimo* (rebel or Nationalist General Headquarters)
CLE	*Colección Legislativa del Ejército*
CTV	*Corpo Truppe Volontarie* (Italian Volunteer Corps)
DOMG	*Diario Oficial del Ministerio de la Guerra*
ECT	*Escuela Central de Tiro* (Central Firing School)
EESM	*Escuela de Estudios Superiores Militares* (Higher Military Studies College)
EM	*Estudios Militares*
EMC	*Estado Mayor Central* (the Spanish army's General Staff)
EPR	*Ejército Popular de la República* (the loyalist or Republican army)
ESG	*Escuela Superior de Guerra* (the Spanish army's staff college)
LCM	*La Correspondencia Militar*
LGP	*La Guerra y su preparación*
MA	*Memorial de Artillería*

MC	*Memorial de Caballería*
MI	*Memorial de Infantería*
NCO	Non-commissioned officer
RCM	*Revista Científico-Militar*
REM	*Revista de Estudios Militares*
RGU 1925	*Reglamento para el empleo táctico de las Grandes Unidades* (Regulations for the Tactical Employment of Major Units) (1925)
RHM	*Revista de Historia Militar*
RTI 1898	*Reglamento para la instrucción táctica de las tropas de infantería* (Infantry Tactical Training Regulations) (1898)
RTI 1908	*Reglamento provisional para la instrucción táctica de las tropas de Infantería* (Provisional Infantry Tactical Training Regulations) (1908)
RTI 1913	*Reglamento Táctico de Infantería* (Infantry Tactical Regulations) (1913)
RTIC	*Revista Técnica de Infantería y Caballería*
UME	*Unión Militar Española* (Spanish Military Union)

List of Tables

Note On Translations, Military Terminology and Sources

The present author translated all the quotations from sources in Spanish. The military terms used in this book do not follow the usage of any specific English-speaking army (for instance, the company-sized cavalry unit is called 'squadron' in the British army and 'troop' in the American army). Nonetheless, it is hoped that their meaning will be readily understood (in case of doubt, see Appendices I and II, and the Glossary). When a Spanish military man is introduced as 'General', this only means that he had general officer rank. The term 'enlisted' is applied to all the military men who are not officers. Capitalization has been used in specific political party or belligerent affiliations in contrast with general ideological orientations (for instance, a Republican is anyone who served in the loyalist side during the Civil War).

Spanish people are formally given both parents' surnames and traditionally put their father's first, but there are no strict rules about the use of the patronymic and the matronymic. However, it is hoped that readers will have no problem to identify people in the text. Arabic terms and names have been transliterated following common usage in Spanish sources. Japanese names are always identified in chapter notes by following the Japanese usage of putting the family name first.

The Spanish military archives were reorganized after the present author finished his research. Therefore, some documents may be currently kept in a different file or even they may have been moved to a different archive.

CHAPTER 1

The Spanish Army Under the Restoration Until 1916

The professional evolution of the Spanish army cannot be studied in isolation from political and social issues affecting Spain's military as a whole.[1] This chapter surveys the background of civil–military relations during the Restoration regime, whose developments led to the military's increasing weight in politics after 1900, and which were influential in the problems surrounding the creation of a general staff. Finally, the influence of career conditions and prospects on the professional core of the army—the officer corps—will be analysed in the last part of the chapter.

Politics, Society, and the Military, 1875–1916

From Political-Military Appeasement to Military Defeat

After the end of the war against Napoleon in 1814, politics in Spain featured the intrusion of army officers. Their usual way to intervene was a sort of (often-bloodless) military revolt called *pronunciamiento* (pronouncement).[2] These interventions did not reflect the political views of the military as a whole. They promoted the interests of political factions and parties, which used sympathizers in the officer corps as spearheads to seize power. These factional interventions ended with the First Republic, when the military endorsed unanimously the *coup d'état* of General Manuel Pavía, who disbanded the parliament on 3 January 1874. Pavía's coup was a highlight in a period of political upheaval after the overthrow of the Spanish Bourbon dynasty, which saw a provisional government

J.V. Herrero Pérez, *The Spanish Military and Warfare from 1899 to the Civil War*, DOI 10.1007/978-3-319-54747-3_1

(1868–1871), a new, fleeting dynasty under Italian-born King Amadeo I (1871–1873), and an equally fleeting republic with four successive presidents and General Francisco Serrano's dictatorship (1873–1874). Moreover, the period was livened up by a separatist insurrection in Cuba (Ten Years' War, 1868–1878), a rebellion to enthrone an anti-liberal Spanish Bourbon line (Third Carlist War, 1872–1876), and a radical federalist revolt with utopian socialist tinges (Cantonalist Rebellion, 1873–1874).[3] The *pronunciamiento* of General Arsenio Martínez Campos on 28 December 1874 restored the monarchy and the son of former Queen Isabel II (ousted in 1868) came to the throne as Alfonso XII.

The first prime minister of the Restoration, Antonio Cánovas del Castillo, set up a parliamentary regime in which two liberal 'dynastic' parties (Conservative and Liberal) would take turns in power after peaceful (and rigged) elections. Cánovas's system was actually oligarchic rule, but it provided stability for the next quarter of a century. The military, after seeing the effects of radical ideologies in the period 1868–1874, also became more apprehensive at anything resembling a threat against the social order and the nation's unity and less willing to intervene in politics. In addition, Alfonso XII's display of both constitutionalist ideas and soldierly demeanour helped to take pressure off civil–military relations. But the concept of the 'soldier-king' also led the army to think that it had an alternative chain of command and thus fostered its sense of institutional autonomy with respect to the civil power.

There were attempts at military reform during the last quarter of the nineteenth century. War Minister General Arsenio Martínez Campos set up the Paid Reserve List in 1883. Its initial purpose was to provide officers for reserve units by paying four-fifths of their current salary to those who retired ahead of schedule and still were young enough for regimental service, but when the military academies could not provide enough subalterns for later colonial conflicts, those non-commissioned officers (NCOs) willing to go overseas were commissioned and allowed to pursue a full career in the reserve list. Thus it eventually became a parallel list for officers promoted from the ranks.[4] This odd arrangement was devised to prevent conflicts with the academy graduates, who went into the Active List. The existence of a separate list for the *chusqueros* (Spanish army vernacular for former rankers) preserved the career prospects of the academy officers.[5] Nonetheless, the latter still displayed mistrust towards the new officer list, while, with poor salaries and professional prospects, the reserve officers in their turn saw themselves as a group discriminated against by social prejudices.[6] For example, the artillery corps prevented the creation of the artillery reserve list until 1908; previously, artillery NCOs had to be commissioned in the infantry or cavalry lists.[7] Moreover, captain was the terminal rank for most *chusqueros*.[8]

The boldest reformist proposals of the period were made by General Manuel Cassola, war minister in 1887–1888. He proposed the abolition of exemption from compulsory military service by paying a fee or purchasing a substitute, the establishment an embryonic general staff, and the disbandment of the staff corps (*Cuerpo de Estado Mayor*), whose duties would be carried out by officers who, after special training, would rotate regimental and staff assignments. Cassola's reforms threatened too many vested interests, whose protection guaranteed the regime's stability, and were rejected as they were. However, their inherent soundness was proved by the fact that most of them were adopted piecemeal over the next fifty years.[9]

The military system of the Restoration was shaken by the *Desastre del 98* (Disaster of 1898): the loss of the last colonial territories in the Caribbean and the Pacific after costly wars.[10] Independence movements had started insurrections in Cuba (1895) and the Philippine Islands (1896). 200,000 troops were sent to Cuba and 25,000 to the Philippines to crush the rebels.[11] By early 1898 the Philippines were virtually pacified, whereas the Cuban rebellion, after harsh counterinsurgency campaigns led by General Valeriano Weyler, had reached a military stalemate.[12] However, the Spanish presence in these territories clashed with the interests of the United States, which declared war on Spain on 25 April 1898. The *casus belli* was the alleged Spanish responsibility for the blowing up of the American battleship *Maine*, when anchored in Havana harbour in previous February (later inquiries pointed to an accidental internal explosion).[13] The war was decided once two Spanish naval squadrons were destroyed in Cavite (Manila Bay) and Santiago de Cuba. American expeditionary forces landed in Puerto Rico, the Philippines and Cuba, but, with the exception of the fight for Santiago de Cuba, there were no major land clashes. An armistice was agreed on 12 August and the peace treaty was signed on 10 December. Spain gave Puerto Rico and the Philippines to the United States and Cuba became an independent state.

The colonial campaigns revealed the flaws in the Spanish military system. The logistical organization was appalling and contributed to the huge losses in Cuba, where 93% of around 45,000 military fatalities died from disease; in the Philippines, the 50,000-strong force (a half of it was natives) suffered 3200 fatalities, a casualty ratio in line with the accepted average for the campaigns of European armies in the tropics.[14] The army maintained its policy of posting overseas only officers who volunteered and since few did it, NCOs and (after shortened training) military academy students had to be commissioned in a hurry. But the high command was not the only culprit for the shortcomings of the army. The dynastic politicians and their supporting social groups were too comfortable with the existing military system to undertake any serious reform, and their criticisms after the defeat were to a great extent a search for military scapegoats to elude their own accountability.

The Repercussion of the Desastre and the Military's Return to Politics

Although the army high command had allowed the neglect of Spain's military preparation during the two previous decades, the inner moral cohesion of the officer corps did not seem to crack after the *Desastre*. Moreover, the cabinet in Madrid was equally accountable for the strategic conduct of the war of 1898.[15] And since the stability of the Restoration regime was dependent to a great extent on not upsetting the corporate interests and status of the military, the army was allowed to carry out an inquiry of its own, which left its top ranks unscathed. With their face saved this way, generals and politicians shelved the subject after 1900.[16] Nonetheless, many officers became resentful of a political system which served the interests of the wealthy oligarchy alone and increasingly used army troops to preserve order against social unrest after 1900.[17] The officer corps only accepted such a task because it saw leftist ideologies and regional nationalisms (such as the Catalonian one) as threats to the social order and the unity of the nation. As a counter, some military essayists advocated the indoctrination of military values in conscripts, who, once released, would spread those values to the rest of the population.[18] Army officers also felt that their service was the victim of the politicians' peacetime budgetary cuts of the period 1900–1910, whereas the politicians believed that poor professionalism turned the military into insatiable wastrels of funds.[19] Thus civil–military relations featured distrust, which had repercussions on the preparation for war. In 1907 a report of the Fourth Military Region (where only 60,000 out of 130,000 reservists reported to their annual review in 1906) complained of the lack of cooperation of the local authorities in issues related to mobilization and reservists, a complaint repeated by the Second and Fifth Military Regions.[20]

The tension between the military and the civil power eventually turned into a tendency of the former to cast into doubt the supremacy of the latter and challenge the government. A major display of this tendency was caused by a cartoon published in the Catalonian satirical weekly *¡Cu-Cut!* in November 1905, which mocked the army's dearth of recent military success.[21] About 200 officers of the Barcelona garrison stormed the weekly's editorial offices and presses and the military bullied the government to pass legislation to put attacks against the army in the press under the jurisdiction of military courts. This pressure bore fruit and a bill (commonly known as the Law of Jurisdictions) which allowed the military to put on trial civilians whose writings allegedly affronted the army or the fatherland was passed on 20 March 1906. The outbreak of the Melilla campaign in the summer of 1909 (see Chap. 3) caused the military's next major involvement in politics. The War Ministry called up reservists in order to bring some expeditionary units to war strength. In Barcelona the order affected 500 reservists who were mostly family men from the city's working-class quarters. This fact and the memories of the suffering caused by the colonial wars of 1895–1898 combined to unleash an anti-war popular protest. This turned into a series of violent riots and strikes in the last week of July, later known as the 'Tragic Week'. The final toll was eight dead in the army and police forces, a hundred civilians reported killed in the bloody street clashes, and five more executed after being sentenced to death by military courts. Besides increasing the mutual mistrust of the military and the working classes, this event reinforced the officer corps' belief that, due to the ruling politicians' incompetence, the army was the only safeguard against social disintegration. The politicians could still keep a rein on the officer corps as long as the top ranks preserved their authority over their subordinates. But when the latter no longer trusted the former, as was to happen in 1917, the government became a hostage of the military.

The Beginnings of the General Staff

The State of the Army: The Problems for Military Planning

After being reorganized following the end of the late 1890s wars, the army force structure on the eve of the campaign of Melilla included, as its standing major units, 14 infantry and one cavalry divisions, and three *cazadores* (light infantry) and three cavalry separate brigades, distributed amongst eight peninsular military regions (territorial commands equivalent to an army corps area). The Balearic and Canary Islands were separate territorial commands with division-sized garrisons and the enclaves in Northern Morocco (Ceuta and Melilla) were guarded by one infantry brigade each. Spain's increased involvement in Morocco brought about a gradual build-up of the army in northern Africa after 1909, including the transfer of several brigades and regiments from the peninsula, the dispatch of expeditionary units (whose number varied according to the intensity of military operations), and the recruitment of native troops. By early 1916, the Melilla command had been expanded to divisional size and reinforced by an attached *cazadores* brigade; the Ceuta command consisted of one infantry and one transferred *cazadores* brigade; and a new territorial command had been set up in the Larache region, garrisoned by an *infantería de marina* (marine) expeditionary regiment, two *cazadores* battalions, and a group of cavalry squadrons.[22]

Most of the home army was only a paper force, because many units were understrength, ill-trained, and ill-equipped. In fact, some regiments seemed to have no reason to exist other than providing assignments for the underemployed officer corps. After 1900, the average peninsular infantry regiment had in peacetime two active (four companies each) and one cadre battalions; its total strength was about 500 officers and men (its war strength was 3000-strong).[23] A report stated in 1907 that such reduced strength made impossible any effective training at battalion and regimental level. The report argued that the disbandment of a number of regiments would increase the strength of the remaining units to 500 troops per infantry battalion and 100 troops per cavalry squadron; these levels of strength would make more feasible realistic unit training.[24]

The financial constraints did not affect personnel only. They caused excessive delays in the construction of fortifications: for instance, Fort Alfonso XII and La Sagueta battery (two strongpoints for the defence of the Pyrenees border) were still unfinished after 30 and 20 years of building, respectively.[25] The state of many army barracks was deplorable.[26] The First Military Region pointed out in 1907 that, besides lacking adequate training grounds near the barracks, the cavalry could not train on rainy days because there were no roofed riding schools.[27] And the replacement of material could take a long time: the engineer corps depot in Granada stored in 1907 sapping tools made in 1848.[28]

Besides its deplorable material state, Spain's army lacked clear guidelines to conduct its professional affairs. The Spanish military policy of the first years of the twentieth century neglected planning for the two kinds of conflict in which the army could be engaged in the near future: a colonial campaign in Northern Africa and intervention in a war between European coalitions. The former perhaps was neglected due to the desire to forget the sad experiences of Cuba and the Philippines, whereas the latter went against the isolationist attitudes of the military. Certainly those generals who were war ministers in this period and had fought in the overseas wars of 1895–1898 (Weyler, Linares, Luque) did not want to change this state of affairs. They preferred to erase memories of the past and adapt their course of action to the neutralist policy of the government, even if this attitude harmed the modernization of the army.[29] The ministers' short terms in office did not favour a steady direction of military affairs either. From 1900 to 1917 there were 25 changes in the ministerial chair, which amounts to an eight months' interval between each change on average (the portfolio changed hands six times in 1905–1906).[30] Although some ministers were in office several times, their alignment with cabinets with different priorities in military policy hampered sound planning (Conservative Party politicians were interested in the organization of the higher planning and command bodies; those of the Liberal Party focused instead on issues related to compulsory military service).[31]

The Political Problem of the General Staff

The problem of short-lived ministers was made worse by the lack of an efficient professional body able to make up for the ministerial turmoil. The issue of the supreme command of the army in peacetime was passed over by the Constitution of 1876 and by the 1878 and 1889 army acts. The terms of the 1889 act assigned the everyday administration and conduct of the army to the war minister. So the minister actually became the commander-in-chief (leaving aside the monarch, who was the nominal supreme commander of the armed forces), but this also implied that organizational authority over the military was tied to political-administrative authority and therefore to the repeated swings of Spanish political life. Nonetheless, the 1878 act set up a *Junta Superior Consultiva de Guerra* (Higher Advisory War Board), whose advisory mission partially resembled the role of a general staff. General Cassola tried unsuccessfully to enlarge the board's role by allocating it some responsibilities for war planning, training, and recruitment.[32]

The War Board disappeared in 1904, when War Minister General Arsenio Linares, who realized the need for a permanent body for war planning, set up at last the General Staff (*Estado Mayor Central*) on 17 July 1904 through an act whose content was developed by a decree on 9 December 1904.[33] The *Estado Mayor Central* (EMC) had executive power and its chief could order territorial commanders to carry out the decisions taken by the EMC (the war minister only intervened when the orders were so important as to need the monarch's signature for their approval).[34] Nevertheless, the EMC soon became entangled in a conflict with the War Ministry apparatus over their respective responsibilities. Who would assume the conduct of operations in wartime, the minister or the chief of general staff? Would the War Ministry become reduced to a sort of administrative management agency for the General Staff as in Germany? Politicians certainly desired the war minister's supremacy, which assured them a tighter control of the army.[35] But the EMC could also be their ally against a war minister's policy. José García Benítez, a reformist military essayist writing under the pseudonym of Comandante Beta, disagreed with the General Staff's role as an instrument for military reorganization, separate from the ministry. It was true that the EMC should provide a continuity of approach to military reform, but actually the argument of the need for such continuity was used by the parliament as a way to hold back the war ministers' reformist proposals on military organization issues.[36]

The concept of an independent body in charge of the reorganization of the army, separate from the traditional bureaucracy of the War Ministry, was not accepted unanimously. It was even argued that the General Staff was unconstitutional, because its head could usurp the role of commander-in-chief. The supporters of the General Staff denied any contradiction with the constitution. The decree of December 1904 defined the EMC as a branch of the army's administrative framework without any organizational link with the other major branch—the War Ministry. Its mission was the preparation for war and the supervision of the staff officers' training, and it was subordinate to the war minister. Therefore the minister's final authority was not lessened by that of the chief of general staff. This meant no change in the spirit of the constitution, which designated the monarch and the war minister as the heads of the army. But this subordination was also the key to the problems affecting the General

Staff. The General Staff, in order to fulfil its function properly, had to limit such subordination. This was necessary in a country like Spain, where even successive war ministers of the same political party in power could support very different views on the same military issue.[37]

Early Organization of the General Staff

Standing for the General Staff did not necessarily mean agreement with the shape it had taken in Spain. Captain Nazario Cebreiros, a highly critical essayist writing under the pseudonym of El Capitán Equis, thought that the *Estado Mayor Central* was badly organized for the tasks expected from a general staff. After a reform in 1906, its structure and responsibilities were as follows:

- Secretariat (staffed by 18 officers): administration, staff corps personnel affairs, staff college (*Escuela Superior de Guerra*), military history, technical studies on communications and signals
- First Section (20 officers): organization and operations (strength and territorial deployment of units, training of officers and enlisted men, allotment of conscripts and reservists, mobilization, war planning, military transport)
- Second Section (nine officers): foreign intelligence
- Third Section (nine officers): ordnance and supply
- Fourth Section (eight officers): fortifications and military buildings
- Fifth Section (two officers): mapmaking and publishing[38]

The bulk of duties usually assigned to a general staff were put together in a single section. Thus only 20 out of 66 officers were employed in true staff work, according to Cebreiros. Excepting those in the Second Section, the rest dealt with administrative matters or technical studies more appropriate for inspectorates or technical staffs.[39] Cebreiros also complained that this mixture of military and technical matters led to the EMC officers coming from too many backgrounds:

> Grouping in the same centre so heterogeneous a personnel [without common training and criteria to deal with military problems] – and all this reinforced by the background of prejudices, rivalries, and convergence of ambitions, which divided the officers [of the different corps] – meant... bringing down the best initiatives...[40]

Moreover, according to Cebreiros, the chiefs of general staff were not above the political struggle because 'there were Conservative chiefs and... Liberal ones; not only this, there were [after the name of contemporary generals] *luquistas*, *linaristas*, *polaviejistas*, *primorriveristas...*'.[41] That is, the appointment and the performance of the chiefs of general staff were not unconnected to the play of political and personal sympathies which weighed so much on the top of the military hierarchy.

Disbandment and Rebirth of the General Staff

The General Staff was suppressed by War Minister General Agustín Luque in late 1912. Luque used the need for slimming down the army bureaucracy as a justification to remove this branch of the military structure. His critics, though admitting that the EMC had become an unworkable organization, reckoned that the solution should be a reform instead of disbandment.[42] But the disbandment also was a drastic way of resolving the political conflict about the General Staff's alleged incompatibility with the terms of the 1876 Constitution and the final accountability of the minister to the parliament.[43] At the same time Luque created a new branch within the ministry, the *Sección de Estado Mayor y Campaña* (General Staff and Field Operations Section). It was headed by a brigadier general and divided into three bureaus (staff corps personnel, staff college, and military attachés; manpower management, ordnance stocks, transport, and communications; mobilization, planning, and operations), manned by 18 staff corps officers.[44]

García Benítez reckoned that the trouble with the EMC was the political-military authorities' desire for 'cure-alls' which solved at once all the problems of the army. Thus he criticized the creation at one stroke of a fully-fledged general staff, instead of pursuing a policy of gradual enlargement from a smaller establishment, better suited to the needs of the Spanish army.[45] His views were a result of his experience in General Ramón Echagüe's 'military bureau'. Echagüe, war minister in 1913–1915, announced a project of military reform to the parliament in December 1914. He set up a special bureau in the ministry in late April 1915 to undertake the preparation in detail of his ambitious reform plans, since no existing ministerial branch was able to do it.[46] García Benítez saw in this bureau a sort of general staff in embryo. Nevertheless, it had allegedly aroused ill-will among those who had an interest in the restoration of a large general staff (especially the staff corps officers).[47]

General Echagüe's project, among others reforms, also created the new *Consejo Superior del Ejército* (Army Supreme Council), which would include former war ministers and provide steadiness in military policy and planning. The solution of professional problems raised by the preparation for war would be the task of the re-born General Staff, turned into an advisory branch without any command and administrative responsibilities. The Conservative and Liberal parties broadly agreed on the contents of the reforms. However, at the time of their discussion in late 1915, Prime Minister Eduardo Dato insisted on passing first a much-contested bill on the reduction of the retirement age for officers. Facing strong opposition on this issue, Dato resigned on 6 December 1915. The fall of his cabinet also paralysed any further discussion of Echagüe's reforms. The reasons for Dato's odd conduct remain obscure, but they had little to do with military policy, given the previous consensus on the reforms.[48]

The next cabinet, headed by Álvaro de Figueroa, Count of Romanones, included Luque as war minister again. And Luque, who disbanded the EMC in 1912, brought it back to life on 24 January 1916.[49] Nevertheless, the new General Staff had only a consultative role (a feature which caused criticism from some sections of the military) and its relationship with military units in peacetime was limited to inspection duties.[50] The EMC was to be headed by a captain general or a lieutenant general (who would perform as commander-in-chief of the field army in wartime or as its chief of staff if the king took command). Its organization was as follows:

- Secretariat (three officers)
- First Section (12 officers): organization, mobilization, and training
- Second Section (eight officers): operations and military communications
- Third Section (five officers): ordnance and war industry
- Fourth Section (five officers): statistics and requisition
- Fifth Section (six officers): fortifications and foreign intelligence[51]

The missions assigned to the Fourth Section reveal a greater concern about the problems of economic mobilization (a consequence of the teachings of the First World War), but uniting intelligence matters and technical studies in the same section still seems an oddity in the General Staff's functional division.

Luque also presented another reform bill in the autumn in order to make up for the EMC's lack of executive power. The mission of giving stability to military policy was assigned to the *Junta de Defensa Nacional* (National Defence Board). It had been created in 1907 and was formed by the prime minister, the war and navy ministers, the army and navy chiefs of general staff, and former prime ministers. The board would set the guidelines on which the General Staff would develop the war minister's initiatives on organizational issues. Once it had finished its work, the General Staff would send the project back to the war minister, who would carry out its implementation. Although the bill did not raise any significant political discussion, Luque's reform did not succeed in being passed due to the adjournment of the parliament in February 1917.[52] And after this date military policy in Spain was to be dominated by the crisis of the *juntas de defensa* (see Chap. 5).

The turbulent early life of the General Staff was hardly surprising given the involvement of the Spanish army's top ranks in politics and the uneasy civil–military relations under the Restoration. Though the dynastic politicians realized the need for a professional body in charge of organizational and planning matters, they were suspicious of anything resembling a core of autonomous military power. But such a fear was unfounded to some degree. The refusal of the Restoration cabinets to make any reform seriously affecting military vested interests was successful in integrating the top of the army hierarchy (the so-called 'political generals') within the regime. As a result, senior generals became so bound to the dynastic parties' politics that they did not worry about preserving enough professional autonomy to keep military policy on a steadier basis. War ministers yielded to political interests and a strong General Staff could not help being an embarrassment in this situation. But at the same time, limiting the power of the General Staff deprived this body of its *raison d'être*. Hence the nature of the Restoration political-military establishment was a major factor in preventing the implementation of reforms which increased the Spanish army's professional effectiveness. Due to the excessive interdependence between the ruling dynastic politicians and the military, only a deep change of Spain's political order could open the way to the necessary reform of the Spanish army. An alternative spur for a reform of the army—a major military defeat, such as the French one in 1870—was too remote, due to Spain's keeping out of the arena of international politics.

The Officer Corps: Social and Professional Background

The Transformation of the Spanish Officer Corps After 1808

The Spanish officer corps' background, like in many other European armies, was mainly aristocratic by the beginning of the nineteenth century. However, the war against Napoleon (1808–1814) broke the traditional structure of the army. Many officers of the old army were lost in the early years of the war. This opened the officer corps to candidates whose proven ability on the battlefield or educational qualifications made up for their lack of noble blood. The campaigns against the independence movements in Spanish America until the late 1820s and the First Carlist War (1833–1840) brought into the ranks of the officer corps more candidates of non-aristocratic origins. Despite the attempts of Fernando VII's absolute monarchy to maintain proofs of noble lineage as a requirement for a commission, the rise of liberalism to power led to their suppression in 1836.[53]

Though a number of aristocrats still pursued the military career, about 60% of the officer commissions were given to candidates coming from, or whose upbringing could be on a par with, a middle or even lower class background by the mid-nineteenth century. This was related with the increasing importance of self-recruitment (that is, the increasing numbers of new officers whose fathers were already military men), which was the most remarkable feature of the social evolution of the officer corps during the nineteenth century.[54] Education in the military academies was not free, but civil higher education was even more expensive and hardly accessible for a military salary. On the other hand, the sons of officers were entitled to advantages in admission and financial benefits. Thus many officers were prone to direct their male offspring towards a military career. The 1897 regulations for military academies both reflected and promote such state of affairs by admitting preferably applicants of military background.[55] Due to the rise in self-recruitment, the officer corps became again a sort of closed, self-renewing caste by the end of the nineteenth century and it members ended up thinking of themselves as a separate class, apart from the rest of society.[56]

Professional Concerns of Career Officers

The life of the average Spanish officer during the last quarter of the nineteenth century was characterized by financial difficulties and professional dissatisfaction. The root lay in the surplus of officers caused by previous wars and political upheavals, since the government was unable to find a suitable way to release redundant officers once the military need disappeared. Few officers were ready to retire prematurely as their pay would be low. In addition, the average officer came from a family which, whatever the father's occupation, could be considered in economic terms part of the wage-earning middle classes, whose sources of income were often jobs in the civil or armed services. The relative backwardness of the Spanish economy limited the chances to make a living in industry or business. So, unless they had substantial private means, officers were very reluctant to give up the safety offered by being on the payroll of the state. Moreover, the officers' living standards improved after a rise in pay in the 1880s. The only solution were long-term policies (for instance, by slowing down promotions) which eased natural wastage. Unfortunately, new conflicts broke out before the benefits of such policies could be reaped, as was the case of the colonial wars of 1895–1898.[57]

Besides the economic issues, the other major concern in an officer's career was promotion. The Spanish army had no uniform policy to promote officers until the late nineteenth century. To simplify, suffice to say that the artillery and the engineer corps—the so-called *cuerpos facultativos* (technical or 'learned' corps)—had adopted the system of *escala cerrada* (closed list, that is, seniority as the only way of promotion); by contrast, the 'general arms' (the infantry and the cavalry) had additionally kept their officer lists open to merit promotion in the system of *escala abierta* (open list). Seniority was slower than selection by merit as a path to high rank, but it prevented arbitrary promotions. This disparity caused bitter disagreements within the service during the nineteenth century. A compromise was reached in 1889: all the officers would be promoted by seniority up to the rank of colonel in peacetime and entitled to merit promotions in wartime (although artillerymen and engineers could exchange war merit promotions for a special pensioned medal—the María Cristina Cross).[58]

The military defeat of 1898 did not encourage a reform of the oversized officer corps due to the dynastic politicians' fear of losing the support of the military. On the other hand, the dynastic parties did not dare for a long time to implement an obvious solution, adopted in other European countries, to the problem of the excess of regular officers: the creation of a reserve of civilians trained as officers during their compulsory military service, who could be called up in wartime and released after the conflict was over. The educated upper and middle classes were the natural suppliers of these officers. But the Spanish ones were virtually exempt from military service until the early 1910s: paying a fee or purchasing a substitute avoided any service with the colours. Thus conscription affected the lower classes alone, which had high levels of illiteracy.[59] A new conscription act passed in 1912 abolished fee payment and substitution: every able-bodied man could be drafted into the army for a three-year term serving with the colours plus 15 years in the reserves. However, the so-called 'quota soldiers' were released after a shortened term upon payment of a fee (2000 or 1000 *pesetas* in exchange for five or ten months with the colours, respectively) and proving a satisfactory level of training. So the main burden of conscription still fell on the lower classes. The 1912 act also provided for the creation of a pool of reservist officers, the *Escala de Complemento* (Complement List), whose members had to satisfy certain educational requirements and pay for their personal equipment, but it was not very successful because most of the would-be middle and upper class candidates preferred to serve as quota soldiers.[60]

The demobilization after 1898 left only 79,000 enlisted men in the colours; they were officered by a cadre 21,500-strong (plus 530 general officers).[61] During the early years of the twentieth century, the ratio of enlisted men to officers was about 4:1 and at certain moments may even have dropped close to 2:1. By contrast, the ratio in other European armies was between 17:1 and 24:1.[62] The effects of such a situation on the preparation of the army and the military careers were appalling. Officer salaries absorbed 58% of the army budget for 1900, which devoted only 9.5% to ordnance. About a decade later, the ratio had risen to 10.5:1, but 35% of the budget was still spent on the salaries of officers (Germany and France devoted only half of this percentage).[63]

Reductions of officer personnel could not be carried out resolutely by the government because they harmed the vested interests of the officer corps and no cabinet wished to fall out with it. The military academies had been closed after 1898, but they opened their doors again in 1902 under the pressure of officers whose sons aspired to a military career. The alleged reason was the shortage of subalterns. Such a shortage was real, but it was due to the eagerness for fast promotions in the early years of an officer's career, which depleted the junior ranks. Spanish subalterns needed nine to 12 years' service to reach the rank of captain (a Prussian subaltern needed from 16 to 18 years) and infantry officers often became majors when they were about 40 years old (most French officers reached this rank between the ages of 44 and 49; the estimate for an American officer commissioned in 1909 were 29 years of service before promotion to major, unless there was a war requiring expansion). But the Spanish officers' careers stagnated afterwards due to the surplus of field officers in the army lists (by contrast, the German army's personnel policy cut down the number of officers suitable for field grade, but their careers speeded up thereafter).[64] This eagerness for promotion also reduced the usefulness of the policy of amortization of vacancies. The government decided in 1899 to eliminate one out of every two vacancies of the same rank (so that there would be one promotion only). Junior officers did not welcome the decision because it meant they must stay longer in the lower paid ranks, and the unrest forced the war minister, General Weyler, to reduce the rate of amortization to 25% of the vacancies.[65] The failure to cut substantially the numbers of underemployed officers (8000 out of 18,000 officers had no active assignment in 1906) also meant that any rise in salaries was out of the question. This could not help fostering the interest of officers not to stay too long in the junior ranks, which therefore caused the need for a steady flow of new junior officers.[66] Such a state of affairs was worsened by the practice of admitting into the academies of the service the sons of military men (mostly officers) who had died or been crippled in war or equivalent service if they had just passed the entrance examination, even though it meant surpassing the limit of vacancies. For instance, although 395 vacancies were offered in the military academies in 1905, 472 new students were actually admitted into them.[67] And 2227 out of the 8685 applicants admitted in the period 1913–1928 were supernumerary (an excess of 34%).[68]

The Officer Corps' Reluctance for Military Reform

The roots of the officer corps' resistance to a significant reform in the way personnel policy was managed did not lie in the alleged fear of a deterioration of the army's military effectiveness. Actually such a reform would have been the basis for an improvement of the army as a whole. The reality was that the Spanish officer corps had become something rather close to one more branch of the state bureaucracy, whose corporate vested interests seemed to be more important than professionalism. Too many officers acted in practice as if their career was merely a way to earn one's living instead of a professional vocation, and any reform meaning cuts in personnel was rejected by arguing that one generation did not have to pay for the mistakes of the previous ones. This situation was tolerated by the dynastic politicians in order not to lose the support of the army.[69]

García Benítez saw this military problem as a derivation of the country's overall problem with the middle classes. Spain's limited economic development did not offer many employment opportunities to these classes. Since the doors of industry and commerce were so narrow, they looked for and hoarded jobs in the branches of the state; the army was no exception. Moreover, as another military essayist, Colonel Modesto Navarro, pointed out, education in the military academies was comparatively short and offered a sure wage at its end, in contrast with university education (which was rather expensive as well). García Benítez concluded that the Spanish army was no more than a productive cow feeding good, educated citizens whose military spirit had become impoverished, if it ever existed.[70]

This self-interested concept of the military career was already noticed before 1900. An article reproduced General José Galbis' earlier comments on the 'advantages' offered by the different corps. An officer of military administration (quartermaster corps) found cushy, danger-free assignments more easily. The military engineer could use his professional education (and thrive) in both military and civil life. The staff corps offered a fast track to the highest ranks. The artillery officer alternated regimental service and quieter assignments in army factories. The cavalry offered as compensation the joys of horsemanship. Finally, the infantry officer found only routine barracks duty, rubbing shoulders with the enlisted ranks, and the prospect of being killed because of his superiors' incompetence.[71] The self-interest also grew partly from the disappointing prospects the Spanish officer faced after being commissioned. For instance, Second Lieutenant Francisco Franco Salgado-Araujo went to an ordinary provincial garrison in his first assignment after graduating at the Academy of Infantry in 1911. He found that he was to be a subaltern in a company which was about 50-strong (the theoretical strength of a platoon was around 80 men). He felt a great disappointment: 'For this reality, I told myself, did I spend three years studying the campaigns of Hannibal, the Great Captain [Gonzalo Fernández de Córdoba], Napoleon, the Franco-Prussian and Russo-Japanese Wars, logistics, tactics of the three fighting arms, and so on?'[72] The poverty of professional life was not made good by the joys of a dazzling social life. Though the military academy educated him to hang out with the select circles of society, an officer could not permit himself to haunt them unless he came from a wealthy family.[73] All the same, many officers conformed to the atmosphere of dreariness.

The bulk of the officer corps drew annual salaries ranging from 2500 (first lieutenants) to 5500 *pesetas* (majors) by the beginning of the century. After deducting the income tax (from 5 to 18% of the basic salary), these wages were enough to cover the essential expenses of the average officer and his family—despite lacking housing benefits and subsidized foodstores. But they were far from allowing classy living standards. A captain at forty, on the eve of promotion to major, earned 3500 *pesetas* per year plus an additional amount of 600 *pesetas* for over ten years in his current rank. A civil servant of an equivalent rank could earn from 3500 to 4000 *pesetas*, whereas a bank clerk earned about 3700 *pesetas* after 20 years in his job. So the average officer had lower middle class earnings while he was supposed to live as a gentleman. Actually the Spanish officer devoted much of his home budget to maintaining such a gentlemanly appearance and behaviour, since he was anxious to display his self-image in front of subordinates and civilians.[74]

Some posts brought extra earnings (for instance, assignments in the regiments in Morocco meant a 50% rise in pay and officers in the Pyrenees border garrisons got better salaries as well). Nevertheless, most officers could not ask frequently for new assignments. On the one hand, the expenses of moving to another garrison were not paid by the army and became a heavy burden for officers who could hardly make ends meet on their salaries. On the other hand, the concession of assignments was subjected to the whim of ministerial authorities at the beginning of the twentieth century: any vacancy could be awarded by the minister's free will, and the senior officers of the branches (*secciones*) of each corps in the ministry were also very influential in the posting of personnel. García Benítez thought that this state of affairs fostered patronage and the loss of self-esteem within the military. An officer who had no patron in high places could even be posted compulsorily to a different garrison in order to create a vacancy needed to satisfy the application of a *protégé*. According to the artilleryman Antonio Cordón, only 'select' officers were assigned to the Horse Artillery Regiment (the so-called 'artillery hussars') due to its aristocratic lifestyle and close relationship with Alfonso XIII (new monarch after 1902). A sensible decree of War Minister General Francisco Aguilera in May 1917 finally regulated the concession of assignments up to the rank of colonel according to the seniority of the applicants. There would be only a few exceptions, such as assignments in the General Staff and War Ministry, teaching staff of the military academies, and military attachés.[75]

Since few officers could ease their financial difficulties within the army, the rest had to look for relief outside it. Officers in unpaid extended leave of absence (*supernumerarios*) and reduced pay situations could work full-time in civil jobs without losing their seniority in the army list. The artillery and engineer officers' technical training fitted easily with the civil economy's demands. However, few infantry and cavalry officers found suitable civil jobs and the rest of them limited themselves to work in part-time jobs (for instance, teaching in private crammers for the military academies).[76] General Luis de Santiago, who inspected the Academy of Infantry at Toledo in the autumn of 1907, commented on the effects of an order issued on 5 October 1907 banning officers in active service and stationed in cities where military academies were sited from teaching in private preparatory schools, unless they became *supernumerario* officers. The result was that teaching in the crammers of Toledo stopped due to the shortage of staff. And this meant a financial loss for the officers, since even full-time teaching in those schools did not make good the loss of army pay.[77]

A Snapshot of the Spanish Officer Corps in 1907

General de Santiago's remarks were included in one of the reports of the inspection review carried out in 1907. They were submitted by the captain generals of the military regions (the general officers commanding these territorial commands) and present an interesting picture of the concerns of the officer corps by this year. The report of the First Military Region noted the sad impression caused by the struggle of most officers to maintain a proper appearance with their pay alone. If a rise in salaries was not possible, the suppression of the income tax for state employees would at least remove the need for usury. The Fourth Military Region included in its report a list of officers who were not recommended for an appointment as commanding officers: 13 out of 19 had got into debt problems.[78] The limited salary maybe was not the only cause of indebtedness for some officers. Gambling proliferated in Spain in the early years of the twentieth century and the officers were not immune to its temptation—as the military airman Ignacio Hidalgo de Cisneros witnessed.[79] The report of the Fifth Military Region reckoned that the root of the indebtedness and poor morality of many officers lay in their formative years in the military academies, most of which lacked compulsory boarding (see Chap. 2).[80]

The report of the First Military Region also denounced that some officers were in the habit of obtaining assignments far from the close supervision of their superiors in order to get better efficiency reports or temporary releases from the service. For instance, a cavalry officer who worked as a military judge had got good reports despite his propensity to get into debt.[81] This situation had been denounced a few years before by Captain Eduardo Gallego, who stated that the lack of any stimulus and supervision caused a deplorable average level of professional proficiency. Reports were filled up year after year with no criteria other than the officer's punctuality and health. And if this procedure made assessing the ability of a regimental officer difficult, the assessment was even more difficult if the officer was serving in a bureaucratic post.[82] It was too easy to obtain temporary releases, which removed the officers from professional service for too long (a staff corps lieutenant colonel in the First Military Region reached this rank despite spending most of his career years in such releases, either for his private business or to fulfil his duties as member of parliament). The report of the Fifth Military Region stated that the *supernumerario* situation, created to have a reserve of officers, had lost its purpose. The *supernumerarios* became so devoted to their private businesses that they forgot their military professional expertise. Even so, they did not lose seniority in their rank, could get the San Hermenegildo Cross (a pensioned award for long service), and were entitled to retirement pay (or a pension for their widows), just like those who had performed active service through their whole military career.[83]

Age and health problems also figured in the review's reports. Those of several military regions observed the difference between the age suitable for every post in the service and the actual age of the incumbent officer. Many young officers held bureaucratic posts in conscription offices and reserve units while officers advanced in years still performed regimental service.[84] This had repercussions for the training of the fighting units. An observer of manoeuvres in 1902 remarked that many infantry company officers were too old and fast combat movements were slowed down due to their inability to command their soldiers from a leading position.[85] The report of the Fifth Military Region is remarkable due to its list of serving officers in unhealthy condition that the careless personnel policy had allowed: a senior administrative officer who held his post for a year though he was paralysed and unable to speak; an infantry lieutenant colonel who could not walk and had been promoted three times without performing regimental service; an infantry colonel, suffering from a hernia, who could not ride; a major unable to walk due to a softening of the marrow; and three 'mad captains' (*capitanes locos*).[86]

The Fifth Military Region also reported several cases of professional inability. A major on temporary release declined to lead the tactical movements of two infantry companies according to the official regulations because he considered himself 'ignorant' (*lego*). Faced with the same demand, several others majors 'started to quake' (*se pusieron a temblar*)—one of them even asked for six months in order to get ready.[87] This region's report found it strange that such cases of ill-health and poor professional proficiency were ignored by their immediate superiors. It attributed such neglect to a misplaced kindness, which gave too many highly positive judgements in the efficiency reports. In the Fifth Military Region this practice was frequent in the infantry, the cavalry and, to a lesser extent, the artillery (by contrast, the engineer and staff corps were sparing with such comments). For instance, officers whose annual reports included rebukes for incompetence saw their ability later described as good. The Sixth and Seventh Military Regions also commented on the lack of reliability of the efficiency reports.[88]

Other territorial commands did not mention this problem. This suggests that their captain generals perhaps consented to such practices and turned a blind eye to officers whose professional performance was less than satisfactory.[89] The cases recorded in the Fifth Military Region were perhaps the oddest ones, but then it seems fair to wonder how many cases of less exaggerated professional decay existed. Surveys of officer files and the army lists (a task which is outside the scope of this study) would perhaps make possible a quantitative analysis of the degree of professional commitment of the Spanish officer corps. According to García Benítez, who took the data from the 1913 edition of the *Anuario Militar de España* (the military yearbook), 850 (46%) out of 1,836 infantry field officers held bureaucratic posts in conscription offices and reserve depots or were on temporary releases.[90]

The consequence of the underemployment and the poor professional and financial prospects was an unsatisfying personal life for the average officer of junior and middle rank. This produced frustration and oversensitiveness to criticisms from the civilians, who thought the military were not worth the money spent in the War Ministry budget. The civilians were partly unfair, but the officer corps—in opposing to personnel cuts—was responsible for the situation as well.[91]

Conclusion

This chapter has shown that the political-military environment in Restoration Spain was not adequate enough for the army to achieve good professional standards. The Restoration regime had bartered the non-intervention of the military in politics for the army's institutional autonomy. Nonetheless, Spain's wars and restless domestic politics of the nineteenth century bequeathed an oversized officer corps to the Restoration regime. Personnel cuts were necessary for a real modernization of the army. But reform was thwarted by the officers' corporate self-interest and its tolerance by the Restoration dynastic political elites, which, rather than cutting away the useless elements of the military, left untouched corporate vested interests to safeguard their own political ones. In consequence the Restoration army's levels of preparation and equipment in peacetime were poor, since regular officer pay consumed a disproportionate part of the military budget.

Nonetheless, despite these deficiencies, the performance of the Spanish army in the overseas wars of 1895–1898 was effective enough. The Philippine rebellion was quashed and the Cuban rebels might well have been eventually defeated if the United States had not intervened. On the other hand, the Spanish military had not a monopoly of inefficiency, mismanagement or unpreparedness for war. 2500 out of almost 2800 American fatalities in the war with Spain died of disease, but most of them died in camps with poor sanitary conditions *in* the United States.[92] In fact, Spain was not alone in suffering an overseas fiasco in the second half of the nineteenth century: Britain (in Afghanistan) and Italy (in Abyssinia) were also unsuccessful.

However, these last countries had failed to expand. For Spain, defeat meant losing the last territories of its former empire. It also alienated the military from politicians, seen by the officer corps as spineless and unable to safeguard the fatherland's territorial integrity and social order. The military's bitterness was increased by a gloomy professional life of mind-numbing barracks routine, slow promotion, and unsatisfactory pay, blamed on the politicians as well, although the officers' dogged defence of their own corporate vested interests contributed to such situation. Given a motive serious enough, the officer corps would step in politics again to implement its own solutions to the problems of the nation and the army.

Notes

1. There is a lot of literature on the Spanish army's political and social history during the late nineteenth and early twentieth centuries. For a sample of helpful works which this chapter's first part is based on, see José Ramón Alonso, *Historia política del Ejército español* (Madrid: Editora Nacional, 1974); Miguel Alonso Baquer, *El ejército en la sociedad española* (Madrid: Ediciones del Movimiento, 1971); Carolyn P. Boyd, *Praetorian Politics in Liberal Spain* (Chapel Hill: The University of North Carolina Press, 1979); Carolyn P. Boyd, *La política pretoriana en el reinado de Alfonso XIII* (Madrid: Alianza, 1990); Gabriel Cardona, *El poder militar en la España contemporánea hasta la guerra civil* (Madrid: Siglo XXI, 1983); Fernando Fernández Bastarreche, *El ejército español en el siglo XIX* (Madrid: Siglo XXI, 1978); Fernández Bastarreche, Fernando: 'El Ejército', in [Luis Suárez Fernández, Demetrio Ramos Pérez, José Luis Comellas, and José Andrés-Gallego (eds):] *Historia General de España y América*, 25 vols. (Madrid: Rialp, 1981–1992), XVI-1. 653–75; Ramón Gómez Martínez, *Constitución y Fuerza Militar (1808–1978)* (Granada: Universidad de Granada, 2010); Pablo González-Pola de la Granja, *La configuración de la mentalidad militar contemporánea (1868–1909): del Sexenio Revolucionario a la Semana Trágica* (Madrid: Ministerio de Defensa, Secretaría General Técnica, 2003); Daniel R. Headrick, *Ejército y política en España (1866–1898)* (Madrid: Tecnos, 1981); Stanley G. Payne, *Politics and the Military in Modern Spain* (Stanford: Stanford University Press, 1967); Stanley G. Payne, *Ejército y sociedad en la España liberal (1808–1936)* (Madrid: Akal, 1976); Carlos Seco Serrano, *Militarismo y civilismo en la España contemporánea* (Madrid: Instituto de Estudios Económicos, 1984); Thilo Jens Wittenberg, 'Mut und Ehre. Die professionelle, ideologische und polistische Entwicklung des spanischen Offizierskorps, im 19. Jahrhundert (1808–1908)' (doctoral dissertation, Albert-Ludwigs-Universität [Freibourg im Breisgau], n.d.). An introductory general history which also addresses political and social issues is Fernando Puell de la Villa, *Historia del ejército en España*, 2nd edition (Madrid: Alianza, 2005).
2. On this form of military intervention, see Miguel Alonso Baquer, *El modelo español de pronunciamiento* (Madrid: Rialp, 1983).
3. For a concise military overview of these and other nineteenth-century conflicts in English, see Geoffrey Jensen, 'The Spanish Army at War in the Nineteenth Century: Counterinsurgency at Home and Abroad', in Wayne H. Bowen and José E. Álvarez (eds), *A Military History of Modern Spain: From the Napoleonic Era to the International War on Terror* (Westport: Praeger, 2007), pp. 15–36.

4. José Miguel Quesada González, *El reservismo militar en España* (Madrid: Ministerio de Defensa, 2014), p. 152.
5. The term *chusquero* alluded to the supposed large amount of *chuscos* (the army issue small loaves) that these officers had eaten during their career in the enlisted ranks; an army vernacular synonym was *oficial de cuchara* (spoon officer): Rafael García Serrano, *Diccionario para un macuto*, 3rd edition (Barcelona: Planeta, 1980), p. 528.
6. For a survey of the Paid Reserve List in the period 1912–1931, see José Fernando García Moreno, 'Servicio militar universal y personal en España. Procedencia social y formación del soldado, clases profesionales y oficial de carrera en el Ejército de Tierra español de 1913 a 1936', 5 vols. (doctoral dissertation, Universidad de Zaragoza, 1985), II. 565–80.
7. Even so, artillery *chusqueros* were assigned as a rule to ammunition supply duties and excluded from any command duty as battery officers: Antonio Cordón, *Trayectoria (Recuerdos de un artillero)* (Paris: Librairie du Globe, 1971), p. 51. But then it may have been a wise policy, if the incompetence of some of them in the Civil War illustrates the average technical expertise of their peers: see the examples in Michael Alpert, *The Republican Army in the Spanish Civil War, 1936–1939* (Cambridge: Cambridge University Press, 2013), pp. 91, 127.
8. García Moreno, 'Servicio militar', II. 578.
9. On Cassola and his reforms, see also Fernando Puell de la Villa, 'El general Cassola, reformista militar de la Restauración', *Revista de Historia Militar* (hereafter *RHM*), 45 (1978), pp. 173–196; Fernando Puell de la Villa, 'Las reformas del general Cassola', *RHM*, 46 (1979), pp. 143–174.
10. For a small sample of literature on these conflicts, see David F. Trask, *The War with Spain in 1898* (New York: Macmillan, 1981); Joseph Smith, *The Spanish-American War: Conflict in the Caribbean and the Pacific, 1895–1902* (London: Longman, 1994); Andrés Mas Chao, *La guerra olvidada de Filipinas 1896–1898* (Madrid: San Martín, 1997); John Lawrence Tone, *War and Genocide in Cuba, 1895–1898* (Chapel Hill: The University of North Carolina Press, 2006); Fernando Puell de la Villa, 'Guerra en Cuba y Filipinas: combates terrestres', *Revista Universitaria de Historia Militar*, 2, 3 (2013), pp. 34–57.
11. The war service of one of the infantry expeditionary battalions is recounted in Mateo Martínez Fernández, 'Un batallón de Infantería en la guerra de separación de Cuba', *RHM*, 35 (1973), pp. 75–103.
12. On Weyler's campaigns, see Valeriano Weyler y Nicolau, *Mi mando en Cuba (10 Febrero 1896 a 31 Octubre 1897). Historia militar y política de la última guerra separatista durante dicho mando*, 5 vols. (Madrid: Felipe González Rojas, 1910–1911); Gabriel Cardona and Juan Carlos Losada Malvárez, *Weyler, nuestro hombre en La Habana* (Barcelona:

Planeta, 1997), Chaps. 7–11. The effectiveness of Weyler's methods still remains a matter of controversy; for a recent positive assessment, see Puell de la Villa, 'Guerra en Cuba y Filipinas', but cf. Antonio Elorza, 'Final del imperio', in Hugo O'Donnell (ed.), *Historia militar de España*, 7 vols. (Madrid: Laberinto and Ministerio de Defensa, Secretaría General Técnica, 2009–2016), IV-1. 355–92.

13. A technical study concluding the accidental nature of the explosion is H[yman] G. Rickover, *How the Battleship 'Maine' Was Destroyed* (Washington, DC: Naval History Division, Department of the Navy, 1976).
14. Figures for Cuba are taken from Tone, *War and Genocide*, p. 9; figures for the Philippines, are taken from Puell, 'Guerra en Cuba y Filipinas', p. 48. The Spanish army was not the only one whose overseas campaigns could be very costly in human lives. Only 25 out of 4,613 French fatalities (about 30% of the expeditionary force) in the 1895 campaign of Madagascar were related to combat, the rest dying from sickness: see Anthony Clayton, *France, Soldiers and Africa* (Oxford: Brassey's, 1988), pp. 80–1; Douglas Porch, *The French Foreign Legion: A Complete History of the Legendary Fighting Force* (New York: Harper Collins, 1991), Chap. 13.
15. Miguel Alonso Baquer, 'La guerra Hispano-americana de 1898 y sus efectos sobre las Instituciones militares españolas', *RHM*, 54 (1983), p. 130; José Antonio Olmeda Gómez, *Las Fuerzas Armadas en el Estado Franquista. Participación política, influencia presupuestaria y profesionalización, 1939–1975* (Madrid: El Arquero, 1988), pp. 97–8.
16. Alonso Baquer, 'La guerra Hispano-americana', pp. 128–9.
17. On the political repercussions of the 1898 debacle, see also Juan Pro Ruiz, 'La política en tiempos del *Desastre*', in Juan Pan-Montojo (ed.), *Más se perdió en Cuba. España, 1898 y la crisis de fin de siglo* (Madrid: Alianza, 1998), pp. 151–260. The best study on the use of the army in the preservation of public order is Manuel Ballbé, *Orden público y militarismo en la España constitucional (1812–1983)* (Madrid: Alianza, 1983).
18. Alonso Baquer, *El ejército*, pp. 196–7; Payne, *Politics*, pp. 89–91. A good example of the military literature on the army as 'the school of the nation' is Joaquín Fanjul, *Misión social del Ejército* (Madrid: Eduardo Arias, 1907).
19. Alonso Baquer, 'La guerra Hispano-americana', p. 143.
20. Archivo General Militar de Madrid, Colección Adicional de Documentos, Legajo 8, carpeta 2 (hereafter AGMM-CAD No. Legajo/No. carpeta), Revista de Inspección al Ejército: Capitanía General, 2ª Región Militar, 1907; AGMM-CAD 8/4, Revista de Inspección al Ejército: Capitanía General, 4ª Región Militar, 1907; AGMM-CAD 8/5, Revista de Inspección al Ejército: Capitanía General, 5ª Región Militar, 1907.
21. Ironically, the cartoonist, Junceda, was son of an officer and, as an orphan, pushed for sitting the Academy of Infantry entrance examination; he not

made the grade: 'Joan García-Junceda Supervia', *Humoristán*, http://humoristan.org/es/autores/junceda/ (accessed 6 September 2016).

22. Ministerio de la Guerra, *Anuario Militar de España 1909* (Madrid: n.p., 1909) (hereafter *AME plus the year of publication*); *AME 1916*. The evolution of the force structure from 1899 to 1936 can be followed in the annual issues of *AME*.
23. [Arturo Sus Sus (ed.),] *Historia de las fuerzas armadas*, 5 vols. (Saragossa: Palafox, 1983), II. 43–4; Eduardo Gallego Ramos, *Proyecto de reorganización y mejora del ejército de tierra* (Guadalajara: n.p., 1905), p. 32. A graduate of the General Military Academy and commissioned in the engineer corps, Gallego (1873–1959) served in the Philippines (1895–1898) and Melilla (1909–1910) and reached the rank of brigadier general.
24. AGMM-CAD 8/5, Revista de Inspección al Ejército: Capitanía General, 5ª Región Militar, 1907.
25. AGMM-CAD 8/5, Revista de Inspección al Ejército: Capitanía General, 5ª Región Militar, 1907.
26. The reports of the 1907 army review, filed in AGMM-CAD 8/1 to 8/8, provide many examples.
27. AGMM-CAD 8/1, Revista de Inspección al Ejército: Capitanía General, 1ª Región Militar, 1907.
28. AGMM-CAD 8/2, Revista de Inspección al Ejército: Capitanía General, 2ª Región Militar, 1907.
29. Alonso Baquer, 'La guerra Hispano-americana', p. 144.
30. Figure for the interval estimated from data in *AME 1935*.
31. Alonso Baquer, *El ejército*, p. 235.
32. Fernando Puell de la Villa, 'Las fuerzas armadas en la crisis de la Restauración. Las Juntas Militares de Defensa', in Mario Hernández Sánchez-Barba and Miguel Alonso Baquer (eds), *Las fuerzas armadas españolas. Historia institucional y social*, 8 vols. (Madrid: Alhambra, 1986), V. 91–2. On the history of the General Staff before the Civil War, see Eladio Baldovín Ruiz, *Historia del Cuerpo y Servicio de Estado Mayor* (Madrid: Ministerio de Defensa, Secretaría General Técnica, 2001), Chap. 8, pp. 208–15.
33. Ministerio de la Guerra, *Colección Legislativa del Ejército 1904* (Madrid: n.p., 1904) (hereafter *CLE plus the year of publication*), Items No. 135, 240 Articles 20 to 36; Alonso Baquer, *El ejército*, p. 237.
34. Puell, 'Las fuerzas armadas en la crisis', V. 92.
35. Alonso Baquer, *El ejército*, pp. 237–9. On a contemporary case of military doctrine affected by malfunctioning of the high command and general staff system, see Douglas Porch, 'The French Army and the Spirit of the Offensive, 1900–14', in Brian Bond and Ian Roy (eds), *War and Society: A Yearbook of Military History* (London: Croom Helm, 1975), pp. 121–2, 124–8.

36. Comandante Beta [pseud. of José García Benítez], *Apuntes para historiar tres años de reformas militares (1915–1917)* (Madrid: Imp. de Bernardo Rodríguez, 1917), pp. 21–5. García Benítez (1872–1948) was a graduate of the General Military Academy and an engineer officer; he held assignments in the Philippines (1895–98), the War Ministry's military bureau (1915), and the General Staff (1916–1918); he reached the rank of brigadier general.
37. Mariano Aguilar Olivencia, 'La opinión pública militar', in Hernández Sánchez-Barba and Alonso Baquer (eds), *Las fuerzas armadas*, V. 156–9.
38. Organization as described in *AME 1912*.
39. El Capitán Equis [pseud. of Nazario Cebreiros], *El problema militar en España. Apuntes para un estudio sincero y al alcance de todos*, 2 vols. (Burgos: Saiz y Comp[a], 1916–1917), I. 51–2. Cebreiros was an infantry officer and, after leaving the army under the terms of the 1931 decree for early retirement, shifted towards very conservative political views, which led to his murder at the outbreak of the Civil War.
40. Equis, *El problema militar*, I. 52.
41. Equis, *El problema militar*, I. 54.
42. Puell, 'Las fuerzas armadas en la crisis', V. 92, 101.
43. Beta, *Apuntes*, pp. 55–6.
44. *CLE 1912*, Item No. 254.
45. Beta, *Apuntes*, pp. 20–2, 62–4, 66–7.
46. Puell, 'Las fuerzas armadas en la crisis', V. 103.
47. Beta, *Apuntes*, p. 64.
48. Puell, 'Las fuerzas armadas en la crisis', V. 103–7.
49. *CLE 1916*, Item No. 22.
50. Equis, *El problema militar*, I. 83; Puell, 'Las fuerzas armadas en la crisis', V. 107.
51. *CLE 1916*, Item No. 22.
52. Puell, 'Las fuerzas armadas en la crisis', V. 92, 107.
53. For a brief overview of the transformation of the Spanish officer corps, see Julio Busquets, *El militar de carrera en España*, revised edition (Barcelona: Ariel, 1984), pp. 48–66.
54. Fernández Bastarreche, *El ejército español*, pp. 8–12, 103–23.
55. According to the regulations, if their entrance examination marks were equal, a military man was admitted instead of a civilian, and a civilian who was son of a military man likewise took precedence over a civilian who was not: *Reglamento orgánico para las Academias militares de Infantería, Caballería, Artillería, Ingenieros y Administración Militar* (n.p: n.p., 1924 [originally issued in 1897]), Article 73.
56. Busquets, *El militar*, pp. 63–6; Headrick, *Ejército*, p. 78; García Moreno, 'Servicio militar', II. 750, 862.

57. Headrick, *Ejército*, pp. 75–6, 85–7; Payne, *Politics*, p. 76.
58. Headrick, *Ejército*, pp. 88–97. The term *facultativo* was applied to state employees whose occupations required an accredited level of higher education, and active list officers of the artillery, engineer, and staff corps did indeed receive a thorough education in technical and scientific subjects. See an idiosyncratic definition in José Almirante, *Diccionario militar: etimológico, histórico, tecnológico: con dos vocabularios francés y alemán* (Madrid: n.p., 1869), *sv* 'Facultativo'.
59. For a general overview of military service until 1914, see Eliot A. Cohen, *Citizens and Soldiers: The Dilemmas of Military Service* (Ithaca: Cornell University Press, 1985), Chap. 2.
60. On the evolution and problems of military recruitment in Spain until the early 1910s, see Fernando Puell de la Villa, *El soldado desconocido: de la leva a la mili (1700–1912)* (Madrid: Biblioteca Nueva, 1996); Quesada, *El reservismo*, Chaps. 2–5. On conscription after 1912, see José F[ernando] García Moreno, *Servicio militar en España (1913–1935)* (Madrid: Servicio de Publicaciones del EME, 1988); Quesada, *El reservismo*, Chaps. 6–7. On the Complement List, see José Miguel Quesada González, 'Origen de la Escala de Complemento', *RHM*, Special Issue (2010), pp. 13–52. See also Headrick, *Ejército*, pp. 101–4; Payne, *Politics*, pp. 99–101; Michael Alpert, *La reforma militar de Azaña (1931–1933)* (Madrid: Siglo XXI, 1982), pp. 49, 95–6; Cardona, *El poder militar*, pp. 8–9.
61. Headrick, *Ejército*, pp. 272, 276–7. During the period 1880–1895, peacetime strength (including the paramilitary police forces) fluctuated between 137,000 and 163,000 troops; these figures are based on official data which, in relation to the enlisted men, provide only theoretical figures and are unreliable until 1891: Headrick, *Ejército*, pp. 272, 277.
62. Fernández Bastarreche, 'El Ejército', XVI-1. 657; Headrick, *Ejército*, p. 75.
63. Fernández Bastarreche, 'El Ejército', XVI-1. 657, 662; Boyd, *Praetorian Politics*, p. 32.
64. Headrick, *Ejército*, p. 76; Boyd, *Praetorian Politics*, pp. 27–8; Paul-Marie de la Gorce, *The French Army: A Military-Political History* (New York: George Braziller, 1963), p. 49; Edward M. Coffman, *The Regulars: The American Army, 1898–1941* (Cambridge, MA: Harvard University Press, 2004), p. 151.
65. Boyd, *Praetorian Politics*, pp. 27–8.
66. Payne, *Politics*, p. 98; Boyd, *Praetorian Politics*, p. 28.
67. José Luis Isabel Sánchez, *La Academia de Infantería de Toledo*, 2 vols. (Toledo: n.p., 1991), I. 246–8.
68. García Moreno, 'Servicio militar', II. 658–9.
69. Boyd, *Praetorian Politics*, p. 27.

70. Beta, *Apuntes*, pp. 18–19; Modesto Navarro, 'Lo moral en lo militar', *Revista de Infantería*, 1, 2 (1901), pp. 71–3. Navarro (1853–1913) was a conscript and later commissioned in the infantry, and became a major general; he took part in the campaigns of Cuba (1875–1878, 1895–1898) and Morocco (1909–1912), and was the head of the infantry section at the Central Firing School (1904–1909).
71. X., 'El espíritu de la juventud militar', *Revista Técnica de Infantería y Caballería* (hereafter *RTIC*), 14, 2 (1903), pp. 32–3.
72. Francisco Franco Salgado-Araujo, *Mi vida junto a Franco* (Barcelona: Planeta, 1977), p. 23. Franco Salgado-Araujo (1890–1975) was a cousin of General Francisco Franco Bahamonde, generalissimo of the rebel or Nationalist side in the Civil War and Spain's head of state (1939–1975); he was his cousin's aide-de-camp and military secretary during much of his professional career from the mid-1920s to the early 1970s and reached the rank of lieutenant general.
73. Aguilar Olivencia, 'La opinión pública militar', V. 147.
74. Boyd, *Praetorian Politics*, pp. 32–4, 310 note 31; Puell, 'Las fuerzas armadas en la crisis', V. 90. For an overview of the social and financial situation of the officer corps based on the contemporary military press, see Aguilar Olivencia, 'La opinión pública military', V. 166–78.
75. Gallego, *Proyecto de reorganización*, p. 28; Puell, 'Las fuerzas armadas en la crisis', V. 94–6; Boyd, *Praetorian Politics*, p. 33; Beta, *Apuntes*, pp. 47–8; Cordón, *Trayectoria*, p. 65.
76. Boyd, *Praetorian Politics*, p. 33.
77. AGMM-CAD 8/1, Revista de Inspección al Ejército: Capitanía General, 1ª Región Militar, 1907. The order was both a reminder and an extension of an article of the 1897 regulations for military academies that banned the latter's teachers from being engaged in such activity; a later order had to extend the ban to any town within a radius of 25 kilometres from a military academy site: *Reglamento orgánico*, Article 11; Isabel, *La Academia*, I. 260.
78. AGMM-CAD 8/1, Revista de Inspección al Ejército: Capitanía General, 1ª Región Militar, 1907; AGMM-CAD 8/4, Revista de Inspección al Ejército: Capitanía General, 4ª Región Militar, 1907.
79. Ignacio Hidalgo de Cisneros, *Cambio de rumbo*, 2 vols. (Barcelona: Laia, 1977), I. 56, 89–90. Hidalgo de Cisneros (1896-1966) took part in the Moroccan campaigns of the 1920s and became the commander-in-chief of the loyalist or Republican air force in the Civil War.
80. AGMM-CAD 8/5, Revista de Inspección al Ejército: Capitanía General, 5ª Región Militar, 1907.
81. AGMM-CAD 8/1, Revista de Inspección al Ejército: Capitanía General, 1ª Región Militar, 1907.

82. Gallego, *Proyecto de reorganización*, p. 18.
83. AGMM-CAD 8/1, Revista de Inspección al Ejército: Capitanía General, 1ª Región Militar, 1907; AGMM-CAD 8/5, Revista de Inspección al Ejército: Capitanía General, 5ª Región Militar, 1907.
84. AGMM-CAD 8/2, Revista de Inspección al Ejército: Capitanía General, 2ª Región Militar, 1907; AGMM-CAD 8/4, Revista de Inspección al Ejército: Capitanía General, 4ª Región Militar, 1907; AGMM-CAD 8/7, Revista de Inspección al Ejército: Capitanía General, 7ª Región Militar, 1907.
85. Domingo Arráiz de Conderena, 'Las prácticas y ejercicios militares de la 6ª Región en octubre de 1902', *RTIC*, 14, 1 (1903), p. 562.
86. AGMM-CAD 8/5, Revista de Inspección al Ejército: Capitanía General, 5ª Región Militar, 1907.
87. AGMM-CAD 8/5, Revista de Inspección al Ejército: Capitanía General, 5ª Región Militar, 1907.
88. AGMM-CAD 8/5, Revista de Inspección al Ejército: Capitanía General, 5ª Región Militar, 1907; AGMM-CAD 8/6, Revista de Inspección al Ejército: Capitanía General, 6ª Región Militar, 1907; AGMM-CAD 8/7, Revista de Inspección al Ejército: Capitanía General, 7ª Región Militar, 1907.
89. But the Spanish army was not the only one where an incompetent or infirm officer could still hold command because his superiors neglected to take any action; a similar situation could be found in the American army: Coffman, *The Regulars*, pp.185–6; Brian McAllister Linn, *The Echo of Battle. The Army's Way of War* (Cambridge, MA: Harvard University Press, 2007), pp. 112–13.
90. Beta, *Apuntes*, pp. 203–16.
91. Boyd, *Praetorian Politics*, pp. 34–5; Aguilar Olivencia, 'La opinión pública military', V. 171.
92. Coffman, *The Regulars*, pp. 12–13; see also Linn, *The Echo of Battle*, pp. 102–3.

CHAPTER 2

Military Corporatism Under the Restoration

External factors (such as financial constraints, or the cabinets' inconsistent military policies) were not the only explanation for the Spanish army's professional weaknesses. The army also suffered from its branches' corporate factionalism. The latter fostered a parochial spirit which, far from promoting cooperation, strengthened distrust and hindered efforts to achieve military modernization. This chapter surveys such corporatism, which took root during military education and affected even the professional elite of the officer corps: the staff officers.

The System of Separate Corps Academies

Spanish Military Education from 1800 to 1893

At the beginning of the nineteenth century the Spanish army trained most prospective infantry and cavalry officers in the regiments, while those of the artillery and engineer corps went to specialist schools.[1] During the war against Napoleon (1808–1814), the Spanish authorities set up several improvised military schools. Three of them (for infantry and cavalry officer candidates) survived after the war and, in addition to the restored artillery and engineer schools, carried on training officers until 1823. A new, single military academy—the *Colegio General Militar* (General Military College)—was opened in 1825. It was criticized because of its adoption of unified teaching for all arms despite their different requirements and a decree dated 5 November 1850 set up a

J.V. Herrero Pérez, *The Spanish Military and Warfare from 1899 to the Civil War*, DOI 10.1007/978-3-319-54747-3_2

system of separate academies for the infantry, cavalry, artillery, engineer and staff corps. Meanwhile, the system of regimental officer cadets was abolished and restored several times until its final abolition after the Third Carlist War (1872–1876).

War Minister General Martínez Campos set up the *Academia General Militar* (General Military Academy) in 1882. The officer candidates took basic officer training for two years and, after passing the second year, went to either the infantry or cavalry academies (where they were commissioned after one year) or followed a preparatory course for the artillery, engineer or staff corps academies. Those who passed the course were commissioned as second lieutenants before joining their specialist school. The General Military Academy aspired to diminish the tense relationship among the branches of the Spanish officer corps. Each branch's academy doubtless impressed an indoctrination of general military values and attitudes in its graduates. But this was not enough to overcome the inward-looking corporatism of the arms of the service, eager to advance their own interests. A general academy, it was hoped, would make easier to develop an *esprit d'armée* that would prevail over the *esprit de corps* and its petty corporate interests. The General Military Academy was not as successful as intended. The syllabus did not reconcile sufficiently each branch's educational requirements. Different syllabi were introduced and the academy eventually became rather a preparatory school for the corps academies. War Minister General José López Domínguez closed the academy in order to cut military expenses in 1893. But the lobbying of the learned corps, eager to preserve their touchy *esprit de corps* and reassume responsibility for all of the whole training process, must also be taken into account. López Domínguez, himself an artilleryman, was doubtless receptive to their arguments.[2]

The Reform of 1893

The decree (dated 8 February 1893) that ordered the closing of the General Military Academy also laid down the new structure of military education.[3] Officers of the combat arms would be once again trained in separate corps academies, sited in Toledo (infantry), Valladolid (cavalry), Segovia (artillery) and Guadalajara (engineers). There would also be a staff college in Madrid. The introduction to the decree pointed out that the system of separate academies allowed the numbers of annual admissions to be settled accurately according to the needs of each corps. The reform also intended to address one particularly unhealthy practice. Many sons of families with some private means (including lots of officers' sons) who aspired to the military academies would join the army as volunteer privates and took advantage of the benefits (higher age limit, exemption from secondary education diploma, subsistence grants, free preparatory studies) for applicants coming from the enlisted ranks. Since the taxpayers' money paid for the studies of those officer candidates who came from the enlisted ranks, volunteering had become a simple (and much abused) way of receiving a free commission. This had corrupted a procedure designed to assist deserving enlisted men. After 1893, the benefits were supposed to be limited to conscripts alone, but in practice the subterfuge carried on as usual.[4]

Applicants should have completed full secondary education and had to prove their proficiency in specific subjects through a system of competitive examinations, with those with the best marks receiving admission. Nonetheless, provided that they got the minimum marks, orphans and brothers of military men who had died on campaign would be admitted as supernumerary students.[5] The duration of the course of studies was either three (in the case of the infantry and cavalry) or five years (in the case of the artillery and engineers). Presence in the academy was only required in the final two years; the subjects in previous years could be studied privately and passed through examination alone. Students in the cavalry, artillery and engineer academies would live outside the academy; those in the infantry academy could be either boarders or external candidates. After passing the third year, the students would be commissioned as second lieutenants. They were promoted to first lieutenants after graduation at the academy (learned corps) or two years of service (infantry and cavalry).[6] The graduates of the learned corps' academies also received a professional certification that was equivalent to a university degree in engineering (industrial engineering for artillery officers and civil engineering for engineer officers).[7] In fact, the engineer corps had recently made a case for accreditation of its commissions as an academic degree entitling its officers to work outside the army as civilian engineers.[8] Since looking for a supplementary job out of the army was not uncommon in the officer corps, such accreditation was very useful indeed.

The Reform of 1893 Under Criticism

López Domínguez's reform was fated to be contentious.[9] The alleged advantage of allowing prospective officers to choose their commissioning branch from the outset was not consistent with their freedom to sit entry examinations in several academies in the same year. If many applicants really wished to become officers in just one specific arm, they should have had no reason to apply for admission in other corps at the same time. In short, the 1893 reform did no more than suppress the intermediate step of the General Military Academy and could therefore still produce officers who were dissatisfied with their arm of service. On the other hand, if this kind of dissatisfaction was as great a problem as the advocates of separate academies avowed, the fact is that the affected officers adapted themselves to their situation remarkably well—or at least did not make a public complaint. One plausible explanation lies in the motivation of applicants for the military career. As has been shown in Chap. 1, the economic security of a job on the state payroll was a major incentive (if not the only one in some cases) to take a career in the army. The effects of such a mindset on the Spanish army's professionalism were deplorable, and the military academies were themselves partly responsible because of their failure to select the right officer material. A decree dated 4 October 1905 forbade applicants to take the entrance examination in more than one academy in the same year, but Antonio Cordón states that in 1911 he took the exams in the infantry, artillery, and engineer academies, all of which admitted him.[10]

What were the faults of Spanish military education at this time? For a start, the military academies allowed the recruitment of applicants who could be as young as 14 years old (instances can be found by glancing through the *Anuario Militar of España*—the army yearbook). The keenness to enter a military academy as soon as possible was a consequence of the closed list, since the younger an officer started his career, the greater his chances of reaching the highest ranks before his retirement.[11] Actually, when the courses of study were shortened because of the 1890s overseas wars, a ministerial order dated 4 July 1896 had to remind the academies of taking into account that, when setting the requirements for new admissions, officer candidates younger than 17 could not be commissioned.[12] Miguel Campins, a graduate of the Academy of Infantry who would later play a prominent role in the restored General Military

Academy, deplored the commissioning of overly young officer candidates, who could be as young as 16 years old and reported at the barracks in immature conditions.[13] Campins also concluded that, due to the excessive youth of their students, the academies of this period were more like secondary schools. The officer candidates were only under the semblance of true military discipline. Hence, for instance, whenever a very important person visited the Academy of Infantry, the student corps felt entitled to ask for a general leave, an interruption of teaching or the release of classmates under arrest. According to Campins, the academies' regulations and the pressure of the students' families made it impossible to counter this behaviour.[14]

An additional cause helped to maintain such behaviour for years: the absence of a boarding regime in most academies. This happened although the 1897 regulations stated that all the officer candidates had to live in a boarding regime until they were promoted to second lieutenants; if they had to carry on their studies in the academy, they would live in extramural lodgings. But the regulations accepted that circumstances could make it permissible for students to live outside the academy, thus providing an opening for non-compliance, and in 1917 the rule would be modified to allow students to live out of the academy if living with their parents, brothers or sisters.[15] The Fifth Military Region's report on the 1907 inspection review stated that those officer candidates living outside the academy walls did not get used to the essential disciplinary habits, thereby paving the way to the poor morality and indebtedness of many officers. The report recommended the implementation of compulsory boarding in all the military academies, since it was reckoned essential for inculcating military customs and ethos deeply and sparing the officer candidates' parents from fears and expenses.[16]

The admission procedures of the military academies were also subjected to criticism in the contemporary military literature. There was no real inquiry about the family and moral background of the applicants, meaning that the selection depended finally on the exam results alone.[17] The problem was exacerbated by the limited scope of the examinations, which could not assess the applicant's specific fitness to be an officer, since they only tested his mathematical knowledge.[18] Moreover, after an educational reform in 1894 lengthened the full secondary education (*bachillerato*) by one year, the military authorities decided to demand only the subjects of its first four years.[19] These academic limitations made admission easier for young teenagers whose later performance in the academy proved their physical or intellectual handicaps (and even their limited basic education: for instance, a memorandum of 1907 noted that spelling mistakes were frequent in the students' work).[20]

The syllabi of the academies were another common target for contemporary critics. According to them, Spanish military education was dominated by theoretical studies, which was, to say the least, of dubious value for junior officers. On the other hand, professional matters such as the essential field service and junior command duties were relegated to a secondary status behind scientific and mathematical subjects.[21] The Academy of Infantry, when it had to shorten the training of officers to one year due to the colonial wars of 1895–1898, reportedly still maintained the teaching of the higher scientific subjects, while paying scant attention to the officer candidates' physical fitness.[22] The emphasis on science was also seen as being detrimental to the indoctrination of military morale and ethos. Colonel Modesto Navarro complained in 1901 about the young officers coming out of the academies: they were well-versed in scientific theory, sometimes to the point of pedantry, but they lacked real military spirit and interest in the everyday details of the profession.[23] A decade later, Ricardo Burguete, a tactical theorist who stressed the importance of moral factors on the battlefield, lambasted an education allegedly oriented to produce cool, rational-minded officers: the heavy mathematical content crushed 'the ardent youthful fantasy, poetical and overwhelming, which makes up the most powerful lever of the warrior'.[24] Within this context, the reader may guess which armies Captain Eduardo Herrera de la Rosa was thinking of when he wrote in his report on the Russo-Japanese War that the officer training in the Japanese army lacked the kind of contents which sapped the energy of

> the unfortunate group of people who, in some countries, come up to nourish the officer corps, to the point of turning them into engineers, doctors in science, architects, astronomers, geodesists, and so on; [these are] very respectable professional qualifications, but they are useless for developing leadership skills, getting used to command duties, deploying large masses of troops, being a good sapper or pontonnier or handling a gun, a rifle or a horse correctly.[25]

In addition, teaching methods were based on rote learning of huge quantities of information from textbooks, with no room for personal work.[26] Rote learning in the military academies has been regarded as prone to producing a tendency to conformity and uncritical acceptance of orders and established procedures.[27] Perhaps this is the origin of the passive obedience which, according to a contemporary military essayist, prevailed among the Spanish officers during the Melilla campaign of 1909.[28] Most professional officers on both sides during the Civil War had passed through the military academies after the reform of 1893. Writing after the conflict, General Alfredo Kindelán concluded that the Spanish officer had 'an extraordinary fear of responsibility, a desire of being given detailed, precise orders which save him from the duty of thinking things over and having his own criteria'. According to Kindelán, as a result of this mental inertia prevailing in the Spanish officer corps, whose members were reluctant to conceive of their own solutions for operational problems, the rebel or Nationalist commanders displayed a deplorable lack of initiative during the war.[29]

Two of the few memoirs recalling with some detail the life in a Spanish military academy at that time were written by the artillerymen Carlos Martínez de Campos and Antonio Cordón. Each of them reflects the bookish nature of their training. Carlos Martínez de Campos, who entered the Academy of Artillery in 1903, recalled those years as endless lessons in the classroom, six hours every day. This burden was lightened by the afternoon gathering with his classmates, in which playing cards often replaced textbooks. Apart from the morning run to be present at the roll call, physical training was reduced to a weekly fencing session, horsemanship during the fourth year and a march to the firing range on some Saturdays. Martínez de Campos did not hesitate to define such a life as absurd for a career soldier.[30]

Cordón also reckoned that the education in the academy was unsatisfactory: 'Training the students as engineers was actually much more valued... than the specific military training.' The principal, more demanding subjects were the scientific ones, while the military matters—artillery firing, tactics, and so on—were considered secondary subjects. Such order of priorities meant that 'the artilleryman, after finishing his academy years, had limited military knowledge and not very deep and complete academic knowledge'.[31] Cordón, like Martínez de Campos, was lodged in extramural accommodation (the boarding regime was not implemented until mid-1914 in the Segovia academy), but he had the opportunity to live for a while in the Academy of Infantry as a visiting student in 1913–1914. He concluded that, in contrast with Segovia, life at the Toledo academy was more military, harder, and nearer to that of the soldier in the barracks.[32]

What the academies seemed to produce was no more than another class of bureaucrats, with no regard for their real proficiency for military command. And too many officers were satisfied with this state of affairs, which provided a salary for life with little effort (especially if promotion was based on seniority alone).[33] They looked like 'military officials' rather than military officers and undemanding peacetime soldiering did not motivate them to hone professional proficiency during their later careers. For instance, the shortage of manoeuvres prevented the officers from putting into practice and improving their training. Thus, after leaving the academy, many officers limited themselves to reading the tactical regulations before an exercise in order to remember the procedures for carrying out the movements.[34] Nonetheless, it must also be said in fairness that some of the deficiencies of the Spanish military academies, such as the prevalence of theory over practical teaching or the emphasis on the sheer cramming of information, were not unique among contemporary armies.[35] And they mirrored those of the contemporary Spanish educational system as a whole, which has little or no regard for the benefits of physical fitness and vigorous sports, also neglected in Spanish society in general.[36]

Attempted Reforms

The idea of a general academy did not die in 1893. On 21 July 1904, War Minister Linares—aware that basic officer training should be as standardized as possible for all the corps—decreed the reopening of the General Military College in Toledo.[37] But this decision was never implemented because of the rapid turnover of cabinets at that time. The cabinet of which Linares was a member left office in December 1904 and, after two short-lived cabinets, War Minister Weyler, through a decree dated 17 August 1905, suspended the implementation of those reforms whose funding depended on a budget plan which had failed to get parliamentary approval.[38] One of those reforms was the General Military College. Nevertheless, this setback did not dissuade Captain Herrera de la Rosa from advocating common training for officers, since it was the best way to instil a common spirit in all the army branches.[39] Nonetheless, a critic of the general academy model pointed out in 1905 that the benefits of a common indoctrination and training would be very small if they were not accompanied by a reform of the promotion system, since the current one allowed the branches of the service different paces of promotion.[40] This fact certainly would continue to cause bitterness among members of the same class.

It does not seem casual that two years after the campaign of Melilla, which exposed some of the deficiencies of the Spanish army, War Minister Luque brought in a reform of the military academies' student selection procedure.[41] Such reform inspired a desire of introducing a more practical orientation in the training of officer candidates and making sure, in particular, that they were fit enough for military life.[42] Henceforward the entrance examination would be the same for all the academies, although each one could determine the relative importance of each exam subject for the applicants' overall score.[43] The new procedure introduced physical tests and balanced the academic requirements by reducing the weight of maths and introducing exams in geography and history (previously a school accreditation could substitute for them and the question paper in Spanish grammar).[44] The minimum entrance age was set at 15, but applicants were no longer required to pass all the entrance examination's question papers at the same time. Applicants who were at least 13 years old could take a few of the question papers of a summon; if they passed, their marks were recorded by the concerned academy and the applicants could take different papers in successive summons until passing all the papers of the entrance examination. Applicants could no longer take the exams after turning 21 (civilians), 24 (enlisted men who had served with the colours less than two years) or 27 (enlisted men who had served more time).[45] Although a step forward, this reform made only a limited effort to raise the entrance age, while the option of passing the exams in the course of several years seems designed to assist sons of military men, besides making it less likely that officer candidates had a homogeneous academic knowledge when they eventually entered the academies (they could have virtually forgotten the subjects of the earliest exams).[46]

Later regulations developed the spirit of the reform in the academies by matching the importance of theory and practice in the officer candidates' marks.[47] This reformist spirit eventually reached their syllabi, which were revised in 1913, and the teaching methods. The new syllabi were based on the following assumptions: each academy could decide the weight of a subject in the grading of students; practical learning would take precedence over theory (exams would be divided into theoretical and practical parts, with a view to remove exams and grade only the students' hands-on performance); military subjects would be as important as scientific and technical ones, and if he failed them, an officer candidate could not proceed to the following year's training; officer candidates would continually prove adequate fitness; and their behaviour would be assessed as well, and such assessment only could be done successfully if boarding was compulsory.[48] Teaching, in its turn, had to adapt to the new approach: it had to be a practical process and limit itself to the tasks an officer would face just after leaving the academy; textbooks would only be used for consultation after the lessons; the latter had to be active and participative, and students would take notes that would form the basis of their learning, supplemented by textbooks; the contents of the syllabi would be simplified, with less mathematics and more history and geography, and would include military moral education to develop soldierly habits and virtues, which only could be guaranteed through boarding; and students had to get an effective physical training.[49]

The reality did not match the expectations. There were no entrance exams in either geography or history, an omission justified by an alleged lack of 'adequate' textbooks to be recommended for the applicants' preparatory studies, and there were few requirements in Spanish grammar.[50] The reformist effort of the early 1910s was derailed by a decree of 30 January 1918 and the abridged courses of the early 1920s.[51] The decree argued that dual grading was no longer necessary because the students' marks in the theoretical and practical parts of the exams usually coincided.[52] The effect was retrogressive. By 1918 teachers and students still had not got rid of the intellectual inertia binding them to the old methods and, once the need for a specific marking of the students' practical proficiency was removed, teaching tended naturally to focus on the theoretical contents again. Moreover, the Spanish military academies' facilities had not been improved enough for an effective practical training. The Academy of Infantry's *Junta Facultativa* (Academic Board) blew the whistle on this retrogression in 1924 and, in the three years to 1927, advocated a return to the dual exams and the enlargement of training facilities in the academies.[53]

With regard to the abridged courses, convened from 1921 to 1923, they were meant for meeting the increased demand of subalterns resulting from the escalation of military operations in Morocco after 1919. They were opposed by the military academies, which demanded at least a rise of the entrance educational requirements in order to focus teaching on the most essential subjects. The reality was quite the opposite, since the official policy facilitated the entrance of sons of military men who had died or were missing in war service. Afterwards the government planned to close temporarily the military academies to new admissions, but it eventually had to convene standard-length courses in 1924 and 1925 for the many applicants who had passed partially the entrance examinations in the previous years.[54] In the end, the most durable effect of the reforms may have been some revival of the regenerating spirit in the military academies, which was most evident in the Academy of Infantry.[55]

Over 30 years after López Domínguez's reform, Spanish military education still fell short of amending its defects, despite some reformist attempts like those described above. Captain Vicente Rojo, a teacher at the Academy of Infantry during the 1920s (and a future chief of the loyalist or Republican General Staff in the Civil War), wrote a memorandum about the flaws of its syllabus dated May 1926, which basically reflected the views of the academy's academic board.[56] He described the syllabus as 'bombast, pedantry, external garb more or less modernist, but without a soul giving it life'. Rojo judged the syllabus' subjects disorganized and poorly balanced, keeping the students in the dark about which contents were really important. In addition, theoretical teaching and practical training were uncoordinated and sometimes at odds. So the faculty actually taught the officer candidates 'to hate books by demonstrating that the latter were good for nothing'. And the academy's graduates eventually joined the 'mass of illiterate officers'.[57] Perhaps the state of affairs was not so bleak in the other academies, but the reader must remind that the infantry branch was the largest one of the service by far, so training in its academy had an effect on the best part of the officer corps.

Thus the officer corps was already a not so united institution for the military academies' new intakes. It should come as no surprise that these prospective officers later developed corporate factionalism and mistrust which damaged seriously the professionalization of the army as a whole. It is also remarkable how much of the regulations on military education became dead letter (for instance, the implementation of boarding). This neglect has been imputed to the disruption in entrance procedures and syllabi caused by the abridged courses of the late 1890s and, at least as a partial explanation, the case does not seem to be off the mark.[58] In any case, training in the academies displayed a much-criticized remoteness from the professional realities of the junior officer's service. These major flaws did not go unnoticed and the period 1911–1913 saw an attempt to remedy some of them, but the reform had miscarried by the early 1920s.

Why was the 1893 model so resilient for more than three decades? The separate education better suited the exclusive corporate spirit of the learned corps than did a general academy, and the learned corps undoubtedly lobbied to thwart any proposed reform in depth. But direct access to the corps academies also meant more opportunities for entry into the officer corps, unlike the bottleneck of a general academy. This was advantageous for families keen to direct their male offspring into a military career, and since the head of many of these families was (or had been) an army officer as well, there also was a 'peer pressure' factor which worked to deter any far-reaching reform. Moreover, the 1893 organization also suited the vested interests of the officers whose careers (and additional incomes) were closely bound to teaching in military academies.[59] Captain Eduardo Gallego wrote of officers who sought to get and remain in teaching assignments at any cost because these were a handy way to increase their earnings.[60] For these officers, the more academies and academic contents the better. One additional problem was that they included the reformist officers most likely to convey the deficiencies of the existing system, but whose professional and financial gain depended on it.[61] As a result, there was limited stimulus for any reform from within. It should come as no surprise to anyone that the two major reforms of military education before the Civil War (which will be tackled in Chap. 6) took place under circumstances out of the ordinary.

The Military Consequences of Corporatism

The Learned Corps' Struggle for Responsibilities

The officers of the artillery and the engineer corps displayed the highest degree of inner cohesion (though most of the evidence used in this chapter concerns the artillery, since they were the largest and most influential of the learned corps). This cohesion stemmed from their being trained entirely in their corps' own academies and the adherence to a career pattern based on the closed list (seniority promotion). A foreign observer, the French journalist Arthur Houghton, provided a portrait of the artillery corps in the 1870s which is also valid for the following decades.

Houghton pointed out that the artillery officers made up a select network (*une espèce de franc-masonnerie aristocratique et sévère*), which turned them somewhat into a separate branch from the rest of the officer corps. Zealous to hold their ground, they supported each other like an aristocratic oligarchy and their manner was distinguished, bordering on haughtiness, with the other corps. Such an attitude was allegedly based on higher professional standards, which—according to Houghton—bore well in comparison with all the European armies.[62]

By the last quarter of the nineteenth century, many artillery officers stood for the abolition of merit promotion (at least within their corps). According to Cordón, during the late 1910s and early 1920s, the junior officers of the corps still thought the closed list a small price to pay for securing fairness within the military and this did not affect the artillerymen's conduct in peacetime or in wartime.[63] In order to satisfy such a desire, the corps' board (*junta*) decided in 1891 to collect in an album the signatures of those artillery officers who agreed to renounce to war merit promotions. All the officers signed in the album and from that year on all the new graduates of the Academy of Artillery did it as well. The heading of the album was as follows:

> The artillerymen who sign this album want to conserve in the corps – and transmit through example to those who will serve in it later – the traditional spirit of honour, union, and fellowship that they received from their forerunners, through which it [the corps] achieved the glory and prestige it enjoys, both for the good of the fatherland and the honour of its members.
>
> And by considering that the closed list is the indispensable condition for the attainment of such high goals, they resolve to maintain it among themselves, offering on their honour to renounce (through the ways allowed by the law) to any promotion within the corps or any general officer vacancy assigned to it which is not conceded due to seniority.[64]

The artillery and the engineer corps looked after their cohesion through their own semi-official *juntas*. The artillery corps organized its *Junta Central de Artillería* (Artillery Central Board) in 1888. This body also assumed the function of watching over the artillery corps' prestige and traditions, and the morals of its officers. Due to this system, comradeship was strengthened and inner resentment disappeared within the corps, which developed a gentlemanly (*hidalgo*) spirit, fond of tradition and hierarchies, and pleased to record the honours and awards of its members. Perhaps, according to Jorge Vigón, an officer and historian of the corps, the artillery became too snobbish, thereby stirring desires for imitation in the rest of the military as well.[65] The Artillery Central Board also meddled in issues which were clearly outside its original field of responsibility. For instance, in 1903 it vetoed the publication in the corps journal (*Memorial de Artillería*) of an article which questioned the artillery's alleged monopoly on the use of machine guns. Even Vigón puts into doubt the competence of the board to judge the matter.[66]

This last example illustrates the mistrustful attitude of the learned corps with regard to issues of technological innovation. Artillerymen and engineers were jealous to preserve their privileged status (due to their academic education) in military technological and scientific matters, while they competed with each other for responsibilities in these issues.[67] Actually General López Domínguez, when he was war minister in the period 1892–1895, strengthened the corporate spirit of self-sufficiency by emphasizing the scientific and industrial responsibilities of the learned corps (artillery, engineer, and staff corps) and even the cavalry (on horse breeding and remount issues).[68] Thus a reformist work of the early 1920s complained about 'the internal struggles between corps, trying to take away responsibilities from each other, not because of a noble desire for a better performance, but due to a spurious interest in holding vacancies [that is, in getting assignments for officers of one's own corps]'.[69]

Indeed, the learned corps often seemed more interested in scientific knowledge than in warfare, an attitude deplored by a number of contemporary military essayists. Captain Gallego, himself an engineer officer and a moderate critic, remarked in 1910 on the extraordinary contrast between the interest given in the Spanish army to the technical and educational institutions (firing schools, laboratories, technical commissions, and so on) and the neglect of the troops' combat training.[70] Two decades later, Brigadier General Emilio Mola, an infantryman, was still criticizing harshly the excessive interest in scientific issues, which reflected and fostered the corporate rivalries:

> I can offer even more proofs of petty corporatism. The fight between the corps to increase each one its importance and extend its influence over the army as a whole has reached anarchical proportions. There are laboratories and special establishments without discernment. A lot of chemistry, a lot of mechanics, and a lot of ballistics! And, above all, too many sages![71]

Moreover, such rivalries often meant the unnecessary squandering of public funds. Gallego pointed out in 1905 that three brand new laboratories (run by the military engineers, the artillery corps and the state civil engineers) had been set up in the period 1898–1899 for a similar task of testing materials. He argued that it would have been more economical and rational to create a single laboratory whose equipment and funding might well have been shared by its three institutional users.[72]

The Corporate Rivalry's Effects on Military Effectiveness

How far the idea of 'the more technical responsibilities, the more prestige' could lead is displayed by a manifesto written by the artillery officers of the Mahón garrison (in the Balearic Islands) in November 1914. They complained that the artillery corps had been deprived of any institutional role in the development of military ballooning; it had no overall responsibility for military motor transport nor involvement in its industrial dimension; and it was excluded from railway issues. They also complained that what they saw as the artillery corps' exclusive scientific authority over the matters of ballistic science had diminished because the infantry and cavalry sections at the Firing School compiled their own firing tables for rifle and carbine. These officers also deplored the fact that the infantry had been granted, thanks to the machine gun, responsibilities for crew-served weapons (which had been a traditional monopoly of the artillerymen). According to them, all this proved that the artillery corps had not received its fair share of responsibilities as the number of new military devices grew.[73] Thus these artillery officers felt aggrieved because the infantry corps was fully responsible for the ballistic studies of its own weaponry. And they lamented the fact that technological innovation did not always mean an extension of their authority over ordnance and its use. Faced with such obstacles, the general arms—and especially the infantry—had to fight hard to overcome the institutional opposition of the more technical-minded corps and develop their own tactical doctrines.

An instance of the way the overlapping of corporate responsibilities could hamper the infantry's own experimental labour was the *Escuela Central de Tiro* (ECT)—Central Firing School. It was formed in 1904 from the Artillery Firing School and divided into four sections (field, fortress, and siege artillery; coastal artillery; infantry; cavalry). The creation of the school had been opposed by the artillery, which wanted to monopolize the 'legal responsibility' for everything related to weaponry and firing matters. The existence of the cavalry section was odd anyway, for the cavalry weapons' firing was similar to the infantry firing. Actually an inspection report of 1907 wondered if it would be a better arrangement to consolidate the infantry and cavalry sections into a 'portable arms school'.[74] Perhaps it was reckoned that the cavalry section provided active assignments for a number of officers.

The ECT itself was an oddity as well because no other country had combined infantry and artillery firing schools: their respective modes of firing were too different and the infantry schools were rather tactical training centres. On the other hand, the ECT did not work as a testing centre either, because this role was performed—even for small weapons—by the Artillery Section of the War Ministry. To add insult to injury, the infantry and cavalry sections' own firing ranges were under the authority of the engineer corps. According to Cebreiros, worse than the organizational flaws was the almost worthless training. The courses had an 'informative' character only. Generals and regimental commanders were exempted from any course at all. Other corps' officers attended the artillery courses but artillery officers did not attend the infantry courses. Moreover, according to Cebreiros again, the teaching staff did not keep its expertise up to date and easily lost track of new doctrines.[75]

Given this situation, it is unsurprising that, as late as 1918, a military essayist, Captain Epifanio Gascueña, complained about the lack of tactical agreement between the fighting arms. The 1913 infantry tactical regulations did not offer a comprehensive picture of combined arms combat, which had become the most usual case, while the 1910 artillery tactical regulations did not define in a clear way the mission of direct support to the infantry.[76] As Gascueña put it:

> Neither of the two regulations considers these more or less lasting groupings of artillery and infantry with a common mission, those frequent cases in which artillery elements are charged with directly supporting an infantry attack with subordination in mission, though not in command, or those exceptional ones in which temporary groupings are formed by making up a tactical unit with subordination of the artillery to the appointed commander for such an unit.[77]

The need for tactical cooperation would not remain a matter of theoretical analysis because the Spanish army was to wage in the 1910s a low-intensity war in the Moroccan Protectorate, where all-arm columns played an important role. During the first phase (1911–1916) the artillery seems to have performed poorly. Vigón blames the dispersal of artillery units in penny packets attached to the columns for this.[78] But it can also be asked if a less corporate attitude to tactical training would have provided a better insight into combined arms operations.

Moreover, the scientific haughtiness of the learned corps may well have been responsible for fostering a fashion for bookish scholarship among the infantry officers, who tried to make up for their lower academic education. This trend seems to have been of no benefit for their professional performance. After the 1909 campaign in Melilla, the General Staff itself complained about the predominance given in the assessment of infantry officers to manifestations of written brainwork—which fostered the production of superfluous works on military theory—over the proficiency displayed in regimental service.[79]

Later scholarship offers somewhat contradictory views about this issue. According to the French scholar Andrée Bachoud, the military authorities fostered an increase in Spanish professional literature after 1909 (at least until the mid-1910s), though the quality of this literature was sometimes poor, since authors seemed more concerned to add up pages than to make a new point.[80] In any case, there were certainly officers who tried to improve their professional lot at any price as prolific authors of works of supposed military value.[81] On the contrary, Andrés Mas Chao has pointed out that non-fighting assignments and scientific training lost prestige among the officers most engaged in the Moroccan war.[82] And Julio Busquets even states that the attitude of anti-intellectual pragmatism displayed by the officers fighting in Morocco was due to their limited military education.[83]

Busquets' assessment is rather questionable. As has been shown above, the criticism of military education recorded in the contemporary professional literature alleged precisely that the academic content was excessive. Perhaps the excess of theory in their academy years was the origin of many officers' dislike of formal professional education for the rest of their careers. In any case, many officers of the fighting arms in Morocco paid small attention to the more cerebral side of military operations in this period. Such a dislike of intellectual effort in warfare could lead them to scorn methodical staff and logistical work, less glamorous than an infantry assault with fixed bayonets, and to boast of relying on *cojones* (testicles, meaning guts) alone.[84]

The corporate rivalries were thus influential in disturbing the Spanish military environment and stood in the way of developments in military doctrine. Corporate rivalries were not unique to the Spanish army and were found in armies all over the world. Nevertheless, some armies overcame the worst effects of them on military doctrine thanks to a type of organization developed in parallel with the mass armies of the late nineteenth century and whose authority stood above all the branches of the military: the general staff.[85] Unfortunately, the development of the general staff in Spain was linked to another case of corporate struggle. This is better understood through a survey of the odd position of the staff corps within the Spanish officer corps.

The Odd Corps Out: The Controversy About the Staff Corps

The Problem of the Reform of the Staff Corps

The war against Napoleon showed the Spanish army's need for highly proficient officers helping the senior commanders in the conduct of operations. The staff corps (*Cuerpo de Estado Mayor*) was born in 1810, after a proposal of General Joaquín Blake. However, as its officers were suspect of liberal leanings, the corps was disbanded during King Fernando VII's periods of absolute power (1814–1820 and 1823–1833) and was not finally restored until 1838. From 1842 to 1893, the staff corps (which recruited prospective officers straight from civil life in the period 1850–1882) reached its highest level of institutional autonomy and specialization. It initially sought prestige through its mapmaking functions, but their officers became more focused on operational matters after 1875.[86]

Although the general arms and the learned corps distrusted each other, they nonetheless had common misgivings about the staff corps. This mistrust was based on the latter's somewhat hybrid nature. The general arms reckoned that its elitist attitudes (based on higher educational standards) too closely resembled those of the traditional learned corps, whereas the artillery and the engineer corps saw the staff corps officers as rivals and parvenus of sorts (because their careers were not ruled by a closed list system). Indeed, the artillery officers even kept a tacit commitment not to join the staff corps.[87] The staff corps certainly did not keep the solidarity with the other learned branches of the service in the defence of the closed list. The bonds linking them on this issue were broken in January 1894, after a staff corps officer, Captain Juan Picasso, accepted a war merit promotion won during the small conflict of Melilla in late 1893. This meant the actual acceptance of the open list by the staff corps.[88]

Such behaviour was born from the staff corps officers' realization, from 1875 on, that the prestige won through their performance on campaign could be more influential than scientific achievement. Thenceforward, they sought to reach general officer rank with first-hand knowledge of the realities of combat and life in the fighting forces, which would support their claims to hold high operational command assignments or high political-military office. Actually some staff corps officers (such as the future General Weyler) had stood out by following this career model, which was in line with General Martínez Campos' policy of fostering the moral unity within the military through participation in the fighting.[89]

Nevertheless, such a policy could be overstretched by self-interested officers. A harsh critic of the Spanish military at the end of the nineteenth century reported that, in the Cuban campaigns of 1895–1898, many army corps and divisional chiefs of staff, instead of performing their essential staff duties, took command of operational columns with the backing of the general officers commanding the major units. Therefore, the staff work was never organized properly in the headquarters. Even if the needs of colonial warfare were different from those of a European regular campaign, a good general staff organization was still required to maintain 200,000 troops in a theatre of operations of over 100,000 square kilometres.[90]

The way staff duties should be performed, either through a separate branch of the military (the staff corps) or through the temporary assignment of officers with special training (the concept of a 'general staff service'), was an issue much discussed during the last quarter of the nineteenth century. By the 1880s a small group of reformist officers drew conclusions about the military events of the 1860–1880 period (especially the Prussian–German victory in the 1870–1871 war against France). One of these conclusions was that an efficient general staff was indispensable for the command of modern mass armies. The success of the Prussian model had convinced the reformists of the need to reform the mobilization system and its territorial organization in order to make effective use of the reserve manpower in wartime. Such a reform meant changing the current organization of the staff corps, focused on its traditional mapmaking role.[91]

General Martínez Campos, the leading military personality of the early decades of the Restoration, was a moderate reformist, who thought that the mapmaking missions of the staff corps (though they had to be entrusted to a scientifically trained branch of it) played only a supporting role and were not essential for the staff work.[92] However, he did not undertake any actual reform of the organization of the staff corps. By contrast, General Cassola, war minister in 1887–1888, was a more resolute character. He reckoned that proficiency in cartography was not the monopoly of any corps and therefore should not determine the role of staff officers. Moreover, he thought that general staff assignments were nobody's monopoly. So, after being appointed war minister, Cassola included among his reformist plans the abolition of the staff corps: from then on, he asserted, staff duties would be performed by especially trained officers of all the other corps, after the Prussian model. These officers would remain in their commissioning corps' lists and alternate tours of regimental service and staff duty. On the other hand, although the traditional, overspecialized mapmaking tasks were discarded, the staff officer's status would be enhanced by turning it into a more influential advisory and executive assignment near the commander.[93]

Although Cassola's reformist proposals were finally defeated by the pressure of the vested interests within the military, the debate about the staff corps and the general staff service remained alive. General López Domínguez tried to solve the issue through a compromise in 1893. He retained the staff corps, whose new members thenceforth would be graduates of the staff college who decided to leave their commissioning branches; the other graduates would remain in their corps as *diplomados de Estado Mayor* (staff graduates).[94] However, this arrangement did not solve the fundamental problem, as will be shown below.

The Training of Staff Officers After 1893

López Domínguez's reform also affected the training of staff officers. A staff college was created at Madrid in 1842, where graduates of the General Military College got additional training. From 1850 to 1882, it admitted officers commissioned in other branches and civilians able to pay for their studies. After 1882, the college only admitted General Military Academy graduates.[95] The staff college was renamed as *Escuela Superior de Guerra* (ESG) in 1893. Requirements for sitting the entrance competitive examination were the rank of first or second lieutenant and at least three years' seniority after commissioning as second lieutenant (including one year of regimental service); if the student wanted to join the staff corps after graduation, he must not be older than 29 years when entering the ESG. Since commissioning officer candidates when they were (or were about to be) 18 years old was not unusual, a Spanish officer could be an ESG student at 21.[96] In the German army, by contrast, most officers were commissioned at 21 and had to fulfil four years of regimental service before applying for admission to the *Kriegsakademie*. The theoretical studies in the ESG lasted three years but the syllabus' military contents seem small in comparison with the academic ones:

First year

i. Elements of astronomy, topography, fortification, military bridging, castramentation
ii. Artillery and engineer ordinances, regimental internal service, foreign tactical regulations
iii. Algebra, spherical trigonometry and analytical geometry, general geometry and general history (optional subjects)

Second year

i. Descriptive and strategic military geography, telegraphic and telephonic signals, railways, ballooning, cryptography
ii. Outline of military and naval organization (in Spain and major powers), military art, staff duties and regulations, survey of floating material
iii. Descriptive geometry, calculus and rational mechanics, chemistry, gunpowder and explosives, ordnance manufacture (optional subjects)
iv. Foreign language (English, German or Arabic)

Third year

i. Survey of Spanish naval artillery and foreign artillery
ii. Political economy, military administration, political and administrative law, international law, survey of medical and quartermaster equipment

iii. Astronomy and geodesy or elements of zoology and botany, mineralogy and geology
iv. Foreign language (English, German or Arabic)

The syllabus' academic orientation repelled many artillery and engineer officers (actually no new student came from these corps in 1896), who thought their scientific 'exclusivity' under threat. This did not benefit the prestige of the ESG.[97]

The German army's *Kriegsakademie*, by contrast, had the following syllabus (though this is the syllabus adopted in 1907, its contents seem not to show significant changes with regard to previous years):

First year

Tactics, military history, fortification, ordnance, medical service, general history, military and international law; (optional subjects) pre-1648 history, mathematics, physics, foreign languages

Second year

Tactics, military history, fortification, topography, communications, naval warfare, general history; (optional subjects) pre-1648 history, mathematics, foreign languages, chemistry, physical geography

Third year

Tactics, military history, staff duties, siege warfare, general history; (optional subjects) pre-1648 history, mathematics, foreign languages, topography, geodesy[98]

Comparing the German syllabus with the Spanish one, it can be seen that the former had a clear-cut core of military subjects centred around tactics and history, whereas all its scientific subjects remained optional. On the contrary, tactical and historical matters look underdeveloped in the Spanish staff college's syllabus.

The teaching of the ESG was judged unsatisfactory in the early years of the twentieth century and the college was reorganized in 1904 by War Minister General Linares. He realized the need to increase the number of officers qualified for operational staff duties and stood for giving priority in promotion to general officer rank to those who had passed the staff college.[99] The new syllabus was as follows:

First year

i. Military and strategic geography, basic notions of geology
ii. Political economy and military administration, hygiene, international law, medical field service
iii. Topography, notions of electricity
iv. Conversation and writing in French
v. Horsemanship, fencing, photography, practical sessions of topography and visits to medical and quartermaster corps centres

Second year

i. Military history
ii. Mathematical algorithm, astronomy, geodesy, meteorology, military industry, technical and practical studies in military communications (optional)
iii. Art of war, grand tactics, military law
iv. Foreign language (English, German, Arabic, Portuguese), fencing
v. Landscape sketching, horsemanship, practical sessions in subjects of (ii) and (iii)

Third year

i. Military history
ii. Use of artillery and fortification in warfare
iii. Staff duties, wargaming, communications as a means of war
iv. Foreign language (English, German, Arabic, Portuguese), fencing
v. Panoramic military sketching, horsemanship, practical sessions in subjects of (iii), staff ride and geographical surveys[100]

The non-military scientific content was substantially reduced (although one cannot help but wonder how far the staff officer's proficiency depended on his skill with the sword). As a whole, this new syllabus bore more resemblance to those of other foreign staff colleges. For instance, in the French army, whose studies for the staff diploma lasted two years, the contents included the following subjects: military history, applied strategy and tactics, staff duties, fortification, ballooning, international law, geology and geography, military law and administration, mobilization, German, Russian (optional), railways, hygiene, medical service, naval warfare, and horsemanship.[101]

The teaching methods of the ESG were reportedly changed as well. Official textbooks were removed; instead the college's teaching staff was supposed to encourage the development of analytical skills and judgement through the writing of essays, group projects, and group debates in the classroom, which allowed the display of the students' personal views. In short, any teaching only based on rote learning had to be banished.[102] Nevertheless, it is not clear how successful this change of methods was. A graduate of the ESG argued in 1921 that the college was still inflexible and unwilling to accept critical views (at least in some subjects). And in the same year, a Spanish officer studying at the French army's staff college commented, via Colonel Juan García Benítez (the military attaché), on the too erudite learning system of the ESG, in contrast with the French institution, where teaching was based on case studies taken from recent campaigns.[103]

Attack and Defence of the Staff Corps

The reforms after 1893 did not stop the antagonism towards the staff corps. One of its most heated critics, Cebreiros, even lambasted the quality of its officers. According to Cebreiros, the root of the problem lay in the system of entrance into the staff corps. After the reform of 1904, applicants for the ESG had to be captains or first lieutenants with two years' regimental service. An officer who got the staff diploma remained in his commissioning branch after leaving the ESG. In order to join the staff corps, an ESG graduate had to apply for admission whenever there were vacancies in its list and before being promoted to the rank of major in his commissioning branch (the career in the staff corps started at the rank of captain).[104] But only the youngest officers (that is, those who could presume a career advance) applied for admission to the staff corps. However, they often graduated with the lowest marks of their class, due to their relative immaturity. Thus the worst qualified students got into the staff corps whereas the best ones—most of them older men with many years of service as captains—preferred to stay in their commissioning branches rather than joining a new officer list and starting again behind younger captains belonging to a previous ESG class and who had been able to apply for a vacancy before. In order to attract the brightest graduates, the staff corps pursued a policy of fast promotions which was only feasible through the creation of as many assignments for staff corps officers as possible.[105]

Cebreiros also pointed out that another result of the coexistence of the staff corps and the *diplomados* was the rivalry between the two groups of staff officers. The staff corps officers wanted to cut the number of *diplomados* to prevent them from occupying staff assignments—if not, the policy of fast careers within the staff corps would be under threat. The fact that assignments near the high command were deemed almost a passport into it also spurred this attitude (which was not unjustified: in 1895, 11 out of the Spanish army's 39 lieutenant generals came from the staff corps, whereas 13 came from the infantry). What made this competition an even more lamentable case of corporate selfishness was the feeble state of the Spanish army. Since the budgets did not allow an effective preparation for war, the term 'staff work' (outside the forces in Morocco at least) and its alleged 'technical' complexity—which the staff corps used as a justification of its own existence—actually concealed more and more red tape. Indeed this institutional self-interest even fostered more bureaucracy.[106] As Cebreiros concluded, '[m]ainly due to the fault of the staff corps, the Spanish army is a lie wrapped in notepaper'.[107] Nonetheless, the reader must bear in mind that Cebreiros was an infantry officer and therefore his criticism suffered from corporate bias as well. So, how justified was the antagonism towards the staff corps?

The military archives keep primary evidence which shows that the staff corps was not very pleased that *diplomados* received the same training as its prospective officers. There is no evidence to find out the extent of such a mood, but the fact that two staff corps officers of very different rank and writing a decade apart proposed similar solutions to cope with the *diplomados* suggests that the mood was shared by large sections of the corps. The first piece of evidence comes from the report on the inspection of the ESG carried out by General Pío Suárez Inclán (who had been promoted from the list of the staff corps) in 1907.[108] It is not clear if the points made in the report are actually General Suárez's own ideas, but it seems plausible that his views were influential in the conclusions of the report. It was certainly a good opportunity to defend his former corps' interests. The report pointed out that it was not a good idea to allocate equal value to all the subjects of the syllabus, because this meant that a student who excelled in general academic matters could get a better place in the graduation list than another who obtained high marks in the essential professional subjects. So the report proposed allocating to each subject a quotient which indicated its relative importance for the overall score of the student. So far, so good.

But after that the report remarked that the ESG students had different motivations in their training as staff officers. The proposed change would not affect the students merely aiming for the diploma, but it would affect those aspiring to join the staff corps since their place in the graduation list settled the precedence for admission and their place in the corps' list. Therefore, the report questioned the policy of teaching the same syllabus to both groups, since it could be deficient for prospective staff corps officers and too demanding for those just looking for an extension of their professional expertise. The solution to this problem put forward in the report was to divide the syllabus into two parts. The first one would be a common course including the essential military matters; the second part would be a more demanding course reserved for those called to perform staff duties permanently (that is, the staff corps officers). Thus more officers would come to the ESG to improve their military education without any need to turn themselves into 'perfect' staff officers after five years of theoretical studies and practice. The implied consequence of this proposal was to render *diplomados* unfit for staff assignments, since only the officers of the staff corps would be fully qualified to hold such posts. Therefore, the corps would be able to justify its enlargement (and new promotions) whenever there were staff vacancies. There is no evidence in the files consulted by the present author about the later fate of the proposal, though it obviously found no favourable reception because there was no significant change in the ESG syllabus.

The second piece of evidence dates from early 1918, so it is contemporary with the political-military crisis caused by the *juntas de defensa* from 1917 on (see Chap. 5). The *Junta Superior de Defensa* (Higher Defence Board) of the staff corps consulted its officers in December 1917 about the need for preserving a corporate organization to perform staff duties. The answer of an officer stationed in Valladolid, Captain Nicolás Benavides, argued for the absolute need for keeping the staff corps instead of replacing it with a general staff service.[109] His first argument was that the common training, the constant performance of staff duties and the cohesion caused by belonging to a single branch provided unity in working procedures and methods. On the contrary, the different corporate background of the *diplomados*—though they carried out similar tasks—would damage the unity of the whole if the general staff service was implemented.

Another of Benavides' arguments was that the staff corps, due to its own corporate organization, limited unhealthy ambition among its members, because all of them had the same legal rights. On the contrary, according to Benavides, the *diplomado* of the general staff service would be placed on a different level in relation to his commissioning branch's fellow officers because of his separate staff duties and the special training for the staff diploma, whereas he would be keen to please his superiors, who could back his career up. Therefore the general staff service would offer more fertile ground for unlimited personal ambition. Finally, Benavides reckoned it very difficult to achieve a good performance in staff duties if these were alternated with regimental service.

Benavides also pointed out several alleged dangers for the survival of the staff corps. One of them was the corps' weak state, which was partly a reflection of the deterioration of the army as a whole, but it was also due to its own neglect and lack of collective effort. Another major danger he saw (and this point shows the degree of ill-will between the two groups of staff officers) were the *diplomados*. Their numbers (and foreseeable increase) and deeds caused and fostered hostility towards the corps in political circles and the press; even the archbishop of Tarragona, who also held a seat in the Senate, had allegedly displayed antagonism! Benavides expressed fears that the political debate could bring about a cut in the number of admissions to the staff corps, which would lead to a consequent increase of *diplomados*. Some political groups and sections of the press—mostly of leftist leanings—also were especially hostile because the corps (according to Benavides) strengthened the unity of the army (the major support of government) and therefore held back their ambitions. Benavides recommended the case for the staff corps to be presented to leading politicians, who could defend its cause in the parliament.

He also recommended the establishment of a secret record of data on the main foes of the corps, especially the *diplomados* (a proposal with somewhat sinister tinges). Benavides concluded with a proposal to turn the ESG into an institution which would graduate staff corps officers only. Those officers who did not want to join the corps would make up a sort of *adjudantur* (adjutant's service) after two years of studies plus one more of practice.[110] These officers would perform ancillary staff duties and become a reserve of staff officers in wartime; they would serve usually within their branches, though they could be called up to work in manoeuvres under the orders of staff corps officers.

The essential coincidence of Captain Benavides' last point about the staff college with General Suárez's views is significant. The ultimate aim of both them was to turn the *diplomados* into second-class staff officers, who would therefore be denied the responsibilities and benefits of being fully qualified staff officers. Thus the staff corps would recover at least the monopoly of the most important staff duties. This monopoly had been broken by the 1893 reform, which gave the same professional qualification to prospective staff corps officers and those who aspired to the diploma alone. Therefore, the staff corps had to remove this equality if it wanted to maintain its *raison d'être*. The fact that a general officer in 1907 and a captain a decade later shared rather similar solutions to this problem makes it plausible to think that corporate self-interest was highly developed in the staff corps and hence that criticism of the staff corps as a whole on this ground was not unfounded. Furthermore, Benavides' paper bear witness to the deep rivalries within the military; his suggestion to spy on the *diplomados* illustrates clearly how far the deterioration of the inter-corps relationship could reach.

The Staff Corps and the Development of the General Staff

What made the rivalry of the staff corps with the rest of the military even more serious were its consequences for the development of the General Staff. According to García Benítez, the staff corps worked against the 'military bureau' set up by War Minister Echagüe; it could not tolerate ideas alien to the corps on issues related to the reorganization of the army. Any crisis within the military caused by a reorganization project meant a crisis for the existence of the staff corps, so the latter was urged to take control of the reform by its *esprit de corps* and instinct of survival. The result was the preservation of the corps without any real improvement of the army.[111] Echagüe's military bureau was kept idle by his successor, General Luque, while the General Staff's rebirth was prepared under the guidance of staff corps officers. According to García Benítez, their aim was to turn the General Staff into 'an organization which saved the hegemony so endangered by that inconvenient military bureau, which had to be destroyed in a hurry, and this [the bureau's disbandment] was carried out to the relief of many people'.[112]

García Benítez, despite his criticism of the attitude of the staff corps, stated that he did not underrate the individual value of its members, whose educational standards were higher than those of the other branches. But this superiority also made more dangerous their corporate selfishness. And since they were facing up to another group with similar professional qualifications (the *diplomados*), the struggle was waged on grounds which damaged the reforms badly needed by the army. The other corporate rivalries seriously affected the effectiveness of the army as well,

> [b]ut these struggles among 'guilds' do not have the [same] importance… as that one waged by the guiding 'guild' against all the guided 'guilds'; and its result is that an army whose general staff functions are still in embryo has got the following organizations to carry them out: the staff corps, a staff college, and a legion of *diplomados*; three sources of unsurpassed material and a very deficient service, the most deficient one of our weak military body.[113]

Certainly, an in-depth reform of the army was not a very realistic goal when the officer corps quarrelled about who was most qualified to direct such a work. That the root of such division lay in corporate factionalism is proven by the fact that there was no criticism (excepting Cebreiros) of the professional proficiency of the staff corps officers as individuals: what displeased the rest of the military were their behaviour and privileges as a group. Nonetheless, a general staff service would not necessarily have been a panacea: the French army had adopted it after 1871, but, according to the historian Douglas Porch, this did not prevent its staff officers from turning into an inward-looking body of careerists, entrenched in routine and the defence of their privileges.[114]

Conclusion

Rivalry between the branches of the service is not unusual in armies. Nonetheless, corporate factionalism and self-interest in the Spanish army were highly developed and turned into serious obstacles to military effectiveness in the early decades of the twentieth century. Much of this corporatism was reflected in military education. The reform introduced by López Domínguez in 1893 had imposed the learned corps' approach to the training of officers, which even seems to have been imitated by the general arms' academies, whose syllabi became more bookish. Moreover, the current organization and working of the academies reveal deficiencies and subterfuges that, despite damaging the quality of officer training, were tolerated because they served vested interests of the officer corps (for instance, the endemic excess of admitted applicants). The reforms of the period 1911–1913 offered a chance of moving towards a new model of military education, but they were ruined by the modification of grading criteria in 1918 and the abridged courses of 1921–1923. The fact that these reforms have been missed by historians, excepting José Fernando García Moreno, may be a sign of their utter failure.

The learned branches, especially the artillery, were eager to enlarge their turf by monopolizing technical responsibilities, including the weaponry of the other corps. Thus the infantry found difficult to test new weapons and their effects on tactics because of the artillery's insistence in controlling all the matters related to firearms. The problem of corporate factionalism perhaps reached its apex in the case of the staff officers. Military reformists had advocated the Prussian–German model of staff officer who alternated tours of staff duties and regimental service in his commissioning branch. The old staff corps lost the monopoly of staff duties after the reorganization of the staff college in 1893, but it managed to survive and thereafter sought to restore its former status through subterfuges. This reinforced the antagonism of the other branches, which perceived the corps as a body of conceited careerists. The consequence of all this was that the Spanish army's professional elite was divided and its role in the modernization was influenced negatively by the needs of corporate self-interest. The contribution of staff work to Spanish military successes in later conflicts perhaps must be seen more as a proof of the staff officers' individual competence than as an evidence of institutional effectiveness.

Notes

1. For a sample of literature on the Spanish military education until the 1890s, see Julio Ferrer Sequera, *La Academia General Militar. Apuntes para su historia*, 2 vols. (Barcelona: Plaza y Janes, 1985), I. 33–188; José Ignacio Muro Morales, *El Pensamiento Militar sobre el Territorio en la España Contemporánea*, 2 vols. (Madrid: Ministerio de Defensa, Secretaría General Técnica, 1993), I. Chaps. 1–6, 10–14; Miguel Alonso Baquer, 'En la hora de las Academias Militares', in Miguel Alonso Baquer and Luis A. Collado Espiga (eds), *Historia de la infantería española*, 5 vols. (Madrid: Ministerio de Defensa, Secretaría General Técnica, 1993–2001), III. 47–79; José Izquierdo, José Ramón Ortiz de Zárate, and Ángel Aparicio, *La Academia General Militar. Crisol de la oficialidad española* (Saragossa: Institución 'Fernando el Católico', 2011), pp. 17–41.

2. Ministerio de la Guerra, *Colección Legislativa del Ejército 1893* (Madrid: n.p., 1893) (hereafter *CLE plus the year of publication*), Item No. 33, Introduction; Daniel R. Headrick, *Ejército y política en España (1866–1898)* (Madrid: Tecnos, 1981), p. 240.
3. *CLE 1893*, Item No. 33, Introduction, Articles 7, 8, 10, 11, 13, 17.
4. *CLE 1893*, Introduction; Julio Busquets, *El militar de carrera en España*, revised edition (Barcelona: Ariel, 1984), p. 80; José Fernando García Moreno, 'Servicio militar universal y personal en España. Procedencia social y formación del soldado, clases profesionales y oficial de carrera en el Ejército de Tierra español de 1913 a 1936', 5 vols. (doctoral dissertation, Universidad de Zaragoza, 1985), II. 588–9, 598, 651–9, 863–4.
5. *CLE 1893*, Item No. 33, Articles 7, 8.
6. Busquets, *El militar*, p. 81.
7. Enrique Ruíz-Fornells and Alfredo Melgar Mata, *Organización militar de España y algunas potencias extranjeras*, 9th edition (Toledo: Rafael Gómez Menor, 1901), p. 119 footnote 2; Busquets, *El militar*, p. 118.
8. Archivo General Militar de Segovia, Fondo del Ministerio de la Guerra, 2ª Sección, División 3ª, Legajo 183 (hereafter AGMS No. Sección/No. División/No. Legajo), Inspección General de Ingenieros, Secretaría to Ministro de la Guerra, 24 May 1892.
9. For a heated and thorough criticism of the Spanish military academies in the early years of the twentieth century, see El Capitán Equis [pseud. of Nazario Cebreiros], *El problema militar en España. Apuntes para un estudio sincero y al alcance de todos*, 2 vols. (Burgos: Saiz y Compª, 1916–1917), II. 41–78.
10. *CLE 1905*, Item No. 200, Article 6; Antonio Cordón, *Trayectoria (Recuerdos de un artillero)* (París: Librairie du Globe, 1971), p. 23.
11. García Moreno, 'Servicio militar', II. 750.
12. *CLE 1896*, Item No. 163.
13. Miguel Campins Aura, 'La Academia General Militar de Zaragoza y sus normas pedagógicas', non-commercial edition (n.p.: [Centro de Ayudas a la Enseñanza de la Armada, 1983]), p. 28. Campins himself was commissioned at 18 after an abridged course of study which lasted just one year. On Campins and his career, see Manuel Touron Yebra, 'El General Miguel Campins y su época (1880–1936)', 2 vols. (doctoral dissertation, Universidad Complutense de Madrid, 1992).
14. Campins, 'La Academia General Militar', pp. 28, 42, 47.
15. *Reglamento orgánico para las Academias militares de Infantería, Caballería, Artillería, Ingenieros y Administración Militar* (n.p: n.p., 1924 [originally issued in 1897]), Article 86. Actually the cavalry, engineer, and quartermaster academies did not implement a boarding

regime until the reform of 1927: García Moreno, 'Servicio militar', II. 677–8.

16. Archivo General Militar de Madrid, Colección Adicional de Documentos, Legajo 8, carpeta 5 (hereafter AGMM-CAD No. Legajo/ No. carpeta), Revista de Inspección al Ejército: Capitanía General, 5ª Región Militar, 1907.
17. Capitán Subrio Escápula, 'Reformas militares', *Revista Científico-Militar* (hereafter *RCM*), 26 (1901), pp. 134–6.
18. Comandante Beta [pseud. of José García Benítez], *Apuntes para historiar tres años de reformas militares (1915–17)* (Madrid: Imp. de Bernardo Rodríguez, 1917), p. 128.
19. *CLE 1894*, Item No. 293; Cordón, *Trayectoria*, p. 23; Busquets, *El militar*, p. 80.
20. AGMS 2/3/123, Revista de inspección, Reformas que se proponen en las Academias y Colegios Militares, 17 June 1907.
21. Critón [pseud. of Ricardo Burguete Lana], *Hágase Ejército. Infantería, Caballería, Artillería. Estudios tácticos al alcance de todos* (Barcelona: Luis Tasso, 1899), pp. 41–4; Efeele [pseud. of Francisco Larrea y Liso]: *El desastre nacional y los vicios de nuestras instituciones militares* (Madrid: Publicaciones de los Estudios Militares, 1901), pp. 291–5; Eduardo Gallego Ramos, *Proyecto de reorganización y mejora del ejército de tierra* (Guadalajara: n.p., 1905), p. 17.
22. Efeele, *El desastre nacional*, pp. 243–4. The contents of this syllabus are listed in Touron, 'El General Miguel Campins', I. 78–81.
23. Modesto Navarro, 'Lo moral en lo militar', *Revista de Infantería*, 1, 2 (1901), pp. 65–6.
24. Ricardo Burguete, *La guerra y el hombre. Psicología de las tropas* (Madrid: Sáenz de Jubera Hermanos, 1911), p. 70. A graduate of the General Military Academy, Burguete (1871–1937) was commissioned in the infantry and fought in Cuba (where won the San Fernando Laureate Cross, the Spanish army's highest award for gallantry), the Philippines (1896–1898), and Morocco (1909–1913), where he later became high commissioner (1922–1924); in the last stage of Alfonso XIII's reign, Burguete displayed an ambiguous conduct which turned a court martial against Republican leaders he chaired in 1931 into an anti-monarchist propaganda success; he went into retirement after the proclamation of the Second Republic and shifted towards leftist and pacifist stances during his last years. Burguete's intellectual thought is analyzed in detail in Geoffrey Jensen, *Cultura militar española. Modernistas, tradicionalistas y liberales* (Madrid: Biblioteca Nueva, 2014), Chaps. 2–4; this work is a revised Spanish edition of Geoffrey Jensen, *Irrational Triumph:*

Cultural Despair, Military Nationalism, and the Ideological Origins of Franco's Spain (Reno: University of Nevada Press, 2002).

25. AGMS 2/8/152, Capitán [Eduardo] Herrera de la Rosa, Impresiones recogidas de la campaña ruso-japonesa con el ejército del general barón Nogui, p. 19, 30 November 1905.
26. Equis, *El problema militar*, II. 58.
27. Corelli Barnett, 'The Education of Military Elites', *Journal of Contemporary History*, 2, 3 (1967), p. 25.
28. Juan Avilés, 'Enseñanzas de la guerra del Rif', *RCM*, 35 (1910), pp. 371–2.
29. Alfredo Kindelán y Duany *Mis cuadernos de guerra* (Madrid: Plus Ultra, c. 1947), pp. 207–8. Kindelán (1879–1962) was commissioned in the engineer corps (1899); a pioneer of the Spanish military aviation, he became chief officer of the army air service (1926); a staunch monarchist, Kindelán left the service in 1931, but he returned to command the Nationalist air force throughout the Civil War; he asked unsuccessfully for the return of the monarchy in the 1940s and finished his career half-ostracized.
30. Carlos Martínez de Campos y Serrano, *Ayer*, 2 vols. (Madrid: Instituto de Estudios Políticos, 1946–1970), I. 28–9. Martínez de Campos (1887–1975), commissioned in the artillery (1908) and a graduate of the staff college (1918), held assignments in the General Staff and as military attaché in Japan and Italy; during the Civil War, he was artillery commander of the Nationalist Navarrese Brigades and Army of the North; after the conflict, he was chief of general staff (1939–1941).
31. Cordón, *Trayectoria*, pp. 29–30. Commissioned in the mid-1910s, Cordón (1895–1969) left the service under Azaña's volunteer retirement decree (1931) and became a Communist; he rejoined the army on the Republican side in the Civil War, held senior staff assignments on the Andalusian and Aragonese fronts and became Ministry of Defence undersecretary (1938) and chief of staff of the Eastern Region Army Group, and was raised to general officer rank in February 1939.
32. Cordón, *Trayectoria*, pp. 25, 37. But even in Toledo a number of students were allowed external lodging.
33. Beta, *Apuntes*, pp. 130, 135; Campins, 'La Academia General Militar', p. 109.
34. Efeele, *El desastre nacional*, pp. 300–6; Gallego, *Proyecto de reorganización*, p. 18.
35. See, for instance, the conclusions of the 1902 and 1923 reviews of British military education: Alan Shepperd, *Sandhurst. The Royal Military Academy Sandhurst and its Predecessors* (Feltham: Country Life Books,

1980), pp. 102–4, 133–4. And rote learning was a long-established (and deplored) feature of the West Point military academy in the United States: Edward M. Coffman, *The Regulars. The American Army, 1898–1941* (Cambridge, MA: Harvard University Press, 2004), p. 147; Jensen, *Cultura militar*, pp. 167–8.

36. Critón, *Hágase Ejército*, p. 46; León Fernández Fernández, *De la enseñanza militar* (Madrid: n.p., 1907), pp. 32–3; Manuel Espadas Burgos, 'La Institución Libre de Enseñanza y la formación del militar español durante la Restauración', in Congreso de Historia Militar (1°. 1982. Zaragoza), *Temas de Historia Militar*, 2 vols. (Madrid: Servicio de Publicaciones del Estado Mayor del Ejército, 1983), I. 498–9.
37. *Diario Oficial del Ministerio de la Guerra* (hereafter *DOMG*), 25 July 1904; Miguel Alonso Baquer, *El ejército en la sociedad española* (Madrid: Ediciones del Movimiento, 1971), pp. 237–8.
38. *DOMG*, 19 August 1905.
39. AGMS 2/8/152, Capitán [Eduardo] Herrera de la Rosa, Impresiones recogidas de la campaña ruso-japonesa con el ejército del general barón Nogui, p. 19, 30 November 1905; another instance of support for a single academy is Miguel A. Moreno y Álvarez, 'Instrucción y reclutamiento de la oficialidad', *RCM*, 36 (1911), pp. 260–1.
40. Capitán Subrio Escápula, 'El Colegio General Militar', *RCM*, 30 (1905), p. 43.
41. *CLE 1911*, Item No. 240.
42. *CLE 1911*, Item No. 240, Introduction.
43. *CLE 1911*, Item No. 240, Articles 1, 28.
44. *CLE 1911*, Item No. 240, Introduction, Articles 2, 25.
45. *CLE 1911*, Item No. 240, Articles 6, 30, 31.
46. García Moreno, 'Servicio militar', II. 668.
47. See, for instance, *CLE 1912*, Item No. 97.
48. García Moreno, 'Servicio militar', II. 708–10.
49. García Moreno, 'Servicio militar', II. 714–18.
50. García Moreno, 'Servicio militar', II. 696–8.
51. García Moreno, 'Servicio militar', II. 693.
52. *CLE 1918*, Item No. 37.
53. García Moreno, 'Servicio militar', II. 720–7.
54. García Moreno, 'Servicio militar', II. 727–31, 733–5, 741–7.
55. García Moreno, 'Servicio militar', II. 751–2.
56. The memorandum is titled 'Crítica de algunos aspectos interesantes del plan de estudios y del régimen escolar de la Academia de Infantería' and reproduced in Alberto Guerrero Martín, 'Análisis y trascendencia de la Colección Bibliográfica Militar (1928–1936)' (doctoral dissertation, UNED [Universidad Nacional de Educación a Distancia], 2015),

pp. 341–57. Vicente Rojo (1894–1966) was commissioned in the infantry (1914), served in Morocco, taught at the Academy of Infantry in the 1920s, passed the staff college in the early 1930s and was a co-publisher of *Colección Bibliográfica Militar*, a successful series of professional literature, from 1928 to 1936; despite his conservative leanings, he was a loyalist in the Civil War; chief of staff in the defence of Madrid in late 1936; as chief of army general staff after May 1937, Rojo planned the main Republican offensives.

57. Rojo's analysis and quotations are taken from Guerrero Martín, 'Análisis y trascendencia', pp. 341–2.
58. García Moreno, 'Servicio militar', II. 693.
59. Equis, *El problema militar*, II, pp. 43–4, 62–3, 71–8.
60. Gallego, *Proyecto de reorganización*, p. 18. As set in the 1897 regulations, the time limit for a tour of teaching duty were six years, or eight if the teacher wrote and got approved a textbook for his subject within the first three years after his appointment (as a comparison a German officer could not hold a teaching assignment for over five years): *Reglamento orgánico*, Article 6; Luis de la Gándara Marsella, *El oficial alemán de la escala activa y de reserva* (Madrid: n.p., 1916), p. 94. The decree of 4 October 1905 reiterated the 1897 rule, later modified by a decree dated 1 June 1911 which set the limit in seven years without further extension; even so, the head of artillery studies at the Segovia academy in 1931 managed to be one of its teachers since 1911, and Vicente Rojo also was a teacher at the Toledo academy from 1922 to 1932: *CLE 1905*, Item No. 200, Article 5; *CLE 1911*, Item No. 109, Article 23; Fernando Puell de la Villa, *Gutiérrez Mellado*Gutiérrez Mellado, Manuel.*Un militar del siglo XX (1912–1995)* (Madrid: Biblioteca Nueva, 1997), p. 71; Guerrero Martín, 'Análisis y trascendencia', p. 167.
61. Jensen, *Irrational Triumph*, p. 68.
62. A[rthur] Houghton, *Les origines de la restauration des Bourbons en Espagne* (Paris: Plon, 1890), p. 13.
63. Cordón, *Trayectoria*, p. 43.
64. The text is taken from Jorge Vigón, *Historia de la Artillería española*, 3 vols. (Madrid: Consejo Superior de Investigaciones Científicas, 1947), II. 134–5.
65. Jorge Vigón, 'Breves notas para la historia de las Juntas de Defensa y de la Dictadura', unpublished manuscript [mimeographed copy filed in Archivo General Militar de Ávila, Fondos del Ejército Popular, armario 47, legajo 73, carpeta 8], p. 3; Vigón, *Artillería española*, II. 134.
66. Vigón, *Artillería española*, III. 240.
67. Equis, *El problema militar*, II. 114. Nonetheless, the artillery officers had debated in the nineteenth century if their corps had to take on a

facultativo (technical) role besides the military (tactical) one: Muro, *El Pensamiento Militar*, I. 421–9.
68. Miguel Alonso Baquer, *Aportación militar a la cartografía española en la Historia contemporánea. Siglo XIX* (Madrid: Consejo Superior de Investigaciones Científicas, 1972), p. 170.
69. A.F.B. y P. de A., *Nuestro ejército. Lo que es y lo que puede ser* (n.p.: S. Fernando, 1923), p. 55.
70. Eduardo Gallego Ramos, *La campaña del Rif (1909). Orígenes, desarrollo y consecuencias* (Madrid: n.p., *c.* 1910), p. 306 footnote 1.
71. Emilio Mola, *Obras completas* (Valladolid: Librería Santaren, 1940), p. 972. Mola (1887–1937) was promoted to brigadier general in 1927 after a successful career in the Moroccan campaigns; director of state security during the last stage of Alfonso XIII's reign; separated from the service by the first republican government, he returned in 1935 under a right-wing cabinet; he became the organizer of the 1936 military rebellion and commanded the Nationalist army in northern Spain until his death in an air crash in June 1937.
72. Gallego, *Proyecto de reorganización*, p. 44.
73. Equis, *El problema militar*, II. 118–19.
74. AGMM-CAD 8/20, Revista de Inspección: Estado Mayor Central del Ejército, 1907.
75. Equis, *El problema militar*, II. 89–100.
76. Epifanio Gascueña, *El principio de la cooperación y enlace de las Armas y nuestros vigentes reglamentos tácticos* (Toledo: Publicaciones del Memorial de Infantería, 1918), pp. 14, 16. Born in 1882, Gascueña volunteered as private in 1900, became a sergeant in 1902, and entered the Academy of Infantry in 1905; after fighting in the Melilla campaign (1909), he passed the staff college; staff corps officer after 1916, he studied in the French staff college (1927–1929); a lieutenant colonel in 1936, the military uprising found him in the loyalist zone; he was unassigned most of the war and surrendered to the Nationalists in April 1939; he was court martialled and dismissed from the service in 1942 after a spell in prison.
77. Gascueña, *El principio de la cooperación*, pp. 16–17.
78. Vigón, *Artillería española*, III. 183.
79. Estado Mayor Central del Ejército, *Enseñanzas de la campaña del Rif en 1909* (Madrid: [Depósito de la Guerra], 1911), pp. 76–7.
80. Andrée Bachoud, *Los españoles ante las campañas de Marruecos* (Madrid: Espasa-Calpe, 1988), pp. 105–6.
81. Jensen, *Cultura militar*, pp. 164–7. A notable instance of this sort of officers (although part of their writings does have intellectual value in their own right) is Antonio García Pérez, an infantry officer and teacher

in his arm's academy; for a biography, see Pedro Luis Pérez Frías, *La vida que fue. Antonio García Pérez—un intelectual militar olvidado* (n.p.: Estudios Especializados, 2015); for an analysis of his thinking, see Jensen, *Cultura militar*, Chap. 5.

82. Andrés Mas Chao, *La formación de la conciencia africanista en el ejército español (1909–26)* (Madrid: Servicio Geográfico del Ejército, 1988), pp. 58, 60–1.
83. Busquets, *El militar*, p. 97.
84. Stanley G. Payne, *Politics and the Military in Modern Spain* (Stanford: Stanford University Press, 1967), p. 154; Gabriel Cardona, *El poder militar en la España contemporánea hasta la guerra civil* (Madrid: Siglo XXI, 1983), p. 36. During the campaign which led the Melilla district forces, under the command of a cavalry general (Manuel Fernández Silvestre), to the disaster of Annual (July 1921), the staff officers played a very small role, if any, in the operational planning and their functions were often reduced to passing on orders: [Juan Picasso González] *Expediente Picasso. Documentos relacionados con la información instruida por el señor general de división D. Juan Picasso sobre las Responsabilidades de la actuación española en Marruecos durante julio de mil novecientos veintiuno*, facsimile edition (Mexico, D.F.: Frente de Afirmación Hispanista, A.C., 1976), p. 311.
85. Michael Howard, *War in European History* (Oxford: Oxford University Press, 1976), pp. 99–101.
86. For a history of the staff corps, see Eladio Baldovín Ruiz, *Historia del Cuerpo y Servicio de Estado Mayor* (Madrid: Ministerio de Defensa, Secretaría General Técnica, 2001), Introduction, Chaps. 1–9. See also Alonso Baquer, *Aportación militar*, pp. 150, 159; Busquets, *El militar*, pp. 189–95; Muro, *El Pensamiento Militar*, I. Chap. 7. The direct recruitment of civilians was due to the shortage of officers applying for the staff college; 47 officers and 74 civilians entered the college in the period 1852–59: Baldovín, *Historia del Cuerpo*, pp. 83–4.
87. Cordón, *Trayectoria*, p. 142.
88. Alonso Baquer, *Aportación militar*, pp. 159–60. In 1921 Picasso (now a major general) headed the commission of inquiry on the Annual debacle, whose proceedings were later known as the *Expediente Picasso* ('Picasso Dossier'); he was an uncle of the painter Pablo Ruíz Picasso.
89. Alonso Baquer, *Aportación militar*, pp. 171, 191. On Weyler's career, see Gabriel Cardona and Juan Carlos Losada: *Weyler, nuestro hombre en La Habana* (Barcelona: Planeta, 1997).
90. Efeele, *El desastre nacional*, pp. 22–3.
91. Fernando Puell de la Villa, 'El general Cassola, reformista militar de la Restauración', *Revista de Historia Militar*, 45 (1978), pp. 180–1.

92. Alonso Baquer, *Aportación militar*, p. 170.
93. Alonso Baquer, *Aportación militar*, p. 170.
94. Alonso Baquer, *El ejército*, p. 196.
95. On the staff college before 1893, see Alonso Baquer, *Aportación militar*, pp. 148–9; Muro, *El Pensamiento Militar*, I. Chaps. 8–9.
96. *CLE 1893*, Item No. 33, Article 27. The students of the intake of 1898 were 20 years old on average: present author's estimate based on data taken from Ministerio de la Guerra, *Anuario Militar de España 1899* (Madrid: n.p., 1899).
97. Equis, *El problema militar*, I. 132–5.
98. Gándara, *El oficial alemán*, pp. 236–7.
99. Alonso Baquer, *El ejército*, pp. 237–8.
100. Equis, *El problema militar*, I. 142–3.
101. Equis, *El problema militar*, I. 109–10.
102. AGMM-CAD 9/1, Memoria del Jefe del Estado Mayor Central, 1907.
103. Enrique Maquieiría, 'Hombres e ideas de la guerra europea', *Memorial de Infantería*, 10, 1 (1921), p. 80; Juan García Benítez, 'Nuevos métodos de enseñanza en la Escuela Superior de Guerra francesa', *La Guerra y su preparación*, 6, 1 (1921), p. 133.
104. *CLE 1893*, Item No. 33, Article 32.
105. Equis, *El problema militar*, I. 137–8.
106. Equis, *El problema militar*, I. 125, 138–9, 159–62. The fear caused among the staff corps officers by the competition of the *diplomados* is voiced in Pío Suárez Inclán, *Organización del Cuerpo de Estado Mayor, 1810–1910* (Madrid: Talleres del Depósito de la Guerra, 1912), p. 192.
107. Equis, *El problema militar*, I. 160.
108. AGMM-CAD 9/1, Memoria del Jefe del Estado Mayor Central, 1907.
109. Archivo General Militar de Ávila, Fondos del Ejército Popular, armario 47, legajo 73, carpeta 7a, Captain Nicolás Benavides to Junta Superior de Defensa del Cuerpo de Estado Mayor, 2 January 1918.
110. The *adjudantur*, in the Prussian–German military system, was a service, separate from staff duties, which handled the administrative affairs in a major command headquarters, thus allowing the staff officers to focus on operational and logistical matters: *Dictionnaire militaire. Encyclopédie des sciences militaires*, 3 vols. (París and Nancy: Berger-Levrault, 1898–1910), *sv* 'Adjudantur'.
111. Beta, *Apuntes*, pp. 64–5.
112. Beta, *Apuntes*, pp. 70–1.
113. Beta, *Apuntes*, p. 72.
114. Douglas Porch, *The March to the Marne. The French Army 1871–1914* (Cambridge: Cambridge University Press, 1981), p. 58.

CHAPTER 3

Firepower and Infantry Tactics, 1899–1913

The development of increasingly effective firearms was transforming warfare in the late nineteenth century and the early years of the twentieth century. This chapter describes and assesses how the Spanish military coped with the new conditions by looking at the evolution of tactical doctrine for the infantry and the performance of this arm in the campaign of Melilla (1909).[1] Although all the tactical arms were affected by this transformation of warfare, the infantry was the largest one and illustrative enough of the military's predicament in this period.

Firepower and the Problem of the Offensive

The Impact of New Firearm Technology on the Battlefield

The Spanish army was not ignorant of the changes that technology had brought to warfare in the last decades of the nineteenth century. Weapons like quick-firing artillery, the magazine rifle, and the machine gun delivered so much firepower on the battlefield as to transform its traditional features.[2] A shrewd observer of technological evolution, Colonel José Marvá, described the change in this way at the beginning of the twentieth century:

> Fire's extraordinary accuracy and destructive energy force the masses to disperse and look for protections which no longer can be found in the metal breastplate or behind the wooden fence. The entrenchments have

J.V. Herrero Pérez, *The Spanish Military and Warfare from 1899 to the Civil War*, DOI 10.1007/978-3-319-54747-3_3

> had to leave their artistic profiles...; the host hides itself in the ground folds, and the skirmishing line disappears from sight...[3]

A distinguished military essayist, Lieutenant Colonel Carlos Banús, elaborated on this point. Due to the increased range of the rifle, infantry troops had to adopt combat order at a distance of 1500 metres or more from the enemy. Moreover, this order had turned from closed ranks, elbow to elbow, into dispersed lines of skirmishers. Thus the soldier was no longer under the close supervision of his officer, and this situation had negative effects on his morale. On the other hand, the advance under fire was slow and, since the distance to cover became longer, the fighting was more prolonged. The longer the battle and the more difficult the advance, the easier it was for the attacker to get demoralized and call off the assault. So the offensive had become a very difficult operation and frontal attacks would be almost impossible—unless combined with enveloping ones.[4]

Nevertheless, the contemporary professional literature, while recognizing the growing importance of firepower, asserted that armies could not remain motionless shooting at each other. Though the value of the bayonet had diminished, the attacking side could not do without the pursuit of a clash if it wanted to complete the work of firepower and defeat the enemy.[5] This meant taking the offensive whenever possible, despite the difficulties caused at the tactical level by the improvement of weapons, in order to impose one's own will on the enemy. And since the major wars from the last third of the nineteenth century to the outbreak of the First World War were won by the side which displayed the most offensive spirit and plans, the military doctrines in vogue all over Europe in this period emphasized the offensive.[6]

The Spanish Army and Its War Experiences, 1873–1898

Although Spain was not involved in any major international conflict after the Napoleonic Wars, the Spanish army had experienced the effects of the growing power of firearms in the Third Carlist War (1872–1876).[7] General Pedro Ruíz Dana observed the same year the war was over how modern firearms gave to the tactical defensive and field fortifications an importance they had lacked before. The Carlist troops in the Basque-Navarrese theatre of operations had learnt to make an extended use of field fortifications, and, according to Ruíz Dana, only a shortage of offensive spirit prevented them from turning their defensive successes into major victories.[8] Actually the Carlist forces suffered many casualties from rifle fire in the combat of Puente la Reina in October 1873, during the beginnings of the regular operations in the North, when they threw themselves in packed battalion columns against the rear guard covering the withdrawal of the government army.[9] So they fought in open order during the battle of Montejurra (November 1873). Moreover, since they fought defensively, the Carlists took advantage of hedges and fences to build earth and stone breastworks. However, the high profile of the works made them easy targets for the government artillery. In December, in the combat of Velabieta, the Carlist defenders dug ditches and raised small breastworks with the removed earth; but the ditches were too wide (1.5 metres) and the breastworks still offered a clear target to the artillery.

In the battles of Somorrostro (February–March 1874) the Carlists' whole battle front was covered by a line of trenches, linked by redoubts, and very low turfs replaced the breastworks in order to reduce the profile. Nevertheless, the artillery still took a heavy toll of the Carlist defenders (almost half of their casualties in the fighting from 25 to 27 March was due to the government cannon).[10] The experience of the combats of February and March proved to the Carlists the need to improve their field fortifications. The trenches became deeper (almost a man's height) and narrower (half a metre): thus very few shells got into the trench and the defenders could avoid the fragments of the rest by crouching. The artillery also had a less clear target since the removed earth was now scattered. This new model trench was used in the battle of Estella (June 1874), but instead of forming long, continuous lines the Carlists dug shorter trenches (15–20 metres); these also had hook-like ends which allowed troops to avoid enemy fire in case of evacuation. Their success led the Carlists to protect the access routes to the territory under their control with lines of this sort of field fortification, which obliged the government forces to carry out major operations of strategic envelopment (the Carlist army did not have enough troops to man permanently all the fortified sectors).[11]

Besides improving their fortifications, the Carlist defenders polished their tactical skills. They occupied only the trenches directly threatened by the enemy. If the shelling was too heavy, they moved in single file, taking advantage of the ground, to other trenches in the flanks and the rear. When the cannonade stopped to allow the infantry's advance, the defenders came back quickly and fired once the enemy troops came within easy range (300–400 metres). These defensive tactics were possible because the attacking infantry, in order not to expose themselves too much to the enemy fire, had to start the advance from a distance of 1000–1500 metres from the trenches.

The experience of this war was taken into account when new infantry tactical regulations were issued in 1881, since they stated: '[B]ecause of the range, the accuracy, the flat trajectory and the speed of fire of the current rifle, fire is the main, almost exclusive way of fighting for the troops in the front line.'[12] However, after 20 years of peace, the memories of those battles had probably faded when the Spanish army again went into action in the colonial conflicts of Cuba and the Philippines after 1895 and the war with the United States in 1898.

Since this last war was fought in far-off islands and the naval superiority of the United States was overwhelming, Spain could not aspire to overall victory. However, it seemed reasonable to think that a successful defence on land could make up for the naval defeats and secure better terms at the negotiating table. The commander-in-chief in Cuba, General Ramón Blanco, followed a strategic defensive to preserve key places, which would be the bases for operations in the field against any American expedition.[13] On paper, there was no shortage of soldiers in Cuba at the outbreak of war: 196,000 troops (including 41,000 local volunteers), of which 34,000 were deployed in the Santiago de Cuba province.[14] However, this force was not all available for operations against an invading force since its numbers included many men in hospitals and a lot of troops on garrison duties. Another handicap was that this colonial army had fought against guerrillas since 1895 and lacked preparation for regular warfare: the Spanish field artillery in Cuba was weak, there were not enough transport means to move large forces, and the troops were unused to fighting against a regular army (though any fighting experience was better than no experience at all).[15]

An American expeditionary corps (17,000-strong) landed in late June 1898 near Santiago de Cuba, a major city and harbour in southeastern Cuba, where a Spanish naval squadron under Admiral Pascual Cervera had taken refuge. The American force did not encounter serious resistance until it reached the external defences of Santiago. About 1700 Spanish troops barred the way to the city from the east. 1200 Spaniards were entrenched on San Juan Hill (Lomas de San Juan); the rest occupied the fortified hamlet of El Caney. On 1 July, the Americans attacked both positions. 6600 troops needed most of the day to drive the stubborn defenders out of El Caney. The other assault force (8400-strong) took a shorter time to seize the trenches on San Juan Hill, but the American troops suffered so heavily (nearly a thousand casualties) that they were unable to carry on. The Spanish troops lost about 600 men; the American casualties amounted to almost 1400 men. After these clashes, the fighting was limited to some skirmishing (the Americans actually reckoned their position so weak as to think of a withdrawal to the beachhead). Meanwhile, Cervera's squadron was destroyed in front of Santiago de Cuba when it tried to escape on 3 July. The Spanish command no longer saw purpose in defending the city and the garrison capitulated on 17 July.[16]

The Debate on the Tactical Lessons of the 1895–1898 Wars

The few Spanish officers who reflected on the combats of the Spanish–American War concluded that one factor in the Spanish defeat had been the lack of offensive spirit and the overconfidence of some commanders in the strength of defence.[17] Such criticism was not limited to military voices. The journalist and essayist Damián Isern also charged the Spanish senior commanders with lack of initiative. General Blanco, for instance, allegedly remained passive when Santiago de Cuba was threatened and failed to reinforce its defenders.[18] As for contemporary military critics, one of them thought that the local command in Santiago suffered from lack of resolution: due to its concern for protecting all the perimeter, it had dispersed its forces instead of keeping them concentrated to go to the aid of the threatened sector.[19] Major Ricardo Burguete, writing under the pseudonym Critón, voiced angrily the feeling that the opportunity to achieve a major success had been missed by the fainthearted Spanish command:

> Who ever saw a campaign start by giving the troops such a passive role as that given to ours? Confined to the perimeter of trenches after three days of gallant defence, after receiving reinforcements [Burguete referred to a column of 3,000 men arrived from other places of the province], watching the symptoms of discouragement and exhaustion in the attacking troops, why was the flag of truce hoisted when the smallest offensive demonstration would have been enough to bring about the withdrawal arranged beforehand by the Americans?[20]

The essayists agreed in criticizing the Spanish senior command's conduct of the operations against the Americans. However, there was no similar agreement when the tactical teachings of the campaigns of the late 1890s as a whole were discussed in the professional literature.

One line of thought stated that the Spaniards had put too much reliance on firepower, which had led them to a lack of offensive drive. Major Ricardo Burguete—who had fought with distinction against the Cuban insurgents, winning the highest award to gallantry—blamed the current tactical regulations for fostering a spirit of excessive caution and hindering the exploitation of success after seizing enemy positions. Though the Spanish troops had rarely been defeated by the Cuban rebels on the battlefield, the former limited themselves to carrying on firing while the latter made good their escape, so the fruits of victory were wasted.[21]

This view was not accepted by other officers who thought the Spanish troops had not displayed lack of aggressiveness—quite the contrary. They achieved tactical victories continuously, since they always took the offensive in the combats in the bush against the insurgents, even if the fire preparation was short. Actually, these essayists said, the nature of the conflicts and the Spanish conscripts' poor training in musketry and skirmishing made the development of real fire tactics like in regular warfare impossible. According to Colonel Modesto Navarro, the usual tactical procedure was to form in line, open a heavy fire (which wasted a lot of ammunition), and charge with the bayonet after a very short time. Any excess of caution or lack of boldness imputed to the troops was a reflection of their officers' qualities: whenever the commanders were bold and aggressive, the troops followed them.[22]

Despite this evident disagreement on tactical doctrine, the Spanish army seemed to push aside any debate in depth about the military teachings of the overseas campaigns. This attitude may have been caused by a desire to forget the painful experience of defeat in the war against the United States—and, by extension, everything related to the wars of 1895–1898. Major Burguete complained that the scorn of irregular warfare and minor tactics allegedly felt by some sections of the military had helped the army to disregard the experience of these conflicts.[23] Nonetheless, such an attitude was not unique to the Spanish officer corps. For instance, the institutional culture of the American army in the nineteenth century focused on regular warfare despite spending far more time in irregular campaigns against American natives. In any case, implementing effective reforms in large bureaucratic institutions is usually difficult, and the task is harder in armies because military officers are instinctively conservative and distrustful of radical innovation.[24]

Bayonet Versus Firepower

The Value of Cold Steel in Modern Warfare

The disagreement on the tactical lessons of the overseas campaigns was linked to a more general debate on the ways to cope with the new conditions on the battlefield. To be more accurate, the issue to be solved was how to carry out the tactical offensive when weapon technology made the defensive stronger. An answer was to emphasize speed and the psychological effects of the bayonet. The foundation of this doctrine was that if defensive fire was always more powerful, then the bayonet charge was the only means through which the offensive proved its superiority. According to this doctrine (which found in Major Ricardo Burguete a very outspoken supporter), the best protections against fire were speed and mobility. The aim of the offensive was to cover the distance quickly, through an advance without interruptions. Thus the enemy could not aim its weapons accurately against a fast-moving target. The light field fortification and the methodical advance in which the soldier took advantage of the roughness of ground must be dismissed; otherwise the troops would be too exposed to enemy fire whenever they made a halt.[25]

In the eyes of offensive-minded essayists, such as Major Burguete and his brother Manuel (who was an infantry officer as well), fire in the offensive was seen more as a liability than as an asset: it delayed the advance and diminished the will to get closer to the enemy. Moreover, since most contemporary armies were made up of short-service conscripts, it was not possible to give them a good military training. If the tactics of gradual advance were used, these troops would end up by staying still behind any cover available and wasting ammunition in a useless, long-range firefight. So they had to be carried to the final assault line as soon as possible. Once there, they would be unable to wait to defeat the enemy through fire, but they would feel a compulsion to charge and destroy the enemy with the bayonet. More training in the use of this weapon would increase the soldiers' offensive spirit and self-trust. Ricardo Burguete even argued that the colonial campaigns (including the Spanish ones in Cuba and the Philippines) had exposed the inferiority of the Europeans in close combat, faced with peoples used to fight with stabbing weapons, because the former usually resort to defensive squares. But he forgot to add that the European soldiers' alleged softness did not prevent them from slaughtering their more robust, offensive enemies through the firepower of modern quick-firing rifled weapons.[26]

A slightly different approach to this doctrine was taken by Captain Ángel Dolla, of the cavalry corps. He argued that firearms were the more effective the closer they were to the enemy. Therefore, in order to increase their effects, they had to be moved forward—thus movement became the decisive factor of the combat. Dolla also argued that the advance provided moral fortitude as a bonus. The more resolute and faster the advance, the greater the morale boost. Moreover, according to Dolla, what had happened in the performance of modern weapons was an increase in their moral effects due to the fear they roused. And this fear increased as the distance from the enemy diminished. The firearms' effectiveness depended more on how long the troops were exposed to their shots than on the volume of their fire per unit of time. So the decision of the fighting had to be achieved as soon as possible through the advance.[27] This argument was rather fallacious, since it underrated the importance of improvements in modern weapons' range and rate of fire. Actually, the effectiveness of the modern quick-firing rifled weapon was based on its ability to inflict damage from longer distances sooner. When weapons could kill in a few minutes as many men as they did in one hour a hundred years before, saying that the longer the troops stood under fire, the more they suffered, was just a truism.

The arguments of the advocates of the doctrine of *offensive à outrance* echoed partly the thinking of the French military essayist Charles Ardant du Picq. He had stated that the soldiers' morale diminishes as they lose confidence in the effectiveness of their weapons to beat the enemy from afar. So if the firearms became more perfect, the advance under fire would have a greater moral effect on the enemy and victory would go to those who marched forward in the most orderly and resolute way.[28] Although the Spanish sources surveyed do not allow one to trace a clear, direct link back to the work of Ardant du Picq, it is plausible to think of it. A Spanish translation of Ardant du Picq's *Etudes sur le combat* had been published in 1883, so he was not a totally unknown essayist. On the other hand, Ardant du Picq's ideas could have been known second-hand. French tactical thinking after 1870 was partly inspired by Ardant du Picq (although essayists usually misread his writings). And there is some evidence suggesting that French military thinking was more likely to be known in

Spain due to reasons of geographical and cultural closeness.[29] Therefore, the Spanish officers were more likely to be influenced by French (and thus by Ardant du Picq's) ideas than by other foreign thinking.

Firepower as the Foundation of Tactics

Facing the doctrine of *offensive à outrance*, there were those who followed ideas more in tune with the official doctrine after 1881. If the strength of the defence lay in firepower, the offensive had to apply even more firepower to overcome the former. The attacking troops had to keep firmly in mind the idea of marching always forward, but this did not mean that fire was an ancillary means which they could do without. Colonel Modesto Navarro expressed the idea as follows:

> In the attack one must have the resolution to advance constantly until arriving, if necessary, at the assault of the enemy lines with the bayonet, but arriving quickly at this situation is less important than arriving in a good condition, that is, after the adversary has been broken physically and morally through long range fighting by means of well chosen and duly spaced out fire from position to position.[30]

Navarro also argued that if speed were indeed so good a protection, the cavalry's role on the battlefield would never have diminished (since horse troops are faster than infantry).[31] According to another advocate of firepower-based tactics, Captain Germán Gil Juste, the bayonet enthusiasts overrated the psychological effect of cold steel (for example, seasoned Spanish troops did not shun hand-to-hand fighting in surprise attacks by Philippine insurgents armed with stabbing weapons). On the contrary, the fear of a hail of bullets and shells, which killed in an impersonal and random way, was as great as that of the bayonet. If the defenders overcame the former, they would also overcome the latter, and therefore the alleged moral superiority of advance per se disappeared. So any assault carried out while the defenders stayed in their positions ran the risk of being wiped out.[32]

For advocates of tactics based on firepower such as Captains Gallego and Gil Juste, the bayonet had lost its decisive role in the combat and assaults with cold steel would have no *raison d'être* in future battles. The combat would be opened by means of firepower, which would also carry out the main attack. Under the cover of artillery fire, the attacking troops would be able to advance to suitable positions, where they would start a shooting duel to achieve the fire superiority deciding the outcome of the fighting.[33] Actually Captain Gil Juste, an infantryman, thought that new artillery ammunition, such as melinite shells, would be able to overcome field fortifications and drive the defenders out of them. Thus tactical manoeuvring would be restored and firepower would be the means to prepare or impede the clash, the combat's supreme act.[34]

However, Gil Juste became less sanguine about the presumed effectiveness of such fire three years later and admitted that the tactical offensive was inappropriate against an entrenched enemy. But this did not mean the impossibility of offensive warfare, as stated by Ivan Bloch, a Russian-Polish financier turned military essayist. The offensive could still be carried out through strategic manoeuvre, which would turn the enemy forces out of their positions and force them into encounter battles. In this way, in having to cope with more unpredictable battlefields, the commanders' genius would still play a fundamental role. On the other hand, battles on unprepared ground offered the quick-firing artillery and magazine rifles many more chances to break the enemy, whose rout would be achieved by the cavalry, which was the shock arm *par excellence* and therefore more effective for the task than the infantry's bayonet attacks. So Captain Gil Juste foretold that cavalry would be the only shock arm in the battles of the future.[35]

This argument of an officer who recognized the growing power of firearms and the need to adapt tactics to this fact is noteworthy because it shows the survival of traditional views about the battle (through the attempt to integrate the traditional tactics of the cavalry, an arm whose role on the battlefield was eroded by technological progress). It illustrates how deeply rooted was the idea of the prevalence of the offensive despite the increased power of the defence. Nevertheless, the idea of strategic manoeuvre evaded rather than solved the concrete tactical problem: how to avoid or limit the effects of defensive firepower in the age of quick-firing rifled weapons.

The Evolution of Infantry Tactical Regulations

The Theoretical Debate on Tactical Regulations, 1899–1913

The rapid changes in warfare and the adoption of the Mauser magazine rifle forced the Tactics Commission to update the Spanish infantry tactical regulations after 1890. But the new regulations were not approved until April 1898.[36] Thus the Spanish troops fought the wars of 1895–1898 following the 1881 infantry tactical regulations. These regulations considered firepower as the principal means to overcome the growing strength of defence. The skirmishing line (called *guerrilla* in the Spanish military terminology) had become the main fighting echelon because of its fire action, dropping its old role as the screen for the assault columns.[37]

The 1898 regulations preserved the essential principles of the previous ones.[38] Fire was 'the main fighting element of the infantry'.[39] Even more, the regulations were based on the conviction that contemporary infantry tactics came down to achieving fire superiority.[40] This was a cause of regret for advocates of shock action like Major Burguete. According to him, the spirit that had inspired the Spanish regulations of the late nineteenth century had been influential in the failures of the overseas campaigns in the late 1890s.[41]

Burguete pointed out that, despite the importance given to firepower, the official regulations did not specify what sort of fire was the most effective, either volleys or individual fire. He thought that the regulations overstated the tactical importance of the possession of ground. The most unwise were the articles recommending a thorough use of ground to get cover during the advance: if they were followed, individualism would kill any collective action since every soldier would advance his own way. It was the task of commanders rather than individuals to decide when to go to ground. The exaggerated concern for protection in the regulations was harmful for the offensive spirit.[42] But Burguete seems to overstate his case as well. The 1898 regulations actually stated that the attacking infantry 'must be more worried about advancing quickly and in an orderly way without heavy losses than about delivering fire of a kind which will not achieve real effects except in the last stages of combat…'.[43]

Burguete blamed the official doctrine's ambivalent spirit for an overcautious approach. Even when talking about the assault, the regulations always bore in mind a possible repulse and the ways to reach safety in the rear, rather than emphasizing the decisive character which the attack must reveal. The 1898 regulations, in the parts dealing with the assault at company and platoon level, certainly did not encourage in junior commanders the idea of pushing far forward. For instance, after the paragraph describing the way the platoon assault had to be carried out, the next one stated that, if the assault was successful, the enemy was to be pursued through fire alone and the platoon would redeploy quickly in order to secure the conquered position against a counterattack. And a third paragraph—which was longer than the first one—explained how the platoon had to retreat in case of failing in the assault. The paragraphs about the company assault were quite similar. According to another essayist, the regulations' warnings about keeping reserves whatever the circumstances were inspired by overcaution, as if the prevailing aim was to avoid setbacks. Thus it was impossible to achieve any decisive success and destroy the enemy, since exploitation would be timid at best.[44]

If Burguete complained of the excessive fear of enemy fire, Captain Gil Juste, on the contrary, thought that the tactical regulations underrated the effects of defensive firepower during the bayonet charge. If the regulations were applied literally, the troops would sometimes stand up in close order in the open at 300 metres from the enemy lines and they were supposed to act in an orderly way in this situation.[45] Another essayist criticized the extent of the 1898 regulations as well. Their 700 pages were compared unfavourably with the approximately 150 ones of the German or the French regulations. The only explanation for this was an excess of detail on matters of no use on the battlefield.[46] But the writers of the regulations argued that the lack of training grounds and major manoeuvres deprived the Spanish officers of the means to exercise their mastery of tactical procedures.[47] That is, it made little sense to replace detailed instructions for tactical evolutions on the battlefield with broad guidelines about the principles inspiring the official tactical doctrine when the officers had few opportunities to put these principles into practice.

Burguete did not limit himself to criticizing the current regulations. He wrote an alternative draft of infantry tactical regulations by 1900 and submitted it to the Tactics Commission. Burguete's regulations were tested by a full-strength company in the *El Rey* Infantry Regiment 1.[48] The basic point of the project was to divide the infantry platoon into four squads arranged in files ten to 15 men deep each. With the squad leaders ahead, the platoon would move in this formation on the battlefield. It would overcome obstacles by increasing the distance between squads, which would find their own way. Any obstacle which the platoon passed through in skirmishing line order could be overcome by this deeper formation and with no loss of the advantages of a closer order. Once it reached a position to rest or shoot, the platoon could deploy in line. The project was inspired by Burguete's ideas of *offensive à outrance* and predominance of speed and shock over fire, explained above.

The commanding officer of *El Rey* Regiment praised the offensive spirit, but he reckoned that officers had to know its limits as well. Anyway, he thought that it was better to avoid any reference to caution in the regulations; otherwise the troops' morale might be harmed.[49]

Burguete's regulations were criticized at length by Colonel Navarro. He argued that they contained an unbalanced approach to warfare by placing so much emphasis on the offensive; the defence remained an important form of warfare and had to be dealt with in the regulations, since it might be forced upon one's own forces. There were also contradictions in the project: despite his cult of the bayonet, Burguete recognized on several occasions the importance of fire, both in the defence and in the attack. So Navarro wondered what the point was in persuading the troops that cold steel was supreme. He agreed that sometimes the troops had to close with the enemy in order to win, but the idea of a clash with the bayonet had become a moral concept rather than a material one. The trick was to convince the enemy that a soldier able to advance under fire would also be able to kill him with the bayonet, if the former waited in his positions for the latter. Fire was the means to break the will to resist such advance. If the destruction of the enemy was not mainly physical (the percentage of casualties had diminished in modern battles), the only tangible proof of having achieved moral superiority and victory was therefore the seizing of ground. The latter thus had a real tactical importance, contrary to what Burguete thought.[50] Navarro pointed out that, despite the talk of uninterrupted movement, Burguete actually adopted successive advances from position to position; this was unavoidable if the enemy had to be broken by fire. Moreover, the extended modern battlefields, the limits of the infantryman's stamina, the ruggedness of terrain, and, of course, the enemy's fire forced troops to make halts during the advance; speed alone was no protection.[51]

Another criticism was that Burguete's deep formations did not offer any substantial improvement when facing modern firearms. Although these would be given more difficult frontal targets, their ballistics made it easier to hit in depth. Moreover, such formations got soldiers used to seeing the man ahead as a sort of shield, whose cover they would be unwilling to leave. Another flaw in Burguete's regulations was the excessive exposure of the squad leaders; morale was harmed when they became casualties.[52]

The sources consulted do not give any information on the official assessment after testing Burguete's regulations. However, a disappointed

Burguete wrote a decade later that his project led 'a misleading, ill-fated official life'.[53] This suggests that his views were too radical to be accepted by the Spanish army at that time.

The Alleged Lessons of the Russo-Japanese War: Field Fortifications and the Rise of Morale Factors

The Russo-Japanese War (1904–1905) gave an opportunity to observe closely the effects of state-of-the-art firearms in regular warfare.[54] Burguete took advantage of this conflict to claim the soundness of his own tactical theories, by stating that they had been adopted successfully by the Japanese army.[55] He praised the aggressive Japanese attitude, while the Russians looked more interested in the possession of positions alone. However, Burguete himself described how 'every inch of conquered ground was a victory [for the Japanese] and a support to carry on advancing. Every soldier—with a sandbag on his back—turned the bit of ground which he conquered into a formidable redoubt…'.[56] This does not seem exactly the same thing Burguete suggested when he talked in his tactical project about advancing without interruptions and the curbing effect of the use of the ground.[57] Indeed, a few years later, Burguete would envisage the offensive combat as the action of a shooting line, from which squads would move forward in order to establish more advanced fire bases. He described this sort of advance as methodical and slow, taking advantage of folds in the ground.[58]

Burguete's writings are also a token of the interest about the use of field fortifications raised by the Russo-Japanese War. Actually some Spanish essayists had already displayed a similar interest around 1900. Given the predominance of defensive firepower, which caused more frequent halts during the advance, Captain Gallego thought that resorting to field fortifications was—besides dispersed fighting orders—the solution to the problem. These fortifications gave a degree of cover against rifle fire and shelling by making use of ground. They could also be used as rallying points if the assault was repulsed or as a line of resistance against counterattacks.[59] Other essayists reckoned that the experiences of the Boer War (1899–1902) endorsed this use of field fortifications and showed that the infantryman had to fight as much with the shovel as with the rifle and had to be trained accordingly in the use of both tools.[60]

Based on the experiences of the Russo-Japanese conflict, several Spanish essayists discussed the importance of using the ground or, more accurately, the light field fortification, in the attack. Two teachers of the Academy of Infantry's faculty wrote that the Russo-Japanese War had proved the poor effectiveness of field artillery against entrenched infantry in contrast with the trials in peacetime. Therefore, the attacking infantry had to rely more on their own fire preparation, but they could not carry this out uncovered within the effective range of artillery. Moreover, once within the range of the enemy's musketry, the advance to the assault line would sometimes become so difficult as to limit it to the night hours (as actually happened in Manchuria). So these essayists concluded the impossibility of any advance on open ground, before the ultimate assault, against any average fortified defence and the need in modern campaigns for resorting to the old techniques of siege warfare in order to approach the enemy strongpoints.[61]

Major Juan Avilés, though recognizing the usefulness of light, fast-built trenches in the defence and the attack, did not infer such a predominance of the field fortification. This was useful for economizing on troops in inactive sectors, but excessive reliance on it could be demoralizing in the long term.[62] A few years later, Avilés (a lieutenant colonel by that time) was to argue that, though they must be used as a rule, field fortifications could (and had to) be discarded whenever this action improved one's own fire: the modern concept of field fortification gave the improvement of one's own fire priority over the protecting role. On the other hand, he also stated that field fortifications did not diminish the offensive spirit of well-trained soldiers, who saw them as a temporary means to achieve fire superiority.[63]

Avilés's arguments on fortification were inspired by the alleged lessons of the Russo-Japanese War, which seemed to endorse the idea that the armies displaying the highest morale and the most offensive attitude would hold the upper hand in any future war, whatever the difficulties they faced on the battlefield. Firepower was now important because of the moral support it provided to the attacking troops, whereas it was underrated in the defence. According to Avilés, the war had taught that the modern rifle did not make impossible the frontal attack in open order and had not substantially increased the casualty figures. Higher rates of fire meant lesser accuracy in practice. The attack could not be successful if it had not been prepared through artillery and rifle fire. But the purpose of this fire was to tire and unnerve the defender, since the latter's fire lost accuracy after two or three hours' firefight.[64] Avilés does not explain why the attacking troops were apparently immune to this phenomenon, unless it was due to their alleged moral superiority.

Another essayist argued in 1907 that fire superiority only became manifest for the attacking side through moral agents: this meant in practice a resolute advance scornful of any danger. Thus the defender, confident within his positions at the start of fighting, would lose heart when he saw his own losses and his inability to stop the attackers; then he would lose fire effectiveness, the casualties he inflicted would diminish and the attacking troops' fire superiority would really be obvious.[65] This argument, which resembles Ardant du Picq's ideas, has a grain of truth (in the last analysis, defeated troops flee to avoid the clash with the enemy, so attacking troops must move forward sometime in order to win), but it portrays fire superiority as the consequence of the advance rather than as a requirement, which in an offensive-minded environment could not help encouraging premature advances.

The influence of the Russo-Japanese War was so strong that praise of morale-related qualities after 1905 spread to those who had endorsed firepower tactics before. One of them was Brigadier General Modesto Navarro, who now expressed ideas which were nearer to the doctrine of the *offensive à outrance* than to fire-based tactics such as those supported by Navarro himself some years before. He explained that the battles in Manchuria had lessened the expectations about the quick-firing artillery's effectiveness. From now on, the artillery's role would be to provide support fire to make easier the infantry's advance, through the temporary paralysis and neutralization of the enemy forces instead of their destruction. This covering fire—despite its diminished material effect—was to stop the trend towards more dispersed formations in combat. A dispersed formation, though less vulnerable, was also more difficult to command, and considerations of the units' inner cohesion and command (so important for morale) must prevail over those related to vulnerability and one's own weapons effectiveness. Facing the danger on the battlefield, the troops needed an orderly grouping. The combat and its outcome were above all human actions of a moral nature. The roots of tactics lay in them.[66]

Even Marvá (also a general at that time) recognized explicitly that, after the experience of the Russo-Japanese War, the bayonet attack and the clash remained decisive against a determined, entrenched enemy since the cannon and rifle fire did not break the latter. Nonetheless, he also carried on arguing about the need to use formations adapted to the ground they fought on and light field fortifications. Moreover, General Marvá warned that, although moral factors were decisive when the material ones were balanced, the latter were becoming more important due to their increasing effectiveness.[67]

Towards New Infantry Tactics: The 1908 Regulations

The transformations of tactics after 1898 led in early 1907 to the creation of a commission under the General Staff in order to write new, updated tactical regulations.[68] The present author's research did not found evidence related to the activities of this commission, so he cannot assess the influence of the Spanish army's own experience on the new tactical doctrine for the infantry. A contemporary essayist, First Lieutenant Verardo García Rey, reckoned that the Spanish officers had become infected with an offensive spirit after the defeat of 1898, which had been reinforced by the alleged lessons of the Russo-Japanese War.[69] There were also Burguete's tactical proposals. Burguete was officially told that some of his ideas had been used for the new regulations. He was doubtful about this point, since he had been present neither in the discussions nor in the writing. But he was sure that only secondary ideas, if any, had been incorporated.[70]

A contemporary source pointed out that the infantry regulations of France (1904), Germany (1906), Italy (1905), and Russia (1907) provided inspiration for the Spanish doctrine.[71] The writings of contemporary essayists suggest that the French regulations were the most important influence. For example, García Rey thought that the Spanish doctrine on the development of combat was similar to the French one: general attacks by subordinate commanders all along the front and exploitation of local success by the high command through a major effort, whereas the German doctrine was based on a thorough attack by the advanced guard combined with an enveloping manoeuvre of the main force. However, the Spanish doctrine attached more importance to firepower than the French one.[72] The Gallic influence is also emphasized by a French essayist, General Louis de Torcy, who stated that the Spanish regulations had taken much of the contents of the French ones approved in 1904.[73]

The new infantry regulations were approved in a provisional way in August 1908.[74] The 1908 regulations actually ordered the advance through a combination of fire and movement by platoons.[75] The troops would reach the main firing position, beyond which any advance was impossible before achieving fire superiority; the halt would be long and the infantrymen would deliver heavy fire.[76] Once the enemy fire looked like weakening, the advance would start again in order to reach the assault line (in which the superiority of fire would also be definitively achieved).[77] The 1908 regulations seemed more offensive-minded than those published in 1898. They called for a thorough bayonet attack if the enemy showed signs of weakness. But if it was not so and the troops were nevertheless very near to the enemy, they must charge with the bayonet all the same.[78] The use of light field fortifications in the attack had to be very limited.[79] Even the defence had an offensive bias: 'By means of manoeuvre, and taking advantage of mistakes and faults *of the enemy, the defence will change its role by taking on an energetic and determined offensive* whenever feasible.'[80]

The regulations' emphasis on the clash and the offensive was a matter for criticism for Gil Juste, whose views were published in a series of articles in *La Correspondencia Militar* in late 1908. The regulations stated that the clash was the infantry's main means of action, but Gil Juste did not believe in its absolute necessity. An enveloping movement could also drive the enemy out of a position. Actually, he added, many defenders already 'beaten' by firepower remained in their positions because of their fear of being shot in the flight; the chance of running away came when the attacking troops charged with the bayonet. Thus the clash was often no more than a confirmation of victory (though it admittedly could be decisive if the enemy was unbroken).[81] Gil Juste asserted that the excessive emphasis on aggressiveness and offensive spirit had been copied from the French regulations. According to him, attacking whatever the place and time was no less harmful than a passive defence, nor did permanent aggression mean real courage. The latter, in any case, could not be inculcated through tactical regulations.[82]

Gil Juste's remarks prompted a reply in the pages of *La Correspondencia Militar*. According to its unidentified author, Gil Juste was saying the same thing as the regulations: whatever the cause, the fact was that the defenders were put to flight once the charge was carried out. So the troops had to be ready for the offensive in order to conquer (a common idea of all the foreign regulations), and this readiness required the strengthening of morale-related factors and their guidance in a positive (that is, offensive) direction. This was the regulations' task.[83]

Gil Juste's answer to this criticism denied that all the foreign regulations resembled each other in emphasizing the offensive. They were different in the ways the offensive had to be undertaken and in the factors which had to be given priority. He stressed that the Spanish regulations had simply copied the French army's doctrine, which had even found opposition in some sections of the French military, although the attachment to Napoleonic tradition and procedures had prevailed. Gil Juste foretold that future experience would prove the French army's mistake.[84]

The 1908 regulations also found severe criticism from Ricardo Burguete, who seems to have had second thoughts on his work of 1900 and whose diatribes against the new infantry regulations show that he had taken firepower more into account by the early 1910s. He criticized the regulations because they forgot the emptiness of the modern battlefield when they ordered troops to fire only against enemies in sight: it would be necessary to answer the enemy's firepower, but actually there were few chances to see any enemy—especially an enemy who had taken up defensive positions. On the other hand, once the troops started firing, the regulations gave no concrete guidelines about fire control and ammunition consumption (they ordered troops simply to fire at will interrupted by pauses, but with no instructions about the duration and rate of the fire periods). In addition, he saw the chapter on fire as proper for a firing range, but not for infantry tactical regulations.[85]

Burguete thought that the regulations also overestimated the moral influence of the officer upon the soldier, whose natural fear would diminish discipline under fire. The soldier had to be told beforehand that the enemy would be invisible. If not, he would find it difficult to aim correctly when ordered, control his ammunition consumption, and stop firing, because he would see no reason to do it and, therefore, to advance.[86] The regulations gave no details about the array of troops on the battlefield beyond echeloning in a skirmishing line, supports, and reserve. Moreover, they were occasionally contradictory: for instance, Article 143's burning desire which inspired the troops to arrive at the clash did not match Article 298's guideline about 'trying' the advance if the enemy looked like weakening. It was supposed that the advance, once started, would be carried out thoroughly; 'attempts' to advance hinted at a not so resolute attitude on the part of the attacker, according to Burguete.[87] He also remarked ironically on the regulations' lack of clear-cut, practical guidelines: it would have been easier to reduce the regulations to a single precept ordering commanders to use troops, weapons, and ground wisely in order to defeat the enemy.[88] This last criticism had already been brought forward in 1908 by Major Luis Bermúdez de Castro, who complained that the highbrow trends within the Spanish military had turned the 1908 infantry regulations into a philosophical essay.[89]

Besides criticisms, the 1908 infantry regulations faced a practical obstacle to their application in the Spanish army. An anonymous article in a professional journal wondered how the troops could be trained in the right implementation of the regulations when the Spanish infantry units were usually so much understrength.[90] By 1905, the official peacetime strength of the average infantry regiment of two active battalions was 517 officers and enlisted men (one battalion in wartime was supposed to be around 1000-strong). A fifth of the strength was headquarters and specialist staff (clerks, musicians, and so on). Therefore, the eight active companies of this undermanned infantry regiment only had about 50 troops each. According to Captain Gallego, absences due to disease and secondments for ancillary duties often reduced the real strength of some companies to two or three men.[91] This situation does not seem to have improved three years later.

The Campaign of Melilla (1909) and Its Consequences

The Military Operations

The 1908 provisional infantry regulations were tested in actual conflict, one year after their approval, in the Melilla campaign of 1909. This campaign was the first step of the Spanish army's involvement in a long, bitter colonial war which was to affect the conduct of the military in Spain's politics during the next two decades.[92] Spain had two enclaves on the northern shore of Morocco in the early twentieth century. They were the seaports of Melilla (which had been under Spanish sovereignty since 1497) and Ceuta. These enclaves had remained as isolated outposts until the early years of the twentieth century, when they provided a foothold to expand Spanish influence in Morocco. In 1907 Spanish mining companies bought from a self-styled regional overlord the rights to exploit iron deposits near Melilla. However, some of the local Moorish native tribes were hostile to the foreign presence and attacked the workers building the railway to the mines in early July 1909. General José Marina, the military commander of Melilla, reckoned his garrison insufficient to undertake punitive operations and protect the mining activities, so he asked Madrid for reinforcements.

The cabinet sent an expeditionary force (including called-up reservists) during the summer. The first troops to arrive were committed prematurely and suffered setbacks in late July, so the military activity was limited to consolidating a defensive perimeter and training the newly arrived forces. Offensive operations started again in September: in a series of limited, cautious advances the Spanish expeditionary army (over 40,000-strong) subdued the nearby tribes and set up a screen of defensive positions and blockhouses around a slightly expanded area under Spanish tutelage. The campaign of Melilla (which was also known as the Rif campaign, after the nearby mountainous region) was over by November. According to the official history of the campaign, the Spanish casualties amounted to 2517 killed and wounded in action (plus 211 fatalities by disease); 1765 Moorish corpses were buried.

The worst setback suffered by the Spanish army during the campaign of Melilla took place in a ravine called the Barranco del Lobo (27 July 1909), when Brigadier General Guillermo Pintos' brigade was mauled by the Moorish natives' musketry. The brigade (six *cazadores* battalions) had to protect the passage of a supply convoy in front of the lowest slopes of Mount Gurugú. Although Pintos needed only to keep the harassing natives at bay, he actually committed the whole brigade to an advance up the ravine. Since Pintos was killed in the early stages of the fighting, it is not clear why he ordered his troops to move forward. The advance of the Spanish brigade turned into a mess as the officers fell under enemy fire from the front and the flanks. The mauled Spanish battalions came to a halt, endured the fire for a while and withdrew in the most orderly way they could, while the artillery fire held the pursuing natives off. The Spanish forces lost about 750 killed and wounded. Estimates of the enemy losses vary from a hundred to 475 killed.

Lessons of the Campaign

The contemporary professional literature admitted that the early performance of the Spanish infantry had been disappointing. According to Lieutenant Colonel Avilés, besides the Spanish expeditionary forces' lack of training and organization, a cause of the setbacks had been the recent introduction of the new tactical regulations, whose spirit the troops still had not familiarized themselves with.[93] General de Torcy wrote that the Spanish tactics in 1909 had been an improvised combination of the old procedures and the new ones, since only the conscripts who had recently joined the colours had been drilled in the 1908 regulations.[94]

The combat of the Barranco del Lobo was analyzed by Avilés and Captain Gallego. According to Avilés, the 1908 infantry regulations had a partial responsibility for the failure of the advance of Pintos' brigade: they were unclear about the consolidation of seized ground, while they were much more explicit about the pursuit. Thus, after the first gains of ground, the desire to exploit this initial success could prevail over the idea of consolidating the positions.[95] Gallego pointed out that the attack formations of Pintos' brigade were too dense. They could only be justified if General Pintos had thought (as Gallego surmised) of throwing the natives out of their positions by the sheer drive of a bayonet charge (which required a closer order than the skirmishing line). However, the attack was lacking in preparatory rifle and artillery fire. The units did not display a capability for manoeuvring and taking advantage of the terrain either: there was too much confidence in sheer bravery.[96] Avilés remarked that the officers tried to encourage their raw troops through a determined, thorough attack, and disregarded the stages set by the regulations; this did no good for the conduct of the combat.[97]

Discussing the performance of the infantry during the whole campaign, Gallego pointed out that the Spanish troops did not follow the basic rule of modern tactics about taking advantage of terrain and light field fortifications. Despite the example of the Russo-Japanese War, the Spanish infantrymen did not know how to use the roughness of the ground nor were the infantry units' sapping tools used during the combats.[98] The 1908 regulations must have been influential in this behaviour, since an officer praised (just before the outbreak of the conflict) how they had prepared the units to rank protection behind fire effectiveness.[99]

Avilés put this problem down to the flaws in infantry tactical training. This had been based (due to the foreign influences and the Spanish army's own colonial experience) on offensive tactics: the clash had been given more importance than firepower. Such state of affairs had been embodied in the 1908 tactical regulations, in which the officers were told to pay much more attention to the troops' morale than to the correct manoeuvring of their units. The Melilla campaign also highlighted the inadequate place given in the regulations to junior leaders within the units. Platoon and squad leaders were to be integrated into the skirmishing line, but they were unable in practice to control their men's fire during the combat. If they took post behind the line, they discouraged the troops. They finally resorted to standing up and walking along the line (thus offering a good target). Peacetime exercises turned the fire preparation of the attack into a few minutes' interlude before the assault. In real war the preparatory fire could last hours. If the troops were not prepared for this situation, they could get demoralized by thinking that the operation was going badly. The troops did not practice the technique of laying down on the ground either (allegedly in order not to damage the uniforms!), so they got disheartened the first time they did it in action.[100]

Besides poor training, the nature of the fighting was another cause of bewilderment for the Spanish soldiers. They had to fight against an irregular enemy able to take the best advantage of the rugged ground, both for movement and shooting. By contrast, the Spanish army moved as on a drilling ground, according to Martínez de Campos (who fought in the campaign as a young artillery subaltern).[101] He witnessed how a divisional commander proved to be distrustful of the official procedures (or their stereotyped implementation, at least). General Antonio Tovar, watching a deployment of troops by the book, warned his subordinates: 'Don't play dirty tricks on me [*No hacerme faenas*]. We are no longer in Carabanchel.'[102] In the aftermath of the Barranco del Lobo, a feature of *La Correspondencia Militar* complained bitterly about the different fighting attitudes of Spaniards and Moors: the gallant displays of the Spanish troops were wrong against the Moroccan natives. Crawling on the ground was an adequate method for fighting the Moors, imitating their 'treacherous' calm.[103]

Avilés concluded that the idea of the offensive had to be understood as an attitude of the commanders' mind and will. The offensive required in practice a superiority in numbers and morale which was not always achieved at the start of the fighting. A temporary defensive posture could be needed in order to wear out the enemy forces. However, the training had given absolute priority to fast, offensive tactics such as those used in the first combats. When these tactics failed at Melilla, the soldiers, accustomed to the success of the simulated offensive, lost heart and confidence in their own tactics. The haste to develop the stages of the attack caused a disorder of the tactical echeloning and was detrimental to any defensive action in the heat of the battle.[104] According to Gallego, such haste was sometimes due to the commanders' impatience, which made them unwilling to spend time in reconnoitring the enemy positions and preparatory shelling.[105]

The General Staff's analysis of the campaign revealed weak points in the implementation of infantry tactical doctrine. The deployments often had not taken advantage of the effectiveness of modern firearms and the small units kept close physical contact unnecessarily. This hindered the capability of these units for small-scale manoeuvre, so important in modern warfare, and the fronts covered were too narrow to allow the units to move with ease.[106] In the fighting against the Moroccan native, the aim had to be victory through fire superiority and skilful manoeuvre. Although the bayonet charge could be essential sometimes, it had to be avoided as a rule since the troops ended up scattered and exhausted.[107] Gallego pointed out that bayonet charges were rare in the campaign and carried out in company or platoon strength only; there never was an actual clash since it was shunned by the enemy.[108]

The Melilla campaign disclosed the weakness of cooperation between the artillery and the infantry. The irregular nature of the fighting gave few teachings, if any, about the performance of the artillery as an individual arm. The concealment and the dispersed fighting order of the enemy did not allow the most effective use of the artillery; many shells were wasted to destroy almost valueless targets. Moreover, there were seemingly no criteria about the operational ratios of artillery to infantry, so the columns were often unbalanced in their strength by arms; for example, a column could be very strong in infantry but weak in artillery. Avilés reckoned that the Spanish commanders' lack of experience in the use of combined arms was a reason for this problem.[109] Ricardo Burguete complained that the artillery had been moved too far forward on many occasions, when good firing positions, suitable for grouping the cannon, were available in the rear. He also pointed out that the Spanish forces lacked liaison means to coordinate the efforts of artillerymen and infantrymen.[110] Although technical issues were doubtless influential, it was also true that the Spanish army's previous experiences had not fostered the tactical integration of both arms. This was tried during the Third Carlist War, but the outcome was not satisfactory because the infantry was very fearful of suffering casualties by friendly artillery fire. The colonial wars at the end of the nineteenth century saw a piecemeal use of the artillery. One or two pieces were attached to the individual columns and often used at point-blank range. Thus the Spanish officers lacked any chance (or saw no need) to develop a doctrine about combined arms.[111]

The Aftermath: The 1913 Infantry Tactical Regulations

The provisional doctrine of 1908 evolved into the official one of late 1913. The 1913 Infantry Tactical Regulations actually emphasized the spirit of the offensive with moral factors in the foreground. This seems somewhat odd after the experience of 1909. So far, the present author's research has not found official evidence to explain the matter. It is plausible that the colonial character of the Melilla operations justified a reluctance to alter the official doctrine, despite the fact that the Moroccan natives' musketry had proved-admittedly on a limited scale-the effectiveness of rifled small arms. But, on the other hand, the campaign could provide arguments to strengthen the role of morale and offensive spirit in the infantry regulations as well.

For instance, Lieutenant Colonel Avilés remarked that, in order to stiffen their fighting spirit, the troops had been exposed by the commanders to the dangers of the firing line with little concern about their vulnerability. They also attacked before the preparatory fire had broken the enemy. Avilés admitted that such a method was dear in casualties during the first weeks, but he claimed that it was economical in the end. For the Spanish soldier thus learnt to fight his enemy and soon obtained a manifest moral superiority which contributed to the conclusion of the war.[112] This last remark is interesting because it shows that Avilés, whatever his criticisms of the official doctrine (see above), shared the image of warfare in terms of morale and psychological factors so frequent in the contemporary professional literature. This vision has been called the paradigm of 'the psychological battlefield' by the historian Tim Travers.[113] Within this paradigm, any concerns about firepower are secondary to maintaining morale and securing a decisive psychological supremacy. Actually Avilés' views on the moral strength of the offensive and the need for accepting losses resembled the spirit inspiring the 1913 regulations.

The new infantry regulations did not offer significant changes in tactical procedures, but any reader could notice a heavier emphasis on morale and the offensive without regard to casualties. The 1908 version of Article 4 talked of the need for infantrymen to have a 'high spirit' *tout court*. In 1913 the regulations stated that the infantry needed 'to be inspired by a *constant offensive and attacking spirit*, which carries them along to reach the enemy and conquer him whatever the cost'.[114]

Fire and movement seemed to be the accepted doctrine in 1908: '[*F*]*ire* is very important and predominant, but the advancing movement, impetuous and overwhelming, is the decisive [means]; the latter will only be possible and fruitful when an effective fire has prepared it and made it possible.'[115] However, the 1913 regulations stated that fire just made movement 'easier' and added: '[O]nly through movement will fire get all its development and its greatest effectiveness.'[116] This article suggests that fire was a useful, supporting means for movement but was assigned a secondary status in relation to the latter, since fire was also dependent on movement for its effectiveness and was not deemed a decisive element. The underrating of firepower also meant a devaluation of tactical defence. The 1913 regulations devoted 89 articles to offensive combat and only 22 to defensive fighting.[117]

The 1913 regulations also included a new article on the need to persist with the attack.[118] This idea was shared by other sections of the military. A commentator on the regulations wrote that 'the determined and forceful offensive is not enough to conquer, if, besides, it is not joined by the features of stubbornness and endurance in attack'.[119] This persistence involved a will to accept great losses, which Captain Gascueña explained this way: 'It is necessary to fight for the decisive victory, even if we must suffer more casualties than the enemy, and one must bear in mind for this... that the attacker suffers during the fight many more casualties than the defender, but he wins in the end.'[120] This was in fact one of the contemporary solutions to the problem of the tactical offensive against increased firepower, which Captain Dolla had already voiced almost a decade before: 'And if we have no other alternative, let us march forward, throwing lives into... the hearth of victory: war is not a sensitive matter...'[121] According to Gascueña, there was no other alternative but to face up to the modern battlefield:

> Protection must also be conceived within a limit. He who thinks that great tactical successes can be achieved in modern warfare without great losses makes a big mistake. He needs to admit resolutely that troops which are not afraid of losses enjoy a huge superiority over all those which look to saving their blood...[122]

Therefore, trenches were only admissible in exceptional cases when preserving conquered ground rather than advancing was appropriate.[123] Actually the Spanish infantrymen would have found it difficult to dig such trenches on their own, because the Spanish army was one of the two armies in Europe (the other was the Turkish army) where the infantry still had not got an issue individual trenching tool by 1915.[124]

Conclusion

This chapter has shown how the Spanish infantry passed from the firepower-based tactical doctrine of the 1890s to a doctrine of *offensive à outrance* in the early 1910s. The Spanish military's experience in the Third Carlist War seemed to instil a healthy respect for firepower, which was reflected in the 1881 and 1898 regulations. The wars of 1895–1898 generated debate but no change in the official doctrine. On the contrary, a foreign conflict, the Russo-Japanese War, probably did more to tilt the balance towards the offensive, which eventually found full expression in the 1908 infantry tactical regulations. And the latter's principles, despite the experience in the campaign of Melilla in 1909, were reasserted in the 1913 regulations. Following foreign examples rather than assessing and learning from its own experiences in colonial wars was consistent with an institutional culture that seemed to feel more comfortable in the realm of regular campaigns than in the quagmires of irregular warfare. Moreover, Geoffrey Jensen has observed that the Spanish military had an additional motive to emphasize even more than other armies the morale-driven offensive: the dearth of artillery.[125] It worth noting that the Austro-Hungarian army of this period, which had the lowest proportion of artillery among the great powers, resorted to the same solution of sorts.[126]

But this does not mean that the Spanish commanders, like the Austro-Hungarians, were doing their best with what they had. The Spanish military performance in the campaign of 1909 (especially during its initial stage) is generally criticized in the historical literature covering this period. One of the usual culprits is the poor professional foresight of the Spanish army, proved by the lack of interest in the experiences of its wars of the 1890s, despite the fact that military involvement in Morocco seemed possible. Thus the Spanish troops sent to Melilla were poorly prepared for what is diversely labelled as a colonial campaign, irregular warfare or guerrilla warfare. But such categorization is based only in the nature of the enemy and has misled later military analysis of the campaign. For the Spanish forces' actual mission was the occupation and defence of a series of positions in the environs of the small coastal enclave, a task whose essence made no difference between regular and irregular warfare. European armies have been censured because they refused to recognize the features of modern combat in the Boer War (another regulars versus irregulars conflict). Such features were also present in the tactics of the Moroccan natives, whose traditional fighting style exploited naturally in the defensive the improved performance of modern rifles.[127]

This state of affairs was unfortunate for the Spanish troops, who had to sort out the same challenge that could be found in a regular campaign: advancing on an empty battlefield in the face of relatively modern firearms. The Moroccan natives eventually yielded to the overall superiority in numbers and firepower and the improved tactical performance of the Spanish forces. But, like other armies, the latter paid heavily in blood whenever they tried to attack with too much alacrity and insufficient fire support. However, this was a lesson that the Spanish military still found hard to learn.

Notes

1. For a descriptive overview of the Spanish infantry's tactical regulations from 1850 to the 1910s, see José Luis Isabel Sánchez, 'La evolución de las técnicas', in Miguel Alonso Baquer and Luis A. Collado Espiga (eds), *Historia de la infantería española*, 5 vols. (Madrid: Ministerio de Defensa, Secretaría General Técnica, 1993–2001), III. 365–403.
2. For a contemporary survey of the evolution of warfare in the nineteenth century, see J[ean] Colin, *The Transformations of War* (London: Hugh Rees, 1912). For a sample of later overviews of the subject, see David Gates, *Warfare in the Nineteenth Century* (Basingstoke: Palgrave, 2001); Hew Strachan, *European Armies and the Conduct of War* (London: Unwin Hyman, 1985), Chap. 8.
3. José Marvá y Mayer, *Ciencias aplicadas al arte militar* (Madrid: Ateneo de Madrid, 1902), p. 91.
4. Carlos Banús y Comas, *Reflexiones acerca de la guerra anglo-boer* (Madrid: Imp. del Memorial de Ingenieros del Ejército, 1902), pp. 63–4. Commissioned in the engineer corps in 1872, Banús (1852–1934) fought against the Carlists (1873–1875), taught at the Academy of Engineers (1877–1890) and was director of the engineer corps laboratory; he won professional and academic renown through his writings on military and technical issues.
5. Antonio Aleixandre, 'Formaciones de combate', *Revista Técnica de Infantería y Caballería* (hereafter *RTIC*), 12, 2 (1901), pp. 73–6.

6. For a sample of overviews of military doctrine and tactics in this period, see Michael Howard, 'Men Against Fire: Expectations of War in 1914', *International Security*, 9, 1 (1984), pp. 45–57; Jay Luvaas, 'European Military Thought and Doctrine, 1870–1914' in Michael Howard (ed.), *The Theory and Practice of War* (London: Cassell, 1965), pp. 82–91; Christopher Tuck, *Understanding Land Warfare* (Abingdon: Routledge, 2014), pp. 58–66. For an analysis of the strategic environment fostering offensive doctrines, see Jack Snyder, *The Ideology of the Offensive: Military Decision Making and the Disasters of 1914* (Ithaca: Cornell University Press, 1984). For a small sample of scholarship in English examining particular armies, see André Bourachot, *Marshall Joffre. The Triumphs, Failures and Controversies of France's Commander-in-Chief in the Great War* (Barnsley: Pen & Sword Military, 2014), Chap. 2; Eric Dorn Brose, *The Kaiser's Army: the Politics of Military Technology in Germany during the Machine Age, 1870–1918* (New York: Oxford University Press, 2001); Azar Gat, *A History of Military Thought: From the Enlightenment to the Cold War* (Oxford: Oxford University Press, 2001), Book II, Chap. 3; Bruce W. Menning, *Bayonets before Bullets: The Imperial Russian Army, 1861–1914* (Bloomington and Indianapolis: Indiana University Press, 1992); Douglas Porch, *The March to the Marne. The French Army 1871–1914* (Cambridge: Cambridge University Press, 1981), Chap. 11; Shelford Bidwell and Dominick Graham, *Fire Power: British Army Weapons and Theories of War, 1904–1945* (London: Allen & Unwin, 1985), Chap. 2.
7. For a detailed military account of this conflict, see Cuerpo de Estado Mayor, *Narración militar de la Guerra Carlista de 1869 á 1876*, 14 vols. (Madrid: Depósito de la Guerra, 1883–1889).
8. Pedro Ruíz Dana, *Estudios sobre la guerra civil en el Norte, de 1872 a 1876* (Madrid: n.p., 1876), pp. 151–4. Ruíz Dana (1822–1891) was a staff corps officer and reached the rank of lieutenant general in 1876; he was director general of cavalry and captain general of Puerto Rico.
9. The following overview of the evolution of Carlist field fortifications is based on Ruíz Dana, *Estudios*, pp. 200–12.
10. Jorge Vigón, *Historia de la Artillería española*, 3 vols. (Madrid: Consejo Superior de Investigaciones Científicas, 1947), III. 123.
11. For a listing of places where the Carlists dug entrenchments, see Federico de Madariaga, 'La Infantería española', *RTIC*, 15, 1 (1904), pp. 6–7.
12. *Táctica de Infantería. Memoria general* (Madrid: Depósito de la Guerra, 1881), p. 16.
13. Miguel Alonso Baquer, 'La guerra Hispano-americana de 1898 y sus efectos sobre las Instituciones militares españolas', *Revista de Historia Militar*, 54 (1983), pp. 134–5.

14. Herbert H. Sargent, *The Campaign of Santiago de Cuba*, 3 vols. (London: Kegan Paul, Trench, Trübner & Co., 1907), III. 157–9.
15. Efeele [pseud. of Francisco Larrea y Liso], 'La guerra con los Estados Unidos', *Estudios Militares* (hereafter *EM*), 18, 1 (1899), p. 294.
16. This summary of the campaign of Santiago de Cuba is based on José Müller y Tejeiro, *Combates y capitulación de Santiago de Cuba* (Madrid: n.p., 1898); Severo Gómez Núñez, *La guerra hispano-americana. Santiago de Cuba* (Madrid: Imprenta del Cuerpo de Artilleria, 1901); Efeele [pseud. of Francisco Larrea y Liso], *El desastre nacional y los vicios de nuestras instituciones militares* (Madrid: Publicaciones de los Estudios Militares, 1901), pp. 54–6; Joseph Smith, *The Spanish-American War: Conflict in the Caribbean and the Pacific, 1895–1902* (London: Longman, 1994), pp. 119–59; Sargent, *Campaign of Santiago de Cuba*.
17. Alonso Baquer, 'La guerra Hispano-americana', pp. 135–6.
18. Damián Isern, *Del desastre nacional y sus causas* (Madrid: n.p., 1899), pp. 490–7.
19. Efeele, *El desastre nacional*, pp. 55–62. In the less important theatre of operations of Puerto Rico the obsession of the Spanish command to defend the capital prevented the garrison from creating a reserve strong enough to counterattack the isolated columns of the American invasion forces: Efeele, *El desastre nacional*, pp. 118–20.
20. Critón [pseud. of Ricardo Burguete Lana]: *Hágase Ejército. Infantería, Caballería, Artillería. Estudios tácticos al alcance de todos* (Barcelona: Luis Tasso, 1899), p. 101. Nonetheless, a Spanish naval officer who witnessed the campaign thought an offensive reaction against the Americans very difficult: Müller, *Combates y capitulación*, pp. 275–6.
21. José Yumrep, 'De la disciplina del fuego', *RTIC*, 11, 2 (1900), p. 432; Ricardo Burguete Lana, *Nuevos métodos de combate. Proyecto de reglamento de ejercicios y maniobras para infantería* (Madrid: [Depósito de la Guerra], 1903), pp. 22–4, 153; Critón, *Hágase Ejército*, pp. 34–5.
22. Eduardo Gallego Ramos, 'El municionamiento de la infantería en el combate moderno', *RTIC*, 10, 1 (1899), pp. 26–7; Luís Fernández de Cordova, Marqués de Mendigorría, 'Táctica del Comandante Burguete', *RTIC*, 13, 1 (1902), p. 116; Modesto Navarro, 'Observaciones sobre la táctica de Burguete', *EM*, 21, 1 (1902), pp. 165, 259.
23. Alonso Baquer, 'La guerra Hispano-americana', p. 144; Ricardo Burguete Lana, *Mi rebeldía: 'Mane-Thecel-Phares'* (Madrid: Librería de 'Fernando Fé', 1904), pp. 310–14.
24. Jensen, Geoffrey: 'The Spanish Army at War in the Nineteenth Century: Counterinsurgency at Home and Abroad', in Wayne H. Bowen and José E. Álvarez (eds), *A Military History of Modern Spain: From the Napoleonic Era to the International War on Terror* (Westport: Praeger, 2007), p. 36.

25. Burguete, *Nuevos métodos*, pp. 25–6; Hilario Hernández Rivera, 'Sobre el espíritu que debe informar un reglamento táctico', *RTIC*, 14, 1 (1903), pp. 143–4.
26. Burguete, *Nuevos métodos*, pp. 15, 26; Manuel Burguete, 'El choque al arma blanca como resolvente único y decisivo en todo combate moderno', *Revista Científico-Militar* (hereafter *RCM*), 26 (1901), pp. 201–3; Critón, *Hágase Ejército*, pp. 32–5. On firearms in the colonial wars, see Daniel R. Headrick, 'The Tools of Imperialism: Technology and the Expansion of European Colonial Empires in the Nineteenth Century', *Journal of Modern History*, 51 (1979), pp. 258–61.
27. Ángel Dolla, 'Resumen de las armas y su aplicación al Arma de Caballería', *RTIC*, 16, 1 (1905), pp. 32–5.
28. Charles Ardant du Picq, *Estudios sobre el combate*, 2 vols. (Madrid: n.p., 1883), I. 160–1.
29. A contemporary survey of foreign professional periodicals mentions 14 French journals, against seven German ones: X., 'La cultura militar en los comienzos del siglo XX', *RTIC*, 15, 1 (1904), pp. 90–6, 139–44. On the influence of Ardant du Picq's ideas on French military thinking, see Stefan T. Possony and Etienne Mantoux, 'Du Picq and Foch: The French School', in Edward Mead Earle (ed.), *Makers of Modern Strategy: Military Thought from Machiavelli to Hitler* (Princeton: Princeton University Press, 1943), pp. 216–18; Azar Gat, *A History of Military Thought: From the Enlightenment to the Cold War* (Oxford: Oxford University Press, 2001), pp. 308–9, 408–9, 413, 417–18.
30. Navarro, 'Observaciones' [1902], p. 292.
31. Navarro, "Observaciones' [1902], pp. 225–34.
32. Germán Gil Juste [or Yuste], 'Formaciones y evoluciones de la compañía y del batallón', *RTIC*, 12, 1 (1901), 355–7. Gil Yuste (1866–1948) entered the Academy of Infantry in 1882 and fought in the Philippines, Cuba, and Morocco; he was commandant of the Academy of Infantry; a retired major general in 1936, he joined the Nationalist side and took charge of its War Secretariat (an embryonic war ministry) from October 1936 to early 1938.
33. Gallego, 'El municionamiento', p. 25; Germán Gil Juste, 'Cómo puede obtener la Infantería superioridad de fuego', *RTIC*, 12, 1 (1901), pp. 259–62.
34. Germán Gil Juste, 'Algo sobre la nueva Artillería de campaña y los combates del porvenir', *RTIC*, 10, 1 (1899), p. 530.
35. Germán Gil Juste, 'Proyecto de reglamento táctico para la Infantería', *RTIC*, 13, 1 (1902), pp. 34–6. Ivan Bloch published a study titled *Technical, Economic and Political Aspects of the Coming War* (1898); it predicted that any future major armed conflict would become a

prolonged stalemate of trench warfare, whose huge human and material costs would exhaust the belligerents or cause a social revolution; on his view of future warfare, see Tohmatsu Haruo, 'Approaching Total War: Ivan Bloch's Disturbing Vision', in David Wolff, Steven G. Marks, Bruce W. Menning, David Schimmelpenninck van der Oye, John W. Steinberg, and Yokote Shinji (eds), The *Russo-Japanese War in Global Perspective. World War Zero. Volume II* (Leiden: Brill, 2007), pp. 179–202.

36. Ministerio de la Guerra, *Colección Legislativa del Ejército 1898* (Madrid: n.p., 1898) (hereafter *CLE plus the year of publication*), Item No. 109. For an overview of the Mauser rifle series, see Neil Grant, *Mauser Military Rifles* (Oxford: Osprey, 2015).
37. Verardo García Rey, *Estudios acerca de la táctica de infantería* (Madrid: Revista Técnica de Infantería y Caballería, 1907), p. 15.
38. Archivo General Militar de Segovia, 2ª Sección, División 8ª, Legajo 527 (hereafter AGMS No. Sección/No. División/No. Legajo), Memoria General sobre el proyecto de Reglamento para la instrucción táctica de las tropas de Infantería, pp. 10–11.
39. Ministerio de la Guerra, *Reglamento para la instrucción táctica de las tropas de infantería*, 3 vols. (Madrid: Depósito de la Guerra, 1898) (hereafter RTI 1898), I. paragraph 147.
40. AGMS 2/8/527, Memoria General sobre el proyecto de Reglamento para la instrucción táctica de las tropas de Infantería, p. 11.
41. Burguete, *Nuevos métodos*, p. 24.
42. Burguete, *Nuevos métodos*, pp. 17–19.
43. RTI 1898, II. Appendix I, Article 1.
44. RTI 1898, II. paragraphs 119, 120, 121, 309, 310, 332; Burguete, *Nuevos métodos*, pp. 20, 23–4; Un oficial del regimiento 'El Rey', 'Nuevos métodos de combate', *RCM*, 26 (1901), p. 316.
45. Germán Gil Juste, 'Los ataques a la bayoneta ante el fusil de repetición y de pequeño calibre', *RTIC*, 12, 2 (1901), pp. 35–7.
46. Un oficial, 'Nuevos métodos', p. 313.
47. AGMS 2/8/527, Memoria General sobre el proyecto de Reglamento para la instrucción táctica de las tropas de Infantería, pp. 9–10.
48. Un oficial, 'Nuevos métodos', pp. 311–12. Burguete's regulations were published under the title *Nuevos métodos de combate. Proyecto de reglamento de ejercicios y maniobras para infantería* and reprinted in 1903 and 1907.
49. Burguete, *Nuevos métodos*, pp. 6–8, 41; Fernández de Cordova, 'Táctica del Comandante Burguete', pp. 115–16.
50. Navarro, 'Observaciones' [1902], pp. 130–1, 168, 196, 321–3.
51. Navarro, 'Observaciones' [1902], pp. 260–5, 289–90.

52. García Rey, *Estudios de táctica*, pp. 50–1; Navarro, 'Observaciones sobre la táctica de Burguete', *EM*, 22, 1 (1903), p. 95.
53. Ricardo Burguete, *La guerra y el hombre. Psicología de las tropas* (Madrid: Sáenz de Jubera Hermanos, 1911), p. 220.
54. On the foreign military observers in the Russo-Japanese War and the 'lessons' that Western armies learnt (or failed to learn), see David Jones, 'Military Observers, Eurocentrism and World War Zero', in David Wolff et al. (eds), *The Russo-Japanese War II*, pp. 135–77; Dominik Geppert, William Mulligan, and Andreas Rose (eds), *The Wars before the Great War: Conflict and International Politics before the Outbreak of the First World War* (Cambridge: Cambridge University Press, 2015), Part 2; Thomas A. Bruno, 'Ignoring the Obvious: Combined Arms and Fire & Maneuver Tactics Prior to World War I' (student research paper, United States Marine Corps Command and Staff College, 2002).
55. Ricardo Burguete, *La Ciencia del Valor. Psicología de la guerra. Aplicación al desarrollo episódico de la batalla de Mukden* (Madrid: Sáenz de Jubera Hermanos, 1907), p. 243.
56. Burguete, *Ciencia del Valor*, pp. 138–9.
57. The Japanese army learnt soon from its heavy losses at the start of the conflict that the so-called 'human-bullet attacks' could not prevail over firepower and modified its tactics accordingly; massed infantry assaults carried out later in the war resulted from incompetence rather than martial prowess: Yoshihisa Tak Matsusaka, 'Human Bullets, General Nogi, and the Myth of Port Arthur', in John W. Steinberg, Bruce W. Menning, David Schimmelpenninck van der Oye, David Wolff, and Shinji Yokote (eds), *The Russo-Japanese War in Global Perspective: World War Zero* (Leiden: Brill, 2005), pp. 179–201; Bruno, 'Ignoring the Obvious', pp. 11–16.
58. Burguete, *La guerra y el hombre*, pp. 214–16.
59. Eduardo Gallego Ramos, 'Trabajos de campaña y herramientas de las tropas de infantería', *RTIC*, 10, 2 (1899), pp. 32–3.
60. Tomás Rodríguez de León, 'La infantería moderna', *RTIC*, 11, 1 (1900), pp. 125–6; Banús, *Reflexiones*, pp. 52–3.
61. Juan Carreras and Ricardo Malagón, *Relaciones entre la fortificación y la táctica* (Toledo: Academia de Infanteria, 1908), pp. 4, 21–3.
62. Juan Avilés, 'Algunas lecciones de la última guerra', *RCM*, 32 (1907), p. 186.
63. Juan Avilés, 'La fortificación de campaña y su elemento fundamental, las trincheras abrigos', *RCM*, 34 (1909), pp. 38–40.
64. Avilés, 'Algunas lecciones', pp. 185–6.
65. M[ariano] Vicente Arcones, 'Estudios sobre el combate de la infantería', *RCM*, 32 (1907), pp. 134–5.

66. Modesto Navarro, 'Avance de las tropas de infantería bajo el fuego de artillería', *RTIC*, 22, 1 (1911), pp. 292, 329, 338, 369–70, 456.
67. José Marvá y Mayer, 'Sitio de Puerto Arturo. Enseñanzas', *RTIC*, 18, 2 (1907), pp. 252–3, 492, 494.
68. *CLE 1907*, Item No. 30.
69. V[erardo] García Rey, 'La nueva táctica', *RTIC*, 20, 1 (1909), pp. 407–8.
70. Burguete, *La guerra y el hombre*, p. 220.
71. Federico de Madariaga, 'Instrucción táctica de las tropas de Infantería', *RTIC*, 19, 2 (1908), p. 495.
72. Verardo García Rey, 'Doctrinas acerca del combate', *RTIC*, 24, 1 (1913), pp. 135, 230, 232.
73. Louis Joseph Gilles de Torcy, *Les espagnols au Maroc en 1909*, 2nd edition (Paris and Nancy: Berger-Levrault, 1911), p. 227. There is a Spanish edition: Louis Joseph Gilles de Torcy, *Los españoles en Marruecos en 1909* (Madrid: n.p., 1911).
74. *CLE 1908*, Item No. 149.
75. Ministerio de la Guerra, *Reglamento provisional para la instrucción táctica de las tropas de Infantería* (Madrid: n.p., 1908) (hereafter RTI 1908), Article 291.
76. RTI 1908, Article 296.
77. RTI 1908, Articles 298, 305, 306.
78. RTI 1908, Articles 310, 311.
79. RTI 1908, Article 320.
80. RTI 1908, Article 381. Italics in the original.
81. Coronel B. [pseud. of Germán Gil Juste], 'Sobre el nuevo reglamento táctico de nuestra Infantería', *La Correspondencia Militar* (hereafter *LCM*), 28 November 1908.
82. Coronel B. [pseud. of Germán Gil Juste], 'Sobre el nuevo reglamento táctico de nuestra Infantería', *LCM*, 2 December 1908.
83. X., 'Sobre el nuevo reglamento táctico de nuestra Infantería', *LCM*, 11 and 14 December 1908.
84. Germán Gil Juste, 'Sobre el nuevo reglamento táctico de nuestra Infantería', *LCM*, 19 December 1908.
85. Ricardo Burguete, *Teoría y práctica de la guerra. Evolución en el arte* (Madrid: Sáenz de Jubera Hermanos, 1913), pp. 178–82.
86. Burguete, *Teoría y práctica de la guerra*, pp. 184–8.
87. Burguete, *Teoría y práctica de la guerra*, pp. 192–4.
88. Burguete, *Teoría y práctica de la guerra*, p. 195.
89. Luis Bermúdez de Castro, 'En contra de la nueva táctica', *LCM*, 31 December 1908.
90. 'La nueva táctica de infantería', *RCM*, 33 (1908), pp. 354–5.

91. Eduardo Gallego Ramos, *Proyecto de reorganización y mejora del ejército de tierra* (Guadalajara: n.p., 1905), p. 32.
92. On the 1909 campaign, see Carlos Hernández de Herrera and Tomás García Figueras, *Acción de España en Marruecos* (Madrid: Imprenta Municipal, 1929), Chap. 8; Servicio Histórico Militar, *Historia de las campañas de Marruecos*, 4 vols. (Madrid: Servicio Histórico Militar, 1947–1981), II. 7–299; Federico Villalobos, *El sueño colonial. Las guerras de España en Marruecos* (Barcelona: Ariel, 2004), Chap. 6; Fernando Martínez de Baños Carrillo, *Aduares y gumias. Melilla, 1909* (Cuarte de Huerva: DELSAN, 2009).
93. Juan Avilés, 'Resumen de la guerra', *RCM*, 35 (1910), pp. 6–7.
94. Torcy, *Les espagnols*, pp. 226–7. Stating, as Stanley G. Payne does, that the Spanish infantry used 'extremely old-fashioned' tactics in the Melilla campaign is wrong in the context of contemporary tactical doctrine; a different matter is their suitability for the battlefield conditions created by modern firearms: Stanley G. Payne, *Politics and the Military in Modern Spain* (Stanford: Stanford University Press, 1967), p. 111.
95. Juan Avilés, 'Enseñanzas de la guerra del Rif', *RCM*, 35 (1910), pp. 51–3.
96. Eduardo Gallego Ramos, *La campaña del Rif (1909). Orígenes, desarrollo y consecuencias* (Madrid: n.p., c. 1910), pp. 152–3.
97. Avilés, 'Enseñanzas de la guerra del Rif', p. 55.
98. Gallego, *Campaña del Rif*, pp. 302–3.
99. Bernardino García Conde, 'El nuevo reglamento en la práctica', *LCM*, 2 July 1909.
100. Avilés, 'Enseñanzas de la guerra del Rif', pp. 98–100.
101. Carlos Martínez de Campos y Serrano, *España bélica. El siglo XX. Marruecos* (Madrid: Aguilar, 1972), p. 96.
102. Martínez de Campos, *España bélica*, p. 97. Carabanchel was the site of a training ground near Madrid.
103. 'Impresiones', *LCM*, 2 August 1909.
104. Avilés, 'Enseñanzas de la guerra del Rif', pp. 113–16.
105. Gallego, *Campaña del Rif*, p. 305.
106. Estado Mayor Central del Ejército, *Enseñanzas de la campaña del Rif en 1909* (Madrid: n.p., 1911), p. 81.
107. Estado Mayor Central, *Enseñanzas*, pp. 82–3.
108. Gallego, *Campaña del Rif*, pp. 304–5.
109. Avilés, 'Enseñanzas de la guerra del Rif', pp. 257–9; Gallego, *Campaña del Rif*, pp. 309–10.
110. Burguete, *La guerra y el hombre*, pp. 244–5.
111. Critón, *Hágase Ejército*, pp. 150, 250–1, 260.
112. Avilés, 'Enseñanzas de la guerra del Rif', pp. 100–1.

113. Tim [H.E.] Travers, *The Killing Ground: The British Army, the Western Front and the Emergence of Modern Warfare, 1900–1918* (London: Unwin Hyman, 1990), pp. 48–50.
114. Ministerio de la Guerra, *Reglamento Táctico de Infantería* (Toledo: n.p., 1914) (hereafter RTI 1913), Article 4. Italics in the original.
115. RTI 1908, Article 260. Italics in the original.
116. RTI 1913, Article 248.
117. RTI 1913, Articles 263 to 352, and Articles 353 to 375, respectively.
118. RTI 1913, Article 5.
119. Emilio González y P. Villamil, 'El nuevo Reglamento táctico de Infantería', *Memorial de Infantería* (hereafter *MI*), 3, 2 (1914), p. 312.
120. Epifanio Gascueña, 'El fuego y la maniobra como medios de combate', *MI*, 3, 2 (1914), p. 321.
121. Dolla, 'Resumen de las armas', p. 70; T.H.E. Travers, 'Technology, Tactics, and Morale: Jean de Bloch, the Boer War, and British Military Theory, 1900–1914', *Journal of Modern History*, 51 (1979), pp. 272–3.
122. Epifanio Gascueña, *Empleo de la fortificación ligera en la ofensiva* (Madrid: Publicaciones de los Estudios Militares, 1915), p. 17.
123. Gascueña, *Empleo de la fortificación ligera*, p. 19.
124. Gascueña, *Empleo de la fortificación ligera*, p. 7 footnote 1.
125. Geoffrey Jensen, *Franco: Soldier, Commander, Dictator* (Washington, DC: Potomac Books, 2005), p. 12.
126. Geoffrey Wawro, *A Mad Catastrophe: The Outbreak of World War I and the Collapse of the Habsburg Empire* (New York: Basic Books, 2014), pp. 60–2.
127. It is remarkable the resemblance of Boer and Moroccan native tactics as described respectively in Federico García Rivera, Introduction to F[ernando] Ahumada, *La Infantería en la Gran Guerra. Su evolución táctica*, 2 vols. (Toledo: Colección Bibliográfica Militar, 1928), I. 16–25; José Díaz de Villegas, *Lecciones de la experiencia. (Enseñanzas de las campañas de Marruecos)* (Toledo: Colección Bibliográfica Militar, 1930), pp. 23–9.

CHAPTER 4

The Machine Gun, 1899–1913

Technology was transforming warfare by the beginning of the twentieth century. Although the military establishment still thought in terms of a battlefield where psychological factors maintained their importance, it nevertheless had to cope with technological innovation. The machine gun provides an instance of such a process. Due to its mechanical production of high volumes of firepower, this weapon symbolized the new kind of warfare. The study of the debate within the Spanish military about the machine gun's importance and tactical role sheds light on the problems of implementing technological innovation.

The Psychological Concept of Warfare

Warfare as an Exercise of Willpower

Captain Dolla, in a lecture published in a professional journal in 1903, defined modern war as a natural phenomenon in the life of nations; it had a passionate, moral nature and affected all the citizens to the highest degree. Because of this extended popular involvement, war disturbed the country's life so much that it could not last long. It must be ended quickly through a vigorous, decisive action aimed at the moral destruction of the enemy (the physical one was an inevitable outcome of the former). The goal was to achieve fast, decisive, moral successes through the physical exertions of the fighting men, which were 'supported and

J.V. Herrero Pérez, *The Spanish Military and Warfare from 1899 to the Civil War*, DOI 10.1007/978-3-319-54747-3_4

encouraged, above all, by great firmness in their hearts and absolute contempt for their own lives'.[1]

Arguments like this endorsed the idea of war and battle as a psychological contest, a clash of wills, which the material factors were subordinated to. Thus Burguete wrote that 'war is a clash of two wills rather than a clash of weapons: the strongest one beats the weakest'.[2] The origins of this concept can be traced back to the works of essayists of the nineteenth century such as Joseph de Maistre, Carl von Clausewitz, and Charles Ardant du Picq, whose thinking was to be used as a justification of the doctrine of the offensive at all costs.[3] Even essayists who realized the increased importance of firepower shared this view: 'Battles are lost from moral reasons, not material ones…', wrote Banús.[4] This concept was also understood as a negation of any revolution, caused by technological progress, which altered the essentials of war (that is, the moral forces), as essayists like Ivan Bloch argued.[5]

Something like moral forces was what, according to a contemporary critic, the Spanish troops in Santiago de Cuba were short of. It was true that they had not collapsed. They made a gallant stand against overwhelming odds, while the Americans often went to ground under Spanish fire. But the latter advanced boldly wherever the danger did not loom large and also did not hesitate to send patrols and small columns far ahead. On the contrary, despite their fine displays under fire, the Spaniards suffered from paralysis outside the firefight. This critic attributed this alleged unwillingness to attack to a lack of 'moral force'; he defined this shortage as a sort of accidental state of mind derived from the lack of strong patriotic and military spirit (this allegedly was a common feature of the contemporary Spanish society and institutions) and of will to win (the Spanish troops fought just enough to satisfy military honour, thinking that they could do nothing to alter the final result of the campaign).[6]

Not that casualties did not matter. The morale breakdown leading to defeat was certainly caused by physical losses. But how many losses were needed to break an army depended on its state of mind or, to use contemporary words, its 'moral energy'. An army could accept defeat after suffering a few casualties, and the next day could keep on fighting and achieve victory despite a heavier loss. Lieutenant Colonel Banús pointed out that modern firearms, despite their improved performance, were no deadlier than the older ones, as comparative percentages of killed and wounded in battle proved. Actually the casualty percentages diminished in more recent wars (for instance, the Franco-Prussian War, the Boer War). Banús put down this divergence between increasing weapon performance and decreasing casualties either to a decline of military spirit in modern armies, or to its insufficient growth in parallel to the difficulties of modern combat. So the reason for defeat was not the power of weapons, but the weakness of troops who could not endure the stress of modern warfare and who therefore broke with a lesser number of casualties.[7]

Actually Banús was aware that the troops had responded to growing firepower by means of dispersion, but he had also thought it immaterial to his point above. The dispersion on the battlefield meant that the percentage of casualties suffered by a whole army per day of fighting declined. But whenever a commander forgot the lesson and ordered his troops to advance steadily across open ground, the latter suffered heavily. The remarkable point is that the soldiers' sensible answer to a more lethal environment was seen as a sign of moral softness.[8]

What modern weapons had created was the 'empty battlefield'. On it, the elements which traditionally lessened the effects of fear and stress on the troops disappeared. Contemporary military essayists like Navarro and Banús recognized that the soldier could no longer see his enemy, could not assess the effects of his own fire, and did not feel the physical contact of his comrades. If battle was a matter of morale, what troops needed in order to keep up their will to win under these conditions was, above all, more moral training in order to foster their courage, discipline, and sense of duty.[9] Thus, Captain Gallego concluded, 'the soldier, far from staying the selfish man of current times, turns into a *being* deserving admiration and respect…'.[10]

Social Evolution and the Problems of Universal Military Service

The military essayists of this period saw in the strengthening of fighting morale the answer to the challenge of firepower. At the same time, however, these essayists denied or doubted the capacity of conscript armies (like the Spanish one) to improve the military virtues of their troops. On the one hand, the terms of service in the colours were getting shorter; on the other hand, there was a belief that the social conditions of civilized, modern nations tended to lessen the military spirit of conscripts. Banús complained of the spreading of anti-militarist doctrines in the working classes, but he also deplored the 'passive' anti-militarism of the bourgeois middle class and the well-off, which evaded conscription by paying a fee or purchasing substitutes. Nevertheless, he reckoned that the alternatives for Spain were not much better (Banús was writing before the 1912 military service act). An all-volunteer army was too expensive a solution, whereas universal military service would just add another group of disgruntled recruits to those already conscripted. Only a revival of patriotism in Spain would allow the development of a reliable conscript army.[11]

Major Juan Calero also argued about the disintegrating effects of class differences for the military units' cohesion if universal military service was implemented. The modern, egalitarian society would never adapt to the gap between officers and enlisted ranks, who had different duties, rights, and functions, and therefore were totally separate. Class differences would make impossible any comradeship among the enlisted ranks; moreover, the educated conscripts would bring the external atmosphere of social and academic indiscipline into the barracks and transmit a rebellious spirit to the uneducated mass. Calero even classified social classes according to their warlike aptitudes: the best soldiers were peasants, followed at some distance by 'craftsmen' (presumably industrial workers), while well-off people were the very worst.[12]

This thinking on the problems of universal military service reflected the Western armies' distrust and pessimism about the nature of industrial, positivist society and its manpower—even in a country like Spain, which started her industrial revolution late and whose army drafted many conscripts from the countryside.[13] Civilization seemed to weaken the qualities of man for war. An infantry officer and teacher at the Academy of Infantry, First Lieutenant Enrique Ruíz-Fornells, had already espoused this idea in 1894. He stated that the development of industry and commerce and the interest in other civil activities, plus the prevailing positivist thinking, had diminished Spain's military spirit.[14] This line of thought also helps to explain the scorn displayed by the Spanish press about the military virtues of the more industrialized United States before the Spanish—American War.[15] The political-military newspaper *La Correspondencia Militar* echoed these views in the months before the outbreak of war, when its articles often referred contemptuously to the Americans as a 'nation of merchants'. One of its contributors even alluded to a lack of manliness when he wrote about the ridiculous war scare in the United States, a nation that had much more wealth and 'more *people* (I hesitate to say *men*) than Spain'.[16] And there was little subtlety in an anonymous feature whose writer stated that war was also waged by means of 'dignity, a sense of self-respect, and a commodity that the Yankees have never had, and therefore it has been impossible for them to store it in factories or banks. And there is precisely more than enough of it in Spain. *Pantalones* [trousers, meaning guts]'.[17]

The defeat in the Spanish—American War might have changed these attitudes about the military virtues of soldiers in an industrial society. However, there is no sign of such a change in the sources surveyed by the present author. There are three plausible reasons for this absence of change. On the one hand, the fighting on land was too brief and limited, and the American expeditionary armies did not win any clear-cut victory (in the war's only major land engagement at San Juan Hill and El Caney, the American troops needed a whole day to expel an enemy one-tenth of its size from a line of weak field fortifications). On the other hand, according to a contemporary writer mentioned above, the Spanish troops suffered from a previous moral decay which offset their military virtues.[18] And the defeat of 1898 could also be seen within a larger context of military decline of the overcivilized European nations, whose scions were no longer good soldiers. This was the stand of Major Calero, who compared the Spanish overseas defeat to European setbacks in other contemporary campaigns against non-European enemies. Although he did not mention them explicitly, Calero was referring to the setbacks of the British army in the Boer War (1899–1902), the Italians in Abyssinia (1896), and the Russians in the war against Japan (1904–1905).[19]

Ricardo Burguete: Searching for Industrial Age Warriors

It has been mentioned above that there was a widespread belief in Europe, related to social Darwinism, that the modern civilized society's soldier was unable to withstand the stress of modern warfare and conform to traditional standards of military virtue.[20] Examples of this pessimistic thinking are provided by that prolific essayist of the first years of the twentieth century, Ricardo Burguete. The influence of anti-positivist irrationalism is very clear in his book *Así hablaba Zorrapastro*, published about 1903. As the title reveals, this work was inspired by Friedrich Nietzsche's *Thus Spoke Zarathustra* and praised the way the German philosopher exalted war and warriors.[21]

According to Burguete, civilization wore out the warlike qualities of man so much as to render conscription harmful for armies. Civilization also turned war into a business for everybody, since technology could turn any man into a mere weapon operator. Despite its importance, however, technology did not always transform war into a matter of calculation. Therefore, military art sought the solution of conflicts in the manliness, habits, and morality of peoples; in other words, in what Burguete called the 'warlike education'. This education had to be acquired at home during the childhood and teenage years. However, civilization despised this sort of education: it showed war as something horrible and paid homage to progress, not to force. Children were punished whenever they displayed willpower, courage, and daring, and taught to loathe weapons and the use of force. Thus they reached the conscription age fully convinced that military service was a stage to pass with resignation and—in case of war—by taking as few risks as possible.[22] On the contrary, what had to be emphasized were the virtues strengthening the character of man. As Burguete wrote later: 'Our cult must be the cult of courage and heroism.'[23] He praised the contribution of the Japanese warrior's code of behaviour (*bushido*) to the Japanese victories in Manchuria in 1904–1905. The *bushido* was, in Burguete's view, a manifestation of the warlike, heroic ethos that true soldiers had to assume.[24] Burguete was one of many foreign military theorists who admired the bravery and self-sacrifice displayed by the Japanese troops in massed infantry assaults. Unfortunately for those infatuated with *bushido*, as was shown in Chap. 3, such a daring was less durable than met the eye.[25]

Contemporary mass armies neither got reliable, 'war-educated' short-service conscripts nor could hope to educate them in the colours. According to Burguete, discipline would not be enough to keep the cohesion of conscript mass armies once in battle. On the modern battlefield, without the close leadership of their officers and NCOs, exposed to the play of inconstant emotions, the conscripts would need very little stimulus to be overtaken by panic and transformed into a crowd in flight. Incidentally, this would also make cavalry an important arm on the battlefield again: their shock action would cause, or take advantage of, the frequent panics in the enemy masses.[26]

Civilization produced smaller numbers of able men for war or made necessary a longer time to train real soldiers. On the other hand, Burguete argued, progress, by improving weapons, emphasized quality rather than quantity. The performance and complexity of modern weapons were increasing continuously, so the soldier should be more robust and skilled as well.[27] However, this was not possible with short terms of conscription (not more than three years). Besides, the law made every man a soldier whatever his aptitudes; since these were no longer thought to be important, the soldier turned into an ancillary element of the rifle, only trained in musketry.[28] Burguete proposed a professional standing army of volunteers matching the standards for modern warfare; compulsory service would be limited to a case of serious threat to the country. However, Burguete then contradicted himself, since he stated that, as a preparation for this contingency, the population would be given a military or warlike education from childhood.[29]

Thus Burguete rejected conscription but he put forward, at the same time, a sort of pre-military teaching whose lack was supposed to render conscription ineffective. So, if the nation's manpower became 'educated' for military service, why could not it be conscripted then? Perhaps Burguete thought that only an invasion of the realm would galvanize his compatriots enough to join the colours willingly. If so, this thinking would reflect once again his low estimation of the modern man's fighting spirit.

In sum, Burguete's ideas embraced two interrelated, contemporary solutions to the problem of the tactical offensive facing the increased firepower of modern artillery and infantry weapons. One of these solutions was the strengthening of the moral and psychological qualities necessary for victory in war (character, sense of duty, 'warlike education', the cult of courage). The other sought to produce a new, much better trained and highly enthusiastic soldier able to overcome the new weapons' effects on morale and to move forward on the modern battlefield (Burguete's ideal of the professional soldier).[30] Many of Burguete's writings displayed views too radical for him to be considered representative of the Spanish army's mainstream thinking. But the evidence shows that Burguete must be valued rather as an extreme point of reference on ideas which were discussed in the military at that time.[31]

For example, ideas about the need for increased moral qualities were certainly widespread in the Spanish military literature. A reason for this might be the influence of the education in the Spanish military academies, which exalted high spiritual values, shaped within a nationalist framework—which, in turn, was based on the military glories of the sixteenth and early seventeenth centuries.[32] The military also sympathized with the idea of the need for 'regenerating' morally a decaying Spanish society: a need which seemed highlighted by the defeat of 1898 and was promoted by contemporary intellectuals and authors such as Ramiro de Maeztu (who corresponded with Burguete) or Joaquín Costa.[33]

Another influence was contemporary widespread social Darwinist ideas, according to which the country whose population had the highest moral qualities would be successful in the struggle for survival among nations.[34]

Burguete was a rather open spokesman for such ideas and, in one of his early works, made explicit mention of Charles Darwin. Burguete compared war to the struggle for existence due to the inequality between the growth of species and the availability of resources, when the imperative of one's own survival obliged the strongest to eat up the weakest.[35] Burguete wrote a few years later: 'A nation cannot live beside other vigorous nations without accepting the struggle for life which the latter impose on the former.'[36] Backward or weaker nations had to disappear in order to allow the natural, selective progress of the rest of the species.[37] Burguete, in interesting contrast with the attitudes towards the Americans before the 1898 war, also compared warlike spirit to the spirit of commercial initiative, since the conquest of markets had become a 'physiological' need of nations.[38] Actually he devoted a whole book (*Morbo nacional. Vida defensiva*) to denouncing the 'defensive', protectionist attitudes of Spain's economy, which could only lead—as in war—to defeat and annihilation.

Among all this literature stressing the psychological and moral factors in warfare, an exception which deserves at least a mention in passing was the ideas of José Marvá, an officer of the engineer corps. He had no doubts on the increase of destructiveness brought by technology. Against this, courage and manliness were of no use. The human elements that could decide battles in the past were no longer useful unless they were complementary to—and based on—the power of modern weapons.[39] A scholar and teacher in issues related to ordnance and military engineering, Marvá was, according to the historian Miguel Alonso Baquer, the most prominent figure in advocating the technological modernization of the Spanish army in the last decades of the nineteenth century. However, after the defeat of 1898, his ideas were the last token of a tradition of reformist, learned thinking born from a scholarly, analytical approach to the military problems. This tradition had virtually disappeared by 1917, and, indeed, Marvá's concerns turned towards social welfare issues from the early years of the twentieth century on.[40]

The Machine Gun: Early Trials and Organizatonial Debates

Machine Guns in the Spanish Army up to 1900

The Spanish army first tested a Gatling machine gun in 1867; as a result, the Artillery Board ordered an extension of the trials by attaching machine guns to an artillery regiment (one machine gun per battery).[41] If the trials were successful, the other regiments would be equipped the same way. However, this plan was cancelled after the appearance of the Cristophe-Montigny machine gun and new trials of this model were carried out. A decision to create six batteries (with six Cristophe-Montigny machine guns each) was finally made in April 1871. But this project was cancelled as well and only a single battery of four machine guns was formed in May 1872 and attached to an artillery regiment. All the same, the military authorities ordered up to 40 Cristophe-Montigny machine guns, which would be made in a Spanish factory (apparently, only 14 guns were completed). The Spanish army carried out trials of another Gatling machine gun in 1874 and compared it with the French model. The Artillery Board judged the machine gun more useful in the defence than in the attack and, if larger numbers of this weapon were ordered, recommended the Gatling model, which offered a higher rate of fire and less encumbrances (actually three Gatling machine guns were used in the Cuban colonial war of 1868–1878).

However, the disappointing role of the French army's machine guns in the Franco-Prussian War generated a wave of mistrust and scorn towards the new weapon. This wave also reached Spain, whose army forgot any idea of field use of the machine gun, and the Cristophe-Montigny unit was disbanded by 1874.[42] The existing weapons were assigned to fortresses, where they were joined by a few Swedish Palkrantz-Nordenfelt machine guns after 1881. Despite the appearance of the Maxim model (the first reliable automatic machine gun) in 1884, the Spanish lack of interest lasted until the outbreak of the last Cuban war. The army purchased 12 Maxim-Nordenfelt machine guns on wheeled carriages and sent them to Cuba in 1896. However, their performance was poor: the Maxims were not adapted to the Spanish issue rifle cartridge and this caused frequent jammings. The problem was not solved until the machine guns were sent back to Spain after the war against the United States. Once they were adapted, eight machine guns were assigned to the fortifications of Ceuta and Melilla and the remaining four to the Central Firing School.

The American army had used a few Gatling machine guns against the Spaniards in the battle of San Juan Hill. According to a Spanish source, the Gatlings performed reasonably well, though their effectiveness was exaggerated by their commanding officer.[43] The other Spanish sources did not mention the machine guns as a significant factor in the outcome of the fighting.[44] All this suggests that the Spanish army was not impressed enough to press forward the procurement of more machine guns.

After 1898 the machine gun seemed to be underrated. A military essayist, Luis de la Gándara, commented sadly by 1906: '[W]e neither have machine guns nor are there reasons to believe that we will own them soon but, on the contrary, there are many reasons to presume that *there is no atmosphere for the present* for the organization of those weapons in our army.' He added that, in his view, budget matters were not the only cause of this situation, but he refrained from mentioning the others.[45]

Corporate Procrastination, 1900–1907

Actually the Spanish army did keep an interest in the machine gun. In January 1902 the Academic Board (*Junta Facultativa*) of the Central Firing School pointed out that the major European armies had carried out experiments with machine guns and that, by that time, Germany, France, Britain, Switzerland and Russia had organized machine gun units. Though there were still doubts about their role on the battlefield, machine guns were to be highly valuable weapons due to their volume of fire (the board mentioned that, after trials in France, one machine gun was reckoned equal to 200 riflemen). The board urged the military authorities to pay serious attention to the issue and suggested its study in two stages. On the one hand, the Central Firing School would carry out the trials concerning the tactical use and firing methods of the weapon; on the other hand, the Artillery Experiment Commission would carry out the technical trials to decide the model to be chosen.[46] The artillery corps' technical board, after studying the proposal, stated several weeks later that experimenting with machine guns was its responsibility through the Artillery Experiment Commission.[47] The artillery technical board got its way, since the Artillery Experiment Commission carried out the trials alone, although these were not undertaken seriously until 1904. In this year several foreign models were acquired after a fact-finding journey of the commission abroad during 1903 and early 1904.[48] But the commission displayed a rather limited enthusiasm for the machine gun. Though the commission admitted the weapon's effectiveness in some particular cases, it stated that the usefulness of the machine gun had diminished after the widespread adoption of quick-firing artillery by all the armies. Indeed, its evaluation of the machine gun continued mainly because of foreign armies' acceptance of the weapon.[49] This last point suggests that the Spanish artillery corps' own assessment of the machine gun might not have justified such trials. The technical board of the artillery corps had actually proposed in 1902 to delay any decision on the machine gun.[50]

The overthorough nature of the trials held back the introduction of the machine gun. The Artillery Experiment Commission alleged that mechanical unreliability of the models was the cause of the slowness in the trials, which therefore delayed the definitive adoption of the machine gun by the Spanish army.[51] This search for mechanical excellence could be reasonable from a technical viewpoint. However, it was also harmful when armies all over Europe had already started equipping themselves with the models available. Indeed, an artillery officer, Captain Pedro Jevenois, urged the army to accelerate the studies to adopt machine guns after the experience of the Russo-Japanese War.[52]

Another problem was the corporate rivalry within the Spanish army. The artillerymen had little enthusiasm for the machine gun, but they had taken charge of its trials anyway. The machine gun was of secondary importance for the artillery corps and this might explain the leisurely attitude of the commission—though it had assumed the sole responsibility for the trials within the army as a whole. On the other hand, the artillerymen prided themselves on their technical knowledge and their responsibilities in ordnance issues. So, in order to preserve their corporate prestige, they were unwilling to accept that other corps took part in the trials.

Even after the adoption of an infantry machine gun, the artillery corps did not want to give up its hold over other corps' weaponry. The Artillery Experiment Commission stated in 1908 that the artillery corps had to assume responsibilities within the organization of machine gun units, since the performance of these weapons depended on their mechanical working and technical ordnance issues were handled by the corps.[53] About the same period (sometime from 1908 to the mid-1910s), the Artillery Experiment Commission allegedly refused to display the newly acquired machine gun model during a course in the infantry section of the Central Firing School unless an artillery officer was in charge of the presentation and explanations.[54]

Who Should Operate the Machine Gun?

One reason for the scepticism and the early ambivalent policy of the army about the machine gun was doubtless the ill-defined status of this weapon during the early stages of its development. Due to their technical characteristics, the first models of machine guns could not be classified neatly as an infantry, cavalry or artillery weapon, or as a new, independent tactical arm. An artillery officer, Lieutenant Colonel José Brull, summed up the hybrid features of the early machine guns: they fired infantry rifle ammunition; they were crew-served weapons and organized in sections and batteries as were the artillery pieces; and they were best suited to a quick and opportune use on the battlefield as the cavalry were. This meant that the infantrymen disliked them because of their similarity in organization and baggage to the artillery and the nuisance they caused for movement. The cavalry also thought that they hindered their mobility and dashing spirit, whereas the artillery disliked the rifle-like performance of their fire and still remembered the French machine guns' fiasco in 1870. Moreover, according to Brull, every arm thought that admitting the machine gun's support within its ranks was a sign of military decline or feebleness.[55] So it is not surprising that, at the beginning of the twentieth century, the teaching on machine guns in the Academy of Infantry formed part of a heterogeneous subject on 'Ordnance'.[56]

But Brull's views about the alleged unwillingness of the tactical arms to accept the machine gun were not shared by other artillerymen. Major José de Lossada was amazed that the artillery officers of foreign armies that had already adopted the machine gun handed it over to the other arms, though it was a weapon whose tactics and technical features made it suitable for the artillery. And Jevenois argued that machine guns should be operated by artillerymen, because if these weapons were assigned to infantry and cavalry units, the latter's mobility would be restrained.[57]

The artillery corps was indeed loath to give up its corporate interests. The Artillery Central Board censored an article written by Major de Lossada for the corps journal because he stated that machine guns were not artillery weapons only. The board stressed that de Lossada's article did not represent the corps' views on the status of the machine gun.[58] On the other hand, if the artillery could not control a new source of firepower that could threaten the corps' traditional role, this was an incentive to keep the new weapon out of the battlefield. Thus the artillery technical board concluded in 1902 that the machine gun would never replace the horse artillery to support cavalry on independent missions. And the board also stated that substituting a machine gun for a certain number of infantrymen armed with magazine rifles did not provide any significant improvement to the firepower of the infantry units, due to the mechanical limitations and cumbersome transport of the early models.[59] Nonetheless, this attitude was not shared by all the artillerymen.[60]

Captain Ruíz-Fornells, of the infantry corps, accepted that all the tactical arms had machine guns adapted to their own needs. On the other hand, since the machine gun was essentially a modified infantry rifle, it was only logical that the infantry were able to decide on its employment.[61] Major Calero—another infantryman—went further by rejecting any idea of the machine gun being considered an artillery piece. The infantry had to take charge of its study. Since it fired the infantry rifle cartridge and followed the infantry's manoeuvres, the machine gun could not be anything but an infantry weapon. Therefore, it must be used in the exclusive service of the infantry, and maybe the cavalry as well.[62]

Calero's last point had been anticipated—at least partially—by Lieutenant Colonel Banús a couple of years before. He concluded from the experience of the Boer War that machine guns did not replace artillery. Their real role was to increase the firepower of the cavalry and the mounted infantry.[63] Such an increase was badly needed on the modern battlefield. The cavalry remained the weapon of opportunity, but they could not create the opportunity by weakening the enemy infantry unless they were supported by horse artillery or their own cavalry machine guns.[64] However, First Lieutenant de la Gándara thought the machine gun so closely integrated within the infantry that no other corps should use it: there must not be machine guns in the cavalry because they could limit the latter's action by forcing the cavalrymen to look for firing positions.[65] Nevertheless, de la Gándara softened his radical view some pages later and reckoned a ratio of one machine gun company per cavalry division enough. A larger scale would cause the underemployment of the weapons, since they would find few opportunities for their use and 'opportunity is a *sine qua non* for them…'.[66]

Perhaps surprisingly, the cavalry showed interest in the new weapon as well. Despite the fact that the increasing firepower of modern weapons was eroding the cavalry's traditional tactical role, at least some cavalry officers seemed eager to adopt the machine gun. The reason was that they saw it as the remedy that would allow them to recover their arm's old status on the battlefield. Since a few machine guns could replace the firepower of many carbines, there would be less need for dismounted fighting. The traditional spirit and tactics of cavalry would be preserved and the *arme blanche* would remain the horseman's main weapon.[67]

An alternative solution to the problem of which tactical arm the machine gun had to be assigned to was the creation of a brand new machine gun corps. An artillery officer, Major de Lossada, had already weighed up an independent machine gun corps in 1903. He ruled it out because the machine gun lacked sufficient shock action, firepower and mobility (the main attributes of infantry, artillery, and cavalry, respectively) and had no other feature overcoming those of the traditional tactical arms.[68] But the idea of independence still found supporting voices in the Spanish military, such as Colonel Modesto Navarro, of the infantry corps. Though admitting their resemblance to the rifle, and therefore their closeness to infantry, Navarro thought that machine guns must become a fourth tactical arm; the combination of several tactical arms was much better than the mixture of different weapons in the same arm, so attaching a machine gun section to every infantry battalion or regiment would go against the prevailing foundations of warfare.[69]

Despite their differences on organization, both de Lossada and Navarro agreed in opposing the integration of machine guns into the basic tactical units (regiment and battalion), a view also shared by Jevenois, who thought that machine guns should be grouped in battalions of three or four batteries under the divisional artillery commander.[70] Navarro reckoned the division the right echelon to integrate a machine gun unit, whose commanding officer should not have more initiative than the divisional artillery commander.[71] De Lossada thought that machine guns would always be a strange element within an infantry or cavalry unit, even if commanded by officers of these arms, so they had to be a temporary attachment only and return under the general officer commanding the major unit afterwards. At any rate, the machine guns had to be used in separate sections.[72] In fact, the Artillery Experiment Commission reached a similar conclusion by 1908 and argued that attaching machine gun sections permanently would hinder the mobility of infantry and cavalry units.[73]

First Lieutenant Ángel García-Pelayo expressed disagreement with these views. A divisional or army corps commander was not in a position to realize the fleeting opportunities for a successful use of the machine gun. Therefore, this weapon had to be integrated into the basic tactical units of infantry and cavalry, whose commanding officers had a closer knowledge of the events on the battlefield.[74] Nonetheless, this did not necessarily mean a better understanding of the machine gun and its tactics, and the way weapons were organized within tactical units was not unimportant. A subaltern commanding a battalion machine gun section had less chance to influence tactical issues than the commander of a regimental company of machine guns.[75]

The Machine Gun and Tactics, 1900–1913

The Problem of Controlling Infantry Firepower

The discussions about the machine gun took place within a more general debate on the effectiveness of infantry fire. Captain Jevenois had small confidence in the infantry's individual musketry: the conscripts' short terms in the ranks, the isolation in dispersed skirmishing lines, and fear led the riflemen to shoot quickly and inaccurately, wasting their ammunition.[76] A proposed solution was collective rifle fire (that is, fire in which the officers announced to their men the targets, sight adjustment, and number of rounds to be shot). However, there was disagreement about the effectiveness of this kind of fire in fighting conditions, especially at medium and long range.

Jevenois stated in 1907 that the Russo-Japanese War had shown the increased importance of collective fire over individual fire.[77] On the contrary, Captain Emilio Rodríguez Tarduchy—an infantryman—was sceptical about the effectiveness of the collective fire proposed by the regulations. He thought that it was based on firing range experiences, far from the real conditions of the battlefield. He also remarked on the great difficulties that officers would find in directing their units' fire: they would be too busy in leading and keeping their men steady, their units would be too dispersed, and the transmission of orders would be too difficult to allow accurate fire direction. Most times fire direction and tactical command of the troops would be the same thing. Collective fire was really individual fire at the same time and the riflemen must be trained accordingly.[78]

It seems plausible that Rodríguez Tarduchy had taken into account the experience of the early Moroccan campaigns. General Rodríguez del Barrio recalled that he never saw the use of collective fire during the years he served in Morocco (1909–1917). Many officers seemed to be ignorant about that kind of fire, although it was included in the firing regulations.[79] But rather than ignorance, perhaps the reason for these officers' conduct was that they thought that collective fire made little sense on the African battlefields. According to Colonel Bermúdez de Castro, the contemporary doctrines of collective fire were useless against the natives. The latter fought scattered on a wide front without forming a regular skirmishing line. Therefore, the riflemen had to rely on their individual marksmanship, since the officers could no longer direct their units' collective fire against clear targets.[80]

However, it seemed difficult to improve the individual musketry training of conscripts when there was a trend to reduce the terms of military service. One solution, proposed by Captain Ruíz-Fornells in 1904, was to increase the numbers of machine guns—which would allow the army to exploit to the full the qualities of accuracy and range of the modern small firearms.[81] A contributor to a professional journal pointed out the greater effectiveness of the machine gun over the riflemen to fire at the range from 600 to 1500 or 2000 metres.[82] Some officers expressed the view that machine gun fire was less effective than rifle musketry in overcoming extended targets. They did not take into account the fact that a machine gun was less vulnerable and easier to control than a line of riflemen, and that the former's capability to hit targets successively through accurate concentrations of bullets made up amply for the 'narrowness' of its field of fire.[83]

The Machine Gun as a Weapon of Opportunity

Although essayists like Ruíz-Fornells envisaged an intensive use of the machine gun as a way to deliver controlled infantry firepower, mainstream thinking did not favour such use by the infantry (nor indeed by any arm). The machine gun was seen as a weapon of opportunity whose use on the battlefield was limited to infrequent favourable moments of the fighting. Even an essayist asserting the machine gun's role in all the stages of battle recommended only careful, occasional use.[84]

There were objective reasons for such a view. In the early years of the twentieth century the machine gun certainly was not a handy weapon to use in the attack. A machine gun in movement—with its crew, its transport horse or mule, and assorted baggage—offered a clearer target than a small group of riflemen taking advantage of the folds of the terrain.[85] The machine gun also generated distrust about its performance because it was exposed to overheating and mechanical breakdowns, and consumed huge amounts of ammunition which was difficult to carry to the firing line.[86] Therefore, it was a logical step to conclude that its effectiveness was only guaranteed for short periods of time (this also meant that its effects were understood as temporary). Thus the machine gun was 'the weapon of the moment' (that is, a weapon of opportunity). According to this concept, the machine gun could not be used continuously in all the stages of the combat; otherwise the machine gun's tactical role would lose its essence and be exposed to failure.[87]

It is noteworthy that the expenditure of ammunition was of such concern that it led de la Gándara to disapprove of any *tactical* use of the machine gun (that is, just like any other accepted weapon on the battlefield), unless it could produce a significant enough outcome.[88] The historian T.H.E. Travers has shown a similar concern in the British army to save ammunition when the essence of the machine gun is precisely to produce high volumes of firepower (which logically consume a lot of ammunition).[89] This suggests that contemporary armies were unwilling to exploit the full capabilities of the machine gun by devoting more effort to solving the problem of the supply of ammunition. In other words, the machine gun—in order to be an effective weapon—needed large amounts of ammunition (which admittedly was not always easy to supply), but instead of seeking ways to overcome supply difficulties, the military rather preferred to solve the problem by restraining the tactical role of this new weapon on the battlefield.

The Battlefield Role of the Machine Gun

Besides the issues mentioned above, the increasing emphasis of the Spanish army on offensive tactics was a significant factor in underrating the military value of the machine gun. Firepower technology had shifted the balance to the side of defence, but, as the first part of this chapter has shown, the Spanish military's mindset was dominated by a psychological concept of warfare. Despite the increased power of firearms (or precisely because of this), the human factors—morale, discipline, and so on—were still exalted as the decisive ones. And within this frame of thought, offensive tactics were the surest way to achieve victory, since they required the highest display of these factors. Machine guns, the latest piece of firepower technology, could not help running into trouble in this atmosphere. The army's initial reaction was the adaptation of the weapon and its qualities to the predominant frame of tactical thinking rather than the other way round.[90]

Captain Herrera de la Rosa, who was an observer attached to the Japanese army during the Russo-Japanese War, reported that the use of the machine gun demanded a certain degree of passivity in the conduct of operations. This, plus the mechanical requirements and the high ammunition consumption, forced the armies to keep the machine gun for the right moment only.[91] So de la Gándara concluded that the machine gun's role 'is reduced to reinforcing [cannon and rifle] fire [and therefore lacks] enough distinctiveness to introduce variations in current general tactics'.[92]

De Lossada acknowledged that machine guns could be effective in the advanced guard (where there were more chances for sudden clashes). But they must be removed from the firing line once the main force of the army went into action, either for attack or defence (de Lossada was more interested in the former).[93] The machine gun was seen as a firepower reserve, which had to be used by surprise in order to achieve the greatest effect in a short time (that is, as a weapon of opportunity). In practice, this meant relegating the machine gun to secondary operations on the fringes of the battle (to deliver firepower when there were not enough troops) and in special kinds of combat (such as mountain and irregular warfare).[94]

Besides keeping the machine gun out of the major clashes on the battlefield, the military underrated its defensive capabilities. Contemporary tactical thinking seemed indeed to be reluctant to envisage the deployment of many machine guns for defence in the firing line. De la Gándara only justified such a deployment if the defensive front was very long, but he also warned about the huge consumption of ammunition it involved.[95]

The Effects of the Machine Gun on Morale

With hindsight, restraints on the use of the machine gun appear surprising at a time when the military essayists were worried by the double-edged effects of contemporary firepower. The increase in individual effectiveness of small firearms after the mid-nineteenth century had been counterbalanced by the dispersal of their operators. This had reduced the officers' control over their men and, in a sense, the collective effectiveness of units (it was more difficult to maintain fire discipline). But the machine gun put back in the commanders' hands the means to produce firepower in a highly disciplined way. Since it was a machine, such a weapon removed the frailties of the human being from the act of shooting.[96]

Captain Ruíz-Fornells acknowledged this fact. He was a rare advocate of the use of the machine gun during all the stages of combat. Ruíz-Fornells stated that it was impossible to exploit fully the effectiveness of rifle fire—even if the riflemen were all marksmen—since the soldiers were too shaken by the ordeal of battle to make the best use of their weapons. On the contrary, the machine gun was essentially a multiple automatic rifle and a nerveless weapon, which substituted a stable mounting for the staggering man as its firing platform. And it needed very few men to produce more effective firepower than a large number of troops armed with rifles.[97]

But these two features of the weapon (its mechanical nature and the reduction of troop numbers it involved) were fated to cause rejection on the part of contemporary military minds, which were attached to an image of warfare based on psychological, human factors. Technological innovation was accepted insofar as it bolstered those elements; those sides of innovation implying an essential change in the established image of warfare were overlooked or underrated. It was admitted, as early as 1900, that the machine gun was 'concentrated infantry' whose fire was more accurate than the 'real' infantry's. However, it lacked the infantrymen's morale. So the machine gun could be superior to any infantry of low morale and poor training in musketry, but it could never be a match for infantrymen who were highly trained and motivated individuals.[98]

On the other hand, the machine gun represented a kind of firepower that was seen as threatening for the foundations of the contemporary offensive tactics. T.H.E. Travers has pointed out the strong concern of the British military establishment with the 'weight' of the attack and, therefore, with numbers and manpower. The 'energy' or drive of an attack was supposed to derive from the latter's 'weight' (that is, the strength of the attacking troops).[99] Thus it was logical to see a weapon that implied a reduction of manpower on the firing line as a weakening of the offensive.

The professional literature shows quite similar concerns in the Spanish army. Captain Dolla pointed out the importance of the advance *en masse*, even in exaggerated numbers. The troops, by pushing each other physically within close-knit masses, would get the advantages of comradeship and discipline.[100] Ricardo Burguete saw in sheer numbers the way to get closer to the enemy on the fire-swept battlefield: 'Against the huge volume of bullets with which the enemy will cover our infantry's front… it is necessary to throw men, and sometimes "human masses"…'.[101] Writing after the Russo-Japanese War, Lieutenant Colonel Banús concluded that decisive success was only achieved, in moral and material terms, through mass.[102] Martínez de Campos recalled a manifestation of this tendency to see fighting value in sheer numbers during the Melilla campaign. He described how, during the night hours, lots of rifles were placed on the stone wall of a Spanish position, ready to fire without warning: 'Nobody thought of replacing them with a few machine guns enfilading the barbed wire. Numbers doubtless gave an impression of greater strength; they gave it to ourselves, at least.'[103]

As regards the machine gun, this thinking meant a failure to understand that the effectiveness of its fire was what gave real fighting value to an infantry unit, and that technology had changed the quantitative relationship between strength and volume of firepower. The authors of a work published in 1908 offer an example of such a lack of understanding: '[S]ome authors define the machine gun as *condensed infantry*; but this can only refer to the intensity of its fire, that is, to the material part, since *concentration* disappears in terms of personnel and actually very few infantry turn out.'[104] These words suggested that, for example, an infantry battalion with a reduced strength and equipped with several machine guns was deemed 'less infantry' than a traditional battalion armed only with rifles. But modern firearms allowed a greater firepower to be produced with the same strength, or permitted a modern unit to deliver the same volume of firepower with a smaller strength, a fact already recognized by Captain Ruíz-Fornells in 1904.[105] Nonetheless, in 1908 the Artillery Experiment Commission did not hesitate to call the idea of condensed infantry a falsehood.[106]

The issue of numbers had a direct influence on tactics. For it linked the thickening of the skirmishing line to the necessary superiority of fire before the assault. That is, in order to win, the attacking side had to achieve moral superiority, which derived in its turn from the superiority of fire. The problem was that there was not a clear way to know when this superiority had been achieved.[107] As a contemporary essayist put it:

> Achieving fire superiority is an issue which almost has to do more with the troops' morale than with the material effects caused by fire. The thing that decides when fire superiority has been achieved is the fact of advancing under fire. If the attacking side makes progress despite the defender's fire, this means that in practice the former has achieved that superiority.[108]

The solution that best conformed to the psychological concept of warfare was to think (as Lieutenant Colonel Banús did) that accumulating the largest number of riflemen for the firefight translated into a superiority of fire—and, therefore, of morale. The fact that the machine gun was very effective against thick lines or groups of riflemen at short range—as de la Gándara admitted—was overlooked, as was the idea of using machine firepower to replace riflemen.[109] Incidentally, Banús reckoned the density of lines necessary for the defence as well, whereas de la Gándara thought that machine guns could save manpower, at least in passive defensive sectors.[110]

The Machine Gun in the Campaign of Melilla, 1909

Procurement

The push for the acquisition at last of the first batch of machine guns for operational purposes after the Cuban campaigns was due to the commercial initiative of the representative in Spain of the French arms manufacturer Hotchkiss, whose name seems to be Esteban Marín, who wrote on 14 August 1907 to War Minister General Fernando Primo de Rivera (uncle of the future dictator of the 1920s). Marín explained that the events in Morocco led him to suppose that the Spanish army would urgently need to acquire machine guns for use in a likely intervention in northern Africa. He offered the opportunity to get 20 Hotchkiss machine guns in the Spanish rifle calibre, which would be delivered within four weeks of being ordered. This could be achieved if the Spanish government made its order within ten days. Marín admitted that the proposed figure was his own guess, but the manufacturer could accept a larger order, if required. He also remembered that the Spanish army had already tested a Hotchkiss machine gun in 1902 and concluded that it had performed better than the Maxim model.[111]

The urgency of the Spanish government was apparently genuine, since a document from the War Ministry (of unidentified origin, but which surely stemmed from the Artillery Section), dated 21 August 1907, records that the war minister gave an order this same day to start the procedures for the direct purchase of 20 Hotchkiss machine guns. Nevertheless, the document's content was a warning against a rash purchase and advised waiting until comparative trials with other models were carried out. If this was not possible, it recommended that a number of officers be sent to test the machine gun in the manufacturer's factory.[112] Urgency prevailed and on 22 August 1907 the Artillery Section informed the captain general of the First Military Region that two officers of the Artillery Experiment Commission were to be sent to France in order to examine the machine guns on offer and to arrange the purchase.[113]

The commission's officers, Colonel Luis de Santiago and Major Luis Esparza, travelled to France to check the mechanical performance of the Hotchkiss machine guns, but they also met army officers to collect information on the French military's ideas about the role of machine guns. The Spanish commissioners' conclusions on the subject (the machine gun was essentially a defensive weapon; it was not condensed infantry; it must be under control of major unit commanders) inspired the Artillery Experiment Commission report of April 1908, and probably much of the Spanish thought after 1908 as well.[114]

The contract of purchase for the first Hotchkiss machine guns was signed on 31 October 1907, with more Hotchkiss weapons being bought by the Spanish army later.[115] But the overall numbers of machine guns were not excessive. In September 1908 there were 12 sections (two machine guns each) for six brigades (including those equipped with the old Maxims). Thus the ratio was one section to three infantry battalions. By early 1910, after the Melilla campaign, there were 22 Sections.[116] Besides the issues about machine guns and tactical doctrine discussed above, the financial factor must also be taken into account. Even if the Spanish army had developed a more favourable attitude towards the machine gun, the continual restraints on ordnance procurement would probably have slowed down any ambitious policy about the machine gun. In 1909 the Spanish army had a ratio of 3.9 artillery pieces to 1000 men; even the armies of countries like Greece (6.9), Portugal (5) or Bulgaria (4.6) had higher ratios.[117] If this happened with a traditional weapon, the slow introduction of a new weapon whose value was still uncertain is not surprising.

The Lessons of the Campaign of Melilla

The brand new machine gun units underwent their baptism of fire during the campaign of Melilla in 1909. Contemporary judgements about their performance in that conflict were broadly positive, with the only exception found by the present author being in the remarks of Lieutenant Colonel Avilés. Avilés stated that the machine gun was not useful against an enemy like the Moroccan irregular fighting man. Restraint prevailed after an initial period of frequent use, according to Avilés, who concluded that the machine gun had to be kept for the defence of special places.[118]

On the contrary, Lieutenant Colonel de Lossada praised the accuracy and rate of fire of the machine gun. These qualities were very helpful against the scattered Moorish fighting array, since the artillery had little effectiveness without clear-cut targets and the musketry of the infantry was poor due to the conscripts' limited training.[119] De Lossada indicated that the machine guns had often fired without interruption during the combats. This went against the Central Firing School's doctrine, which based machine gun tactics on surprise.[120] Actually, the first regulations for machine gun units, put into effect in July 1909, stipulated occasional use and surprise. But in practice things were rather different. Captain Federico Medialdea—an instructor of the Central Firing School sent to observe and report on the machine guns' performance—pointed out that the weapons used to go into action in the early stages of offensive combats and stayed in their fire positions until the end, sometimes firing without any interruption.[121]

The official doctrine about use as a weapon of opportunity was not followed in the defence either. Medialdea stated that the machine guns' performance was excellent *despite* their constant use.[122] And according to de Lossada, riflemen usually did not open fire when there were machine guns available.[123] The field commanders sometimes had ordered the machine guns to intervene even if the latter lacked suitable targets, because their moral effect made up for their small material effectiveness in such cases. The Spanish soldiers got used to the rhythmic shooting of the machine gun and highly valued its covering fire. On one occasion Medialdea witnessed how the riflemen, advancing under heavy enemy fire, looked back from time to time and wondered why the machine guns did not open fire. Such behaviour had a logical foundation. Whenever the machine guns swept the positions whose fire was more dangerous, the enemy's shooting diminished perceptibly. Then the Spanish skirmishing line could advance with a greater resolution, though the soldiers still did not see their adversaries.[124]

In its post-mortem about the campaign, the General Staff discarded any notion of using the machine gun units as an independent fighting arm. Their employment according to the tactical rules of the artillery was rejected as well. But other conclusions were rather contradictory. The General Staff still supported the idea of using surprise as the base for machine gun tactics. At the same time, however, it concluded that the machine gun had to be used frequently, since it had turned out to be a powerful support of musketry.[125] Frequency and surprise are requirements difficult to match on the battlefield, and this contradiction reveals hesitant thinking about the tactical role of the machine gun. However, the experience of Melilla did not seem to lead to significant alterations to the prevailing views about the machine gun. Proof of this is the almost total neglect of the weapon in the 1913 infantry regulations: there were only five articles, and these were dedicated to fighting *against* machine guns.[126]

In his survey of the machine guns in the Melilla campaign, Captain Medialdea made a point that is useful in understanding contemporary attitudes towards the machine gun. He advised against the general use of sweeping fire on the enemy positions due to the huge expenditure of ammunition it involved, even though it was of benefit to morale.[127] This appears an odd conclusion when the military were so eager to sustain the offensive spirit of the troops. For it was contradictory to dispense with the machine gun when this weapon actually bolstered the morale of the attacking troops. But it was not contradictory within the terms of the psychological concept of warfare ('the psychological battlefield'), which placed too great a weight on the human-centred factors to the detriment of firepower (a more mechanical factor).

Conclusion

How does the Spanish army's early controversies and policy on the machine gun compare with those of other European armies? There was nothing really exceptional in them. The views expressed by the Spanish military in official sources and the professional literature did not diverge from those generally held in other armies.[128] Nor were the Spanish officers more reluctant to adopt the machine gun than other foreign colleagues. For instance, armies generally failed to implement the lessons of the Russo-Japanese War on the use of this weapon because of factors which can be also found in the Spanish case.[129] In the highly respected German army all the arms had rejected the new weapon and its integration in their units on the eve of the Russo-Japanese War in 1904, and it was 1909 before the German line infantry started to receive machine guns.[130] The most idiosyncratic feature of the Spanish case is the debate over what corps should be responsible for the weapon and therefore what corps should not be. Given that the machine gun met a good deal of opposition within each corps, the source of such contention could only be the corporate rivalry to hoard technical and tactical responsibilities which afflicted the Spanish army. It led corporate zealots to claim their own corps' control over a weapon they could actually dislike.

Apart from this, the way the Spanish military coped with the machine gun was essentially similar to the approach of other armies. The machine gun was a kind of firepower which, in being produced by machines, stressed the quantitative side of warfare, based on non-human, tangible factors (for instance, technology). These were susceptible of being measured and treated as physical, mechanical processes unaltered by human emotions. While the crewmen executed the mechanical movements needed to fire the weapon and feed it with ammunition, their emotions would not change the accuracy and the number of rounds fired per minute by the machine gun on a fixed beaten zone. Thus it was possible to develop mathematical models and statistical tables about events on the battlefield (how many men crossing the beaten zone would be hit, how many machine guns were needed to seal off a certain stretch of the front line). For the military establishment of the early twentieth century, accepting the increasing importance of mechanical firepower meant accepting a relative decline of the qualitative, traditional military virtues (discipline, self-sacrifice, and so on) that had been valued until then as the decisive factors in warfare and in which professional officers had grown up.[131] For many officers, if not all of them, this was a difficult change of mind, a change which could only be stimulated by the actual experience of combat.

Notes

1. Ángel Dolla, 'Concepto e importancia de la Caballería en abstracto', *Revista Técnica de Infantería y Caballería* (hereafter *RTIC*), 14, 1 (1903), pp. 376–7.
2. Ricardo Burguete Lana, *Mi rebeldía: 'Mane-Thecel-Phares'* (Madrid: Librería de 'Fernando Fé', 1904), pp. 72–3.

3. Isaiah Berlin, *The Hedgehog and the Fox: An Essay on Tolstoi's View of History* (London: Weidenfeld and Nicolson, 1988), pp. 51–4; Michael Howard, 'The Influence of Clausewitz', in Carl von Clausewitz, *On War*, Michael Howard and Peter Paret edition (Princeton: Princeton University Press, 1976), pp. 27–44; Stefan T. Possony and Etienne Mantoux, 'Du Picq and Foch: The French School', in Edward Mead Earle (ed.), *Makers of Modern Strategy: Military Thought from Machiavelli to Hitler* (Princeton: Princeton University Press, 1943), pp. 206–17, 227–8; Azar Gat, *A History of Military Thought: From the Enlightenment to the Cold War* (Oxford: Oxford University Press, 2001), pp. 296–309.
4. Carlos Banús y Comas, *Reflexiones acerca de la guerra anglo-boer* (Madrid: Imp. del Memorial de Ingenieros del Ejército, 1902), p. 55. Clausewitz, in Book IV, Chap. 4, of his most famous work wrote that '[t]he decision rests chiefly on the state of morale, which, in cases where the victor has lost as much as the vanquished, has always been the single decisive factor': Carl von Clausewitz, *On War*, Michael Howard and Peter Paret edition (Princeton: Princeton University Press, 1976), p. 231.
5. Burguete, *Mi rebeldía*, pp. 77–80. On Bloch, see Tohmatsu Haruo: 'Approaching Total War: Ivan Bloch's Disturbing Vision', in David Wolff, Steven G. Marks, Bruce W. Menning, David Schimmelpenninck van der Oye, John W. Steinberg, and Yokote Shinji (eds), The *Russo-Japanese War in Global Perspective: World War Zero. Volume II* (Leiden: Brill, 2007), pp. 179–202; Antulio J. Echevarria II, *Imagining Future War: The West's Technological Revolution and Visions of Wars to Come, 1880–1914* (Westport: Praeger, 2007), pp. 43–5.
6. Efeele [pseud. of Francisco Larrea y Liso]: *El desastre nacional y los vicios de nuestras instituciones militares* (Madrid: Publicaciones de los Estudios Militares, 1901), pp. 163–7. Navarro also mentioned reduced morale as a reason of the defensive attitude in Santiago de Cuba, in addition to a local inferiority in numbers and the shortage of field artillery: Modesto Navarro, 'Observaciones sobre la táctica de Burguete', *Estudios Militares* (hereafter *EM*), 21, 1 (1902), p. 166.
7. Banús, *Reflexiones*, pp. 55–62, 65.
8. Martin van Creveld, *Technology and War: From 2,000 BC to the Present* (London: Brassey's, 1991), pp. 173–4.
9. Modesto Navarro, 'Lo moral en lo militar', *Revista de Infantería*, 1, 2 (1901), pp. 15–18; Banús, *Reflexiones*, p. 65.
10. Eduardo Gallego Ramos, 'El municionamiento de la infantería en el combate moderno', *RTIC*, 10, 1 (1899), p. 28. Italics in the original.

11. Carlos Banús y Comas, *El arte de la guerra a principios del siglo XX* (Madrid: Memorial de Ingenieros, 1907), pp. 394, 399–402, 407, 412, 415–16; Banús, *Reflexiones*, p. 65.
12. Juan Calero Ortega, 'Ideas sobre organización militar', *RTIC*, 15, 2 (1904), pp. 452–4.
13. See, for instance, similar views in the American army in Brian McAllister Linn, *The Echo of Battle: The Army's Way of War* (Cambridge, MA: Harvard University Press, 2007), pp. 110–11.
14. Enrique Ruíz-Fornells, *La educación moral del soldado* (Toledo: Viuda e hijos de Juan Peláez, 1894), p. 50. Ruiz-Fornells is an officer who deserves further attention from the historians of the Spanish army, since he played a part in matters (the 1908 infantry regulations, the 1925 major unit regulations, Azaña's reform) that shaped it; for a study of his general intellectual outlook, see Geoffrey Jensen, *Cultura militar española. Modernistas, tradicionalistas y liberales* (Madrid: Biblioteca Nueva, 2014), Chap. 6. On the contemporary perception of this Spanish military 'decline', see also Jose Álvarez Junco, 'La nación en duda', in Juan Pan-Montojo (ed.), *Más se perdió en Cuba. España, 1898 y la crisis de fin de siglo* (Madrid: Alianza, 1998), pp. 448–9, 455–62.
15. Damián Isern, *Del desastre nacional y sus causas* (Madrid: n.p., 1899), pp. 371–2; Stanley G. Payne, *Politics and the Military in Modern Spain* (Stanford: Stanford University Press, 1967), p. 81; John Lawrence Tone, *War and Genocide in Cuba, 1895–1898* (Chapel Hill: The University of North Carolina Press, 2006), pp. 254–5.
16. Claudio, 'El miedo yankee', *La Correspondencia Militar* (hereafter *LCM*), 24 February 1898. Italics in the original.
17. 'Saetas', *LCM*, 12 March 1898. Italics in the original. At the onset of the war American soldiers also held scornful views of their Spanish opponents, which would be changed by the experience of combat: Albert A. Nofi, 'From "Dagoes" to "Nervy Spaniards". American Soldiers' Views of their Opponents, 1898', *Strategypage*, http://www.strategypage.com/articles/default.asp?target=spaniard/spaniard.htm (accessed 7 May 2015).
18. Efeele, *El desastre nacional*, pp. 163–7.
19. Calero, 'Ideas sobre organización', pp. 454–5.
20. T.H.E. Travers, 'Technology, Tactics, and Morale: Jean de Bloch, the Boer War, and British Military Theory, 1900–1914', *Journal of Modern History*, 51 (1979), pp. 266–7.
21. Ricardo Burguete, *Así hablaba Zorrapastro* (Valencia: F. Sempere, c. 1903). See also Jensen, *Cultura militar*, Chaps. 2–4.
22. Burguete, *Así hablaba Zorrapastro*, pp. 46, 54–62, 64.

23. Ricardo Burguete, *Preparación de las tropas para la guerra (nuevo Vegecio). Estudios de ética militar* (Madrid: Imp. de Fortanet, 1905), p. 128.
24. Burguete, *Preparación*, p. 128. It is worth noting that General José Millán Astray, founder of the offensive-minded Spanish Legion (an elite force of Spanish and foreign volunteers), wrote years later that he had been inspired by *bushido* to shape the legionary fighting spirit: José Millán Astray, foreword to Nitobë Inazo, *El Bushido (el alma del Japón)* ([Madrid:] n.p., 1941), pp. 9–10. On Millán Astray's thinking, see Jensen, *Cultura militar*, Chap. 7.
25. After the war, asked why fresh recruits were chosen to attack a position doggedly defended by the Russians, a Japanese officer answered that soldiers who had attacked while under machine gun fire once would not do it for a second time: Günther Kronenbitter, 'The Perception of the "Wars before the War" in Austria-Hungary', in Dominik Geppert, William Mulligan, and Andreas Rose (eds), *The Wars before the Great War: Conflict and International Politics before the Outbreak of the First World War* (Cambridge: Cambridge University Press, 2015), p. 194.
26. Burguete, *Así hablaba Zorrapastro*, pp. 48–9. On crowd-like behaviour in battle, see John Keegan, *The Face of Battle* (London: Jonathan Cape, 1976), pp. 172–4.
27. Burguete, *Así hablaba Zorrapastro*, pp. 46, 49–50.
28. Burguete, *Mi rebeldía*, pp. 94–7.
29. Burguete, *Así hablaba Zorrapastro*, p. 51.
30. Dennis E. Showalter, 'Army and Society in Imperial Germany: The Pains of Modernization', *Journal of Contemporary History*, 18 (1983), p. 587; Travers, 'Technology, Tactics, and Morale', pp. 272–4.
31. A line of argument standing against the all-volunteer army and for universal military service (but with some privileges for well-to-do conscripts) is found, for instance, in Pío Suárez Inclán, *El problema del reclutamiento en España* (Madrid: El Trabajo, 1905), pp. 44–57, 81–98.
32. Gabriel Cardona, *El poder militar en la España contemporánea hasta la guerra civil* (Madrid: Siglo XXI, 1983), p. 23.
33. Andrée Bachoud, *Los españoles ante las campañas de Marruecos* (Madrid: Espasa-Calpe, 1988), p. 140.
34. Travers, 'Technology, Tactics, and Morale', pp. 282–3. For an examination of the contemporary debate over the relationships of biological determinism with war, see Paul Crook, *Darwinism, War and History: the Debate over the Biology of War from the 'Origin of Species' to the First World War* (Cambridge: Cambridge University Press 1994).
35. Ricardo Burguete, *¡La guerra! Cuba (diario de un testigo)* (Barcelona: Maucci, 1902), pp. 41, 111.

36. Ricardo Burguete, *El problema militar (España ante los grandes imperios del porvenir)* (Palma de Mallorca: Francisco Soler Prats, 1905), p. 12.
37. Ricardo Burguete, *Morbo nacional. Vida defensiva* (Madrid: Imp. de Fortanet, c. 1905), p. 9.
38. Burguete, *El problema militar*, p. 10.
39. José Marvá y Mayer, *Estudio histórico de los medios de ataque y defensa* (Madrid: Revista Técnica de Infantería y Caballería, 1903), pp. 391–2. Marvá (1846–1937) spent many years in teaching and scientific research assignments; after 1900, though he still remained in active service until the 1910s, his career was dominated by civil appointments related to social welfare issues.
40. Miguel Alonso Baquer, 'José Marvá (1846–1937), entre la tecnología militar y las reformas sociales', in José Luis García Delgado (ed.), *Economía española, cultura y sociedad: homenaje a Juan Velarde Fuertes*, 3 vols. (Madrid: EUDEMA, 1992), II. 210–11, 219–20.
41. This summary of the history of the machine gun in the Spanish armyfrom 1867 to 1898 is based on Luis de la Gándara Marsella, *Ametralladoras de campaña en el Ejército español* (Madrid: R. Velasco, 1906), pp. 3–7; Luis de la Gándara Marsella, *Estudios acerca de ametralladoras* (Madrid: Imprenta de Eduardo Arias, 1908), pp. 247–51; Luis de la Gándara Marsella, *Estudios acerca de ametralladoras*, 2nd edition (Madrid: E. Dossat, 1910), pp. 287–9. An infantry officer, de la Gándara (1880–1928) taught at the Academy of Infantry and the Central Firing School and took part in the writing of new infantry regulations in the 1920s. For an overview of the technical development of the machine gun before the First World War, see George M. Chinn, *The Machine Gun: History, Evolution, and Development of Manual, Automatic, and Airborne Repeating Weapons, Volume I* (Washington, DC: Bureau of Ordnance, Department of the Navy, 1951).
42. Archivo General Militar de Segovia, 2ª Sección, División 1ª, Legajo 9 (hereafter AGMS No. Sección/No. División/No. Legajo), Ministerio de la Guerra, Artillería, Comisión de Experiencias, Informe sobre 'Las experiencias efectuadas con diversos modelos de ametralladoras' (primera parte), 29 April 1908.
43. Lorenzo Cabrera Mac-Kintosh, *Ametralladoras* (Madrid: Imp. de Jaime Ratés, c. 1911), pp. 33–4. A senior American general later depicted the self-promoting commander of the Gatling unit as 'a pestiferous, immodest ass': Edward M. Coffman, *The Regulars. The American Army, 1898–1941* (Cambridge, MA: Harvard University Press, 2004), p. 161.
44. Some in the United States shared this view: Linn, *The Echo of Battle*, p. 98. But cf. Peter Smithurst, *The Gatling Gun* (Oxford: Osprey, 2015), pp. 41, 44; Donald H. Dyal (ed.), *Historical Dictionary of the Spanish*

American War (Westport: Greenwood Press, 1996), *sv* 'Gatling gun'; Tone, *War and Genocide*, p. 278.
45. Gándara, *Ametralladoras de campaña*, pp. 7–8. Italics in the original.
46. AGMS 2/1/7, Artilería, Escuela Central de Tiro, Junta Facultativa, Informando sobre la necesidad de experimentar ametralladoras y cañones automáticos de pequeño calibre, 10 January 1902.
47. AGMS 2/1/7, Artillería, Junta Facultativa, Informe sobre 'Necesidad de experimentar ametralladoras y cañones automáticos de pequeños calibres', 27 February 1902.
48. AGMS 2/1/9, Ministerio de la Guerra, Artillería, Comisión de Experiencias, Informe sobre 'Las experiencias efectuadas con diversos modelos de ametralladoras' (primera parte), 29 April 1908.
49. Comisión de Experiencias de Artillería, *Resumen de los trabajos realizados por la Comisión de Experiencias de Artillería durante el año 1905* (Madrid: Memorial de Artillería, 1906), pp. 24–5. The present author used a copy filed in Archivo General Militar de Madrid, Colección Adicional de Documentos, Legajo 8, carpeta 15 (hereafter AGMM-CAD No. Legajo/No. carpeta), Memoria, Revista de Inspección del Ministerio de la Guerra, 1907.
50. AGMS 2/1/7, Artillería, Junta Facultativa, Informe sobre 'Experiencias con una ametralladora Hotchkiss', 20 December 1902.
51. Comisión de Experiencias de Artillería, *Resumen de los trabajos realizados por la Comisión de Experiencias de Artillería durante el año 1906* (Madrid: Memorial de Artillería, 1907), p. 50. The present author used a copy filed in AGMM-CAD 8/15, Memoria, Revista de Inspección del Ministerio de la Guerra, 1907.
52. Pedro Jevenois, 'Consecuencias tácticas de la guerra ruso-japonesa', *EM*, 26, 1 (1907), p. 422. Jevenois (1878–1941) was an observer in the Russo-Japanese War; he joined the Nationalist side in the Civil War and was divisional commander and artillery commander of the Army of the South.
53. AGMS 2/1/9, Ministerio de la Guerra, Artillería, Comisión de Experiencias, Informe sobre 'Las experiencias efectuadas con diversos modelos de ametralladoras' (segunda parte), 29 April 1908.
54. El Capitán Equis [pseud. of Nazario Cebreiros], *El problema militar en España. Apuntes para un estudio sincero y al alcance de todos*, 2 vols. (Burgos: Saiz y Comp^{a}, 1916–1917), II. 116.
55. José Brull, 'Las ametralladoras en los campos de batalla', *Memorial de Artillería* (hereafter *MA*), 60, 1 (1903), pp. 556–7.
56. Gándara, *Ametralladoras de campaña*, pp. 30–1.
57. José de Lossada, 'Moderna Artillería', *MA*, 57, 2 (1900), p. 292; Pedro Jevenois, 'Ametralladoras', *MA*, 59, 1 (1902), p. 177.

58. 'Crónica interior', *MA*, 60, 2 (1903), pp. 65–6; Jorge Vigón, *Historia de la Artillería española*, 3 vols. (Madrid: Consejo Superior de Investigaciones Científicas, 1947), III. 240. Some details on this incident given in Vigón do not coincide with those provided by *Memorial de Artillería*, but no other sign has been found to think that Vigón was writing about a different case. In any case both sources display the artillery's corporate attitude.
59. AGMS 2/1/7, Artillería, Junta Facultativa, Informe sobre 'Necesidad de experimentar ametralladoras y cañones automáticos de pequeños calibres', 27 February 1902.
60. Jevenois, 'Ametralladoras', pp. 30–1; Lossada, 'Moderna Artillería', p. 292.
61. Enrique Ruíz-Fornells, 'Fuegos de la Infantería', *EM*, 23, 1 (1904), pp. 23–4.
62. Calero, 'Ideas sobre organización', p. 451.
63. Banús, *Reflexiones*, pp. 70–1.
64. Eduardo Gallego y Ramos, 'Misión de la Caballería en las guerras antiguas y modernas', *EM*, 20, 1 (1901), p. 164; see also Jevenois, 'Ametralladoras', pp. 30–1.
65. Gándara, *Ametralladoras de campaña*, pp. 20–5.
66. Gándara, *Ametralladoras de campaña*, p. 89.
67. Eliseo Sanz Balza, *La ametralladora en caballería* (Madrid: El Trabajo, 1906), pp. 6, 11–12.
68. José de Lossada Canterac, *Ametralladoras* (Segovia: Tipografía de la Academia de Artillería, 1903), p. viii. Commissioned in the artillery corps in 1879, de Lossada was assigned in the Academy of Artillery (1896–1911); promoted to brigadier general in 1919, he was head of the artillery section at the War Ministry (1920–1921).
69. Modesto Navarro García, foreword to Gándara, *Estudios* [1908 edition], pp. xiii–xviii.
70. Jevenois, 'Ametralladoras', p. 177.
71. Navarro, foreword to Gándara, *Estudios* [1908 edition], pp. xix–xxii.
72. Lossada, *Ametralladoras*, pp. viii, 84, 177.
73. AGMS 2/1/9, Ministerio de la Guerra, Artillería, Comisión de Experiencias, Informe sobre 'Las experiencias efectuadas con diversos modelos de ametralladoras' (segunda parte), 29 April 1908. See also 'Ametralladoras Hotchkiss 1907', *MA*, 65, 1 (1908), pp. 33–4.
74. Á[ngel] G[arcía-]Pelayo, 'Las ametralladoras en campaña', *RTIC*, 16, 1 (1905), pp. 431–2, 472–4.
75. Shelford Bidwell and Dominick Graham: *Fire Power: British Army Weapons and Theories of War, 1904–1945* (London: Allen & Unwin, 1985), pp. 22–3, 55.
76. Jevenois, 'Ametralladoras', pp. 23–4.

77. Jevenois, 'Consecuencias tácticas', p. 423.
78. Emilio R[odríguez] Tarduchy, *El tiro de la Infantería. Ensayo crítico independiente* (Burgos: n.p., 1913), pp. 22–3. Rodríguez Tarduchy, after leaving the service in 1931, was a founder and the first chairman of the Spanish Military Union (see Chap. 5).
79. Ángel Rodríguez del Barrio, 'El Regimiento de Infantería. El Pelotón de combate. El Batallón de ametralladoras', *La Guerra y su preparación*, 11, 2 (1926), pp. 4–5. Rodríguez del Barrio was involved in the start of the military conspiracy of 1936, but he withdrew soon because of serious illness (and perhaps lack of confidence in the plot's success); he died from cancer a few weeks after the Civil War broke out.
80. Luis Bermúdez de Castro, 'Táctica para el combate en Marruecos', *Memorial de Infantería*, 3, 1 (1914), pp. 29–30.
81. Ruíz-Fornells, 'Fuegos', p. 24. The German army had drawn at that time the conclusion that it was easier to teach conscripts to produce huge volumes of fire with a machine gun than to turn them into marksmen: Bidwell and Graham, *Fire Power*, p. 22.
82. Juan Laverón Agut, 'Conferencia sobre ametralladoras', *EM*, 27, 1 (1908), pp. 101–2.
83. Eladio Rodríguez, Pascual Torras, and José Muñoz, *Conferencias sobre ametralladoras* (n.p.: n.p., 1908), p. 32; Bidwell and Graham, *Fire Power*, pp. 28–9.
84. Laverón, 'Conferencia', pp. 100, 105–6.
85. Eric Dorn Brose, *The Kaiser's Army: the Politics of Military Technology in Germany during the Machine Age, 1870–1918* (New York: Oxford University Press, 2001), pp. 95–6, 144. However, de Lossada and Jevenois thought that the size of contemporary machine guns allowed them to go unnoticed and reach easily the most advanced points of the firing line: Lossada, 'Moderna Artillería', p. 292; Jevenois, 'Ametralladoras', pp. 29–30.
86. Niemand, 'Crónica general', *Revista Científico-Militar* (hereafter *RCM*), 25 (1900), p. 98; Rodríguez, Torras, and Muñoz, *Conferencias*, p. 33.
87. Lossada, *Ametralladoras*, p. 176; Rodríguez, Torras, and Muñoz, *Conferencias*, pp. 32–3.
88. Gándara, *Estudios* [1908 edition], p. 255.
89. T.H.E. Travers, 'The Offensive and the Problem of Innovation in British Military Thought, 1870–1915', *Journal of Contemporary History*, 13 (1978), p. 536.
90. Tim Travers' comments about the British army are applicable to the Spanish case for the most part: Tim [H.E.] Travers, *The Killing Ground: The British Army, the Western Front and the Emergence of Modern Warfare, 1900–1918* (London: Unwin Hyman, 1990), pp. 68–71.

91. AGMS 2/8/152, Capitán Herrera de la Rosa, Impresiones recogidas de la campaña ruso-japonesa con el ejército del general barón Nogui, p. 88, 30 November 1905.
92. Gándara, *Estudios* [1908 edition], p. 255.
93. Lossada, *Ametralladoras*, pp. 86–95.
94. Victor Martín García and Francisco Gómez Souza, *Estudios de arte militar*, 3 vols. (Madrid: Tip. de la Revista de Archivos, Bibliotecas y Museos, 1910), II. 299–301; Navarro, foreword to Gándara, *Estudios* [1908 edition], p. xviii.
95. Gándara, *Estudios* [1908 edition], pp. 254–5.
96. Keegan, *Face of Battle*, pp. 228–30.
97. Ruíz-Fornells, 'Fuegos', p. 20; a similar argument is found in Jevenois, 'Ametralladoras', p. 34.
98. Niemand, 'Cronica general', p. 98.
99. Travers, 'The Offensive and the Problem of Innovation', pp. 541–3.
100. Ángel Dolla, 'Resumen de las armas y su aplicación al Arma de Caballería', *RTIC*, 16, 1 (1905), p. 38.
101. Ricardo Burguete Lana, *Nuevos métodos de combate. Proyecto de reglamento de ejercicios y maniobras para infantería* (Madrid: [Depósito de la Guerra], 1903), p. 26.
102. Banús, *El arte de la guerra*, p. 46.
103. Carlos Martínez de Campos y Serrano, *Ayer*, 2 vols. (Madrid: Instituto de Estudios Políticos, 1946–1970), I. 60–1.
104. Rodríguez, Torras, and Muñoz, *Conferencias*, p. 32. Italics in the original.
105. Ruíz-Fornells, 'Fuegos', p. 20.
106. AGMS 2/1/9, Ministerio de la Guerra, Artillería, Comisión de Experiencias, Informe sobre 'Las experiencias efectuadas con diversos modelos de ametralladoras' (segunda parte), 29 April 1908.
107. Travers, *Killing Ground*, pp. 70–1.
108. Francisco Sigüenza, 'La artillería en relación con la infantería', *MA*, 67, 2 (1910), p. 263.
109. Banús, *El arte de la guerra*, pp. 45–6; Gándara, *Estudios* [1908 edition], p. 263.
110. Banús, *El arte de la guerra*, p. 46; Gándara, *Estudios* [1908 edition], p. 262.
111. AGMS 2/1/73, Marín to Ministro de la Guerra, 14 August 1907. 'Esteban Marín' is the present author's reading of the signature in the letter, the only indication of the representative's identity.
112. AGMS 2/1/73, Ministerio de la Guerra, [Sección de Artillería?, memorandum on the acquisition of 20 Hotchkiss machine guns], 21 August 1907.
113. AGMS 2/1/73, Ministerio de la Guerra, Sección de Artillería to Capitán General de la 1ª Región, 22 August 1907.

114. AGMS 2/1/73, Artillería, Comisión de Experiencias, Memoria redactada por el Coronel D. Luis de Santiago y Comte. D. Luis Esparza como resultado de la Comisión que les fue conferida por R.O. de 22 de Agosto p°.p°. y que trata de la ametralladora Hotchkiss en particular y de las apreciaciones de los mismos acerca del empleo esta clase de armas en general, 20 September 1907.
115. Documents concerning the purchase are filed in AGMS 2/1/73.
116. Gándara, *Estudios*, 2nd edition, pp. 290–3.
117. Vigón, *Artillería española*, II. 299 note 215.
118. Juan Avilés, 'Enseñanzas de la guerra del Rif', *RCM*, 35 (1910), pp. 289–90.
119. [José de Lossada Canterac,] Conde de Casa-Canterac, 'Ametralladoras', *EM*, 29, 2 (1910), p. 402.
120. Casa-Canterac, 'Ametralladoras', pp. 402–6, 414.
121. Federico Medialdea Muñoz, *Las ametralladoras en la campaña del Rif (1909)* (Madrid: n.p., 1910), pp. 54, 56. For an operational account of the machine gun units in the campaign, see Medialdea, *Las ametralladoras*, pp. 91–108. Medialdea (1880–1926) was an instructor at the Central Firing School (1908–1914) and the Academy of Infantry (1915–1920); he served in later campaigns in Morocco (1924–1926), where he died from heart disease.
122. Medialdea, *Las ametralladoras*, p. 61.
123. Casa-Canterac, 'Ametralladoras', 402.
124. Medialdea, *Las ametralladoras*, pp. 56–7. This behaviour resembles British Major N.R. McMahon's 'tabloid' firepower tactics, based on the distinctiveness to the ear of machine gun fire: Bidwell and Graham, *Fire Power*, pp. 51–2.
125. Estado Mayor Central del Ejército, *Enseñanzas de la campaña del Rif en 1909* (Madrid: [Depósito de la Guerra], 1911), pp. 113–14.
126. Ministerio de la Guerra, *Reglamento Táctico de Infantería* (Toledo: n.p., 1914), Articles 457 to 461.
127. Medialdea, *Las ametralladoras*, p. 57.
128. See, for instance, Travers, *Killing Ground*, pp. 62–71; Brose, *The Kaiser's Army*, pp. 93–8, 142–5; Coffman, *The Regulars*, pp. 161–2. For an overview of the introduction into military service of the automatic machine gun, see Paul Cornish, *Machine Guns and the Great War* (Barnsley: Pen and Sword Military, 2009), Chap. 1.
129. See Daniel J. Kenda, 'Lessons Learned from the Use of the Machine Gun during the Russo-Japanese War and the Application of Those Lessons by the Protagonists of World War I' (Master's dissertation, United States Army Command and General Staff College, 2005).
130. Brose, *The Kaiser's Army*, pp. 97, 145.

131. Travers, *Killing Ground*, p. 77. It is noteworthy, by contrast, that many Spanish military essayists of the sixteenth century (especially its second half) saw no conflict between soldierly virtues and the use of new technology (firearms); on the contrary, such a transformation of warfare demanded the best qualities from contemporary soldiers: José Antonio Maravall, *Antiguos y modernos. Visión de la historia e idea de progreso hasta el Renacimiento*, 2nd edition (Madrid: Alianza Editorial, 1986), pp. 542–50.

CHAPTER 5

The Officer Corps and Military Reform, 1916–1936

This chapter surveys the most relevant political events in relation to the army and its reform. It deals initially with the political-military crisis of 1917–1923, unleashed by the officer corps' zeal to safeguard its vested interests. It goes on to survey military policy during General Primo de Rivera's dictatorship, whose final failure dragged the monarchy down. Finally, it provides an overview of the military reforms and policy under the Second Republic, which failed to implement a solid liberal democracy and repair the fractures the Spanish society had developed.[1]

THE *JUNTAS DE DEFENSA*

Morocco and the Growth of Professional Dissatisfaction

After 1909 military operations in Morocco's northernmost regions (which became a Spanish protectorate after November 1912) reintroduced war merit promotions and rekindled the officer corps' internal dispute about promotion.[2] The creation of the Spanish army's Moroccan regular troops (*Fuerzas Regulares Indígenas*, commonly known as *regulares*) in 1911—which were to carry the burden of fighting after that year—improved the prospects of officers hoping to speed up their careers through courage and skill on the African battlefields.[3] Their promotions came at a price in blood, since it was assumed that native troops required leaders who were less daunted than usual by the perils of the firing line (for instance, the Spanish officers and NCOs of *regulares* operating in

J.V. Herrero Pérez, *The Spanish Military and Warfare from 1899 to the Civil War*, DOI 10.1007/978-3-319-54747-3_5

the Melilla district from early 1920 to the eve of the disaster of Annual suffered an average of 70% casualties).[4] These officers were known under the name of *africanistas* (Africanists) and their ambition became conspicuous.[5] Social culture factors may explain to some extent their eager pursuit of promotion, as in the French colonial army, whose officers' practices were markedly *arriviste*. In contrast with their British colleagues in colonial garrisons, the French and Spanish officers' careerist zeal did not seem restrained by a code of gentlemanly social behaviour that prevented the individual pursuit of professional promotion from getting out of hand.[6]

Nonetheless, relatively few officers volunteered to fight in Morocco, though the military supported the Spanish involvement. Most officers carried on dull careers in the undemanding routine of home garrisons and resented the Africanists' rapid promotion on the grounds of their wartime experience, feeling the latter a threat to their rights as tenured state employees, which had been secured by the 1878 army act (and thus advancing, on the other hand, the process of professionalization of the Spanish officer corps).[7] Merit promotion was certainly open to abuse, but promotion based only on the seniority principle was not a good solution either. The latter, however, was most advantageous for those officers whose professionalism withered on the vine.

On the other hand, merit promotion in wartime was initially regarded more as a reward for campaign hardships than as recognition of proven ability for command, a stance which partially explains the Spanish army's liberality in war rewards: 236,718 war merit promotions and decorations were reportedly awarded from 1909 to 1917.[8] Thus officers were promoted in the campaigns of 1909 and 1911 because they had been slightly wounded in either an arm or a leg. The General Staff's analysis of the 1909 campaign admitted that combats were assessed according to the blood toll for one's own side rather than by the military advantages obtained over the enemy.[9] Rules for a more selective awarding of war merit promotions were introduced after 1912 and seem to have been partially successful, but the previous experiences had made the peninsular officers so sensitive that even a smaller number of cases of favouritism could still produce a lot of resentment. In addition, promotion policy was not the only source of concern for the Spanish officers in the mid-1910s. The First World War worsened the officers' financial situation after 1914 because it unleashed inflation and salaries lost half of their buying power.

Rise and Fall of the Juntas de defensa, 1916–1922

Infantry officers of the Barcelona garrison, inspired by the example of the lobbying boards of the artillery and engineer corps, began to organize their own *Junta de Defensa de Infantería* (Infantry Defence Board) after mid-1916 in order to defend what they considered their legitimate professional interests. Their example had spread to the rest of the army by early 1917, although the Barcelona infantry board (chaired by Colonel Benito Márquez) was to play a leading role. In order not to upset the dissatisfied officer corps further the military high command did not interfere with this process (promoted by junior and middle-ranking officers) until May 1917. Then a new prime minister, Manuel García Prieto, and his war minister, General Francisco Aguilera, attempted to stem a movement that threatened the discipline of the army. Márquez and the other members of the Barcelona infantry board were ordered to disband the *junta* and arrested after they refused to do it. The other existing *juntas* all over Spain demonstrated widespread support to their comrades.[10] On 1 June 1917 a new Barcelona infantry board, in addition to threatening the liberation of their arrested colleagues by force, submitted a manifesto which contained, wrapped in vague demands for political reform, the basic claims of the insubordinate officers (especially promotion based on seniority alone). The government released the arrested officers because it did not want to lose the army's loyalty in a time of social unrest (the memory of the Russian Revolution in early 1917 was fresh).[11]

In November 1917 García Prieto chaired again the cabinet, which included the first civilian war minister in Spanish history, Juan de la Cierva. De la Cierva promised to satisfy the professional demands of the boards if they ditched any sort of reformist political programme of their own. The ready approval of the *juntas* proved that corporate self-interest was their main reason to defy the civil authority. It also frustrated the political ambitions of Márquez, who resigned as chairman of the Infantry Higher Board in December and was eventually dismissed from the army in March 1918. De la Cierva's military policy took shape in the 1918 army reform act. Thereafter, seniority would be the rule for all promotions. In exceptional cases, a war merit promotion would be awarded after a thorough inquiry and parliamentary approval. This new system was demoralizing for the officers fighting in Morocco.[12] De la Cierva also provided more command slots by enlarging the peacetime army to 16 infantry and three cavalry divisions. The act was accompanied by a substantial increase in salaries (which were raised again in 1920). Depending on rank, officer pay was raised from 11 to 50% in 1918 (the average pay per hour of a Spanish skilled worker rose 86% in the period 1914–1920).[13] Nonetheless, the zeal of the military to preserve its vested interests did not disappear after de la Cierva's reform, as the dynastic politicians' need of military support to keep the Restoration regime afloat meant that the army's inner corporate conflicts became major political issues.

The strength of the *juntas* wore out because they were unable to bridge the internal divisions of the officer corps. Despite the capacity of the army to influence the decisions of the dynastic politicians, its traditional corporate factionalism remained alive and well—as the 1919 debate about the staff corps showed.[14] Moreover, the escalation of military operations in Morocco after 1918 intensified the hostility between peninsular officers supporting the boards (*junteros*) and Africanists, who were also gaining popular appeal thanks to the successes in Morocco from 1919 to early 1921 and the *panache* of the brand new Spanish Legion (a volunteer force created in 1920, which bore the brunt of the fighting together with the *regulares*).[15] By contrast, the small-minded corporatism of the *juntas* was more evident day after day, while the public became fed up with the political turmoil they caused.

The position of the defence boards started crumbling after the debacle of the Spanish army in the eastern district of the Moroccan Protectorate (the so-called 'disaster of Annual') in the summer of 1921.[16] Major General Manuel Fernández Silvestre, appointed commander of the Melilla district in early 1920, increased notably the pace of advance towards the heart of the Rif (the mountainous region in the centre of the Protectorate), but he had overstretched his forces by late spring of 1921.[17] When Fernández Silvestre realized the vulnerability of his position, he ordered a hasty evacuation of the main advanced camp in Annual on 22 July. The harassment of the natives, led by their new chieftain Abd el-Krim, turned the poorly prepared withdrawal into a flight.[18] The tribes at the rear areas rose in arms and the whole eastern district collapsed. About 8000 Spanish troops were slaughtered in a few days (Fernández Silvestre stayed alone in Annual and committed suicide or was killed).[19] By August, the only area under Spanish control was the city of Melilla itself.

The Africanists blamed the *juntas* for the decreasing fighting spirit of the officer corps, which had allegedly been influential in the Spanish misfortunes in Morocco. Unable to give a convincing answer to this charge, the already weakened *junteros* fell into discredit and lost their power to face up to the government. Thus Prime Minister José Sánchez Guerra was finally able to disband the defence boards on 14 November 1922. But when the turmoil caused by the inquiry about the political responsibilities for the Moroccan disaster of 1921 brought about an opportunity to restore the supremacy of civil power (by starting a regeneration of the parliamentary regime), the officer corps could not help seeing this as a threat to its corporate interests. Therefore, the government found no support in the military to face up to the bloodless *coup d'état* of Lieutenant General Miguel Primo de Rivera, captain general of the Fourth Military Region, on 13 September 1923, which imposed an alternative solution—an authoritarian regime—to replace the decaying Restoration political system.

The Decline of Monarchy and the Second Republic, 1923–1936

Primo de Rivera's Dictatorship and the Struggle Against Corporatism, 1923–1930

The army initially agreed to Primo de Rivera's military authoritarian regime, which seemed to bring political stability and social order back to Spain and kept military corporate affairs out of the hands of the civil power.[20] However, Primo de Rivera was not a watchdog of traditional corporate interests and his military policy led him to a clash with sections of the officer corps. With regard to Morocco, Primo de Rivera was initially inclined to stop any further military advance.[21] This policy angered the Africanists. Fortunately for the dictator, Abd el-Krim attacked successfully the French protectorate in April 1925. The following July Spain and France reached an agreement to crush Abd el-Krim's rebellion through combined military action. This turn of events allowed Primo de Rivera to undertake a resolute war effort that finished the Moroccan conflict in July 1927. The death toll for the Spanish army from 1909 to 1927 may have been around 20,000 European soldiers (10,000 of them in 1921) and several thousands of native troops.[22]

The victories in Morocco brought a shower of promotions on the most outstanding officers of the war. It was possible because Primo de Rivera suppressed the control procedures for war merit promotions and introduced selective promotion in both wartime and peacetime. Primo de Rivera also tried to reduce the excess of officers, but he linked cuts in personnel to political loyalty. Thus the artillery and engineer corps lost more than the infantry and the cavalry—the most loyal branches. The dictator's desire to keep the army and military policy under his control was also behind the decision to disband the General Staff on 14 December 1925. The official reason was the difficulty in adapting this body to the Spanish military administrative framework. The oddly named *Dirección General de Preparación de Campaña* (General Directorate for Preparation of Field Operations), fully integrated within the War Ministry and whose head was a direct subordinate to the minister, took charge of the EMC's functions.[23] The reform was essentially limited to cutting the General Staff's autonomy, which supports the contemporary view that Primo de Rivera's decision was a follow-up of the dismissal of his last chief of general staff, Captain General Weyler, who had allegedly used the post to hinder the dictator's policy.[24]

The artillery corps was the most conspicuous military opponent of Primo de Rivera, who wanted to curb its haughty *esprit de corps*. Primo de Rivera ordered in 1926 that artillery officers could no longer exchange war merit promotions for the María Cristina Cross and accepted those awarded after 1925. The result was a strong protest by all the artillery officers, but the dictator did not yield and they were suspended without salary in September 1926. The artillerymen came back to their duties in November after signing a humiliating document of obedience to the government. In early 1929 artillery officers were the military backbone of a failed *coup d'état* against Primo de Rivera and the dictator decided to dismiss all the officers of the corps from the service. They were admitted back after signing a declaration in which they swore loyalty to Primo de Rivera's current government.

The authoritarian, uncompromising ways of Primo de Rivera's military policy gradually alienated too many sections of the officer corps. The disaffection of the army eventually led the dictator to resign on 29 January 1930. The position of the monarchy, which in accepting the suppression of the constitutional regime in 1923 had linked its destiny to Primo de Rivera's dictatorship, was weak. A cabinet chaired by General Dámaso Berenguer (former high commissioner in Morocco in

1919–1922) tried unsuccessfully to restore the constitutional monarchy. An attempted republican *putsch* in December 1930 ended in total failure, but it was seen as a demonstration of the military's growing alienation from the monarchy.[25] Local elections on 12 April 1931 were won by republican candidate lists in most provincial capitals, an unequivocal result which signalled the nation's real political mood. Alfonso XIII abdicated and left the country to allow the formation of a republican government. After the political failure of Primo de Rivera, the military (though most officers had no republican feelings) adopted a passive attitude and did not stand either for or against the change of regime.

The Second Republic and Azaña's Military Reform, 1931–1933

The Second Republic was proclaimed on 14 April 1931. The first republican war minister was Manuel Azaña, a left-of-centre politician. An admiring scholar on the French Third Republic's political-military policies before the First World War and especially their efforts of 'republicanization' of the armed forces in the 1900s, he sought to safeguard the supremacy of civil power over the military.[26] Azaña's interest in military policy lay essentially in the political aspects, such as putting civil–military relations on a basis suitable for a contemporary democratic state (for instance, the 1906 Law of Jurisdictions was revoked).[27] Nonetheless, he also tackled the professional problems and undertook a deep reorganization of the army.

Azaña did not hesitate in tackling the root problem which had hindered military reform for five decades: an oversized officer corps. In order to reduce the numbers of officers, Azaña issued a decree of voluntary retirement dated 25 April 1931.[28] Applying within a thirty-day deadline after the publication of the decree, an officer could retire from the service ahead of schedule and still draw his current rank's full basic pay (with the exception of officers who were already near the statutory retirement age). About 8000 out of 20,500 officers (almost 40%) took advantage of the decree, which meant a substantial reduction of the officer corps at last.

The promotion system would be based on a combination of seniority and selection. Captains had to pass a compulsory course before promotion to major. The course marks would change the order of seniority for promotion; thus the best officers in the course would be at the top of their academy class segment within the field grade ranks. Colonels would also be required to pass another course before being proposed for promotion to general officer rank. The reserve list was incorporated into the active list, but NCOs would have to pass more demanding trials than hitherto (including a period of study in the military academies) before being commissioned.[29] On the other hand, in order to improve the professional and social status of the career NCOs who lacked the ability for a commission, Azaña created a regular non-commissioned officer corps. Prior to this, the career of the NCO was subject to the approval of successive fixed-term enlistments.[30]

The General Staff was restored. The peacetime army was reduced to eight infantry divisions, two mountain brigades, and one cavalry division, which would be supervised by three new army general inspectorates. Azaña's reform suppressed the territorial organization in military regions, the office of territorial captain general, and the ranks of captain general and lieutenant general.[31]

Azaña also looked for fresh military leaders. The historian Miguel Alonso Baquer classified the Spanish military elite of 1931 in groups based on Morris Janowitz's categories (heroic leaders, managers and technicians) and introduced a fourth group of his own: the 'humanist' officers. These were either colonial soldiers with political and administrative experience as native affairs officers or reformist-minded military academy teachers.[32] By 1917 the career models leading to the top ranks were the heroic and the managerial ones. But Primo de Rivera's dictatorship had granted all the advantages to the heroic model (which Africanists excelled at) by 1930. Azaña sought to change the situation in order to favour technicians and humanists, but they were too small a pool to choose from.[33] Accordingly, the republican regime lacked a reliable army leadership to manage the dissatisfaction of the conservative sections of the officer corps with the reformist policies of the period 1931–1933.

The soundness of the reforms was damaged by the arrogant, abrasive character of Azaña, whose own rhetorical fluency played a dirty trick on him. During a political meeting in June 1931 Azaña spoke of the need for grinding down the dominance of the traditional Spanish oligarchy as he had ground down other obstacles to Spain's progress. There was no mention of the military, but since he had carried out a major reform as war minister, it was easy for his adversaries to present Azaña's policy as a deliberate, sinister plan to destroy the army. Azaña was not wise in his choice of words, but only hostile political partisanship can present his actual achievement as a whole as destructive for the army.[34] In fact, General Francisco Franco Bahamonde, autocratic ruler of Spain and nemesis of the republic Azaña aspired to build, reportedly said three decades later that the early retirement decree 'was not badly planned nor so faulty as was said at that time; it had the sectarian purpose of desiring to remove from the army's ranks the officers with monarchist ideas; but this was not achieved because... most of us stayed on'.[35] Despite the contemporary criticism voiced by some conservative sections of the army, Azaña's reforms were sound from a professional point of view and remained essentially unchanged during the later right-wing cabinets (1933–1935). The reforms were the foundations for an improvement of the army's effectiveness in the long term, once the natural wastage of the retired personnel made available financial resources for equipment and training. Unfortunately, the outbreak of the Civil War in 1936 wrecked these prospects.

From Reform to Civil War, 1933–1936

The leftist leaning of the cabinets during the early life of the republic did not arouse the military's enthusiasm for the new regime. Despite introducing reformist policies, the republican government could not prevent the social strife fostered by the radical left. This unrest and the concession of home rule to Catalonia could not help nourishing the distrust of lots of officers whose political instincts were conservative. Nevertheless, the army was still unwilling to get involved in conspiracies, as it was proved by the lack of support to General José Sanjurjo's *putsch* on 10 August 1932.

The results of the parliamentary elections of November 1933 gave an overall majority to the right, poised to correct what they saw as excesses of the previous republican cabinets. When a coalition of right-wing parties named *Confederación Española de Derechas Autónomas* (Spanish Confederation of Autonomous Rightist Groups) or CEDA later entered the cabinet, the left started a general revolutionary strike on 6 October 1934. The strike quickly failed all over Spain, except in the northern province of Asturias, an important mining area with strong working-class organizations. The revolutionaries seized the weapon stockpiles of arm factories located in the province and the strike turned into an armed uprising (6–19 October 1934). War Minister Diego Hidalgo Durán appointed Major General Francisco Franco (then commander of the

Balearic Islands), who found himself in Madrid by chance, as his personal adviser. Franco became a de facto chief of general staff, since Hidalgo considered the incumbent chief, General Carlos Masquelet, politically unreliable. About 15,000 troops—including *regulares* and Legion units sent from Morocco—and 3000 members of the police forces were employed to crush the rebellion. The government forces lost about 300 killed. The toll of defeat for the rebels were around 900 killed and 15,000 imprisoned.

The leader of the CEDA, José María Gil Robles, became war minister in May 1935. Gil Robles tried to turn the army into an instrument to prevent the left from coming back to power. He chose officers relegated under Azaña for key assignments (for instance, Franco became chief of the General Staff), promoted all the army *subtenientes* (the highest NCO rank, which was suppressed at the same time) to second lieutenants, and prepared an ambitious plan of rearmament (also justified by an increasingly tense international environment in the Mediterranean). The CEDA leader could only carry out his projects partially because of the political crisis in the second half of 1935, which eventually led to new elections. On the brink of leaving office in the last days of 1935, Gil Robles played for a while with the idea of staging a military-backed *coup d'état*, but the conservative generals—despite their political sympathy—still refused to join any action outside the republican legality.

The right and the left (grouped in the Popular Front coalition) finished virtually equal in the elections of February 1936. But the disproportionate electoral system, mob coercion, and later irregularities distorted the outcome and resulted in an ample majority of parliamentary seats for the left, which appointed Azaña as president of the republic in April. The new president and the moderate sections of the Popular Front were unable to hold into check the revolutionary rants of the radical left. The latter thought that the moderation of the early republican cabinets had made possible the conservative backlash of 1934–1935. In their turn, many on the right adopted more extreme attitudes as well, since they saw the heated leftist rhetoric as the announcement of an impending revolution. The result was a polarization of Spain's politics, which generated an escalation of political violence on the streets.[36] The Popular Front's victory and policies led finally a section of the officer corps to think of seizing power through a military *putsch*.

Members of the *Unión Militar Española* (Spanish Military Union) or UME formed the first cells of the 1936 military conspiracy.[37] Created in 1933 as a kind of underground professional association which, initially, was supposed to have no definite political identity, the UME had become more of a networking channel for conservative officers by early 1936. Only relatively small numbers of officers (perhaps no more than a thousand, including those who had left the service) were actively engaged in the conspiracy under the direction of General Emilio Mola as senior organizer (Sanjurjo, exiled in Portugal, was the formal leader). Nevertheless, many, if not most, officers were ready to join the uprising if their superiors rebelled or the plotters could get rid of the commanders loyal to the government. The murder of the conservative political leader José Calvo Sotelo by leftist policemen acting on their own on 12 July 1936 (as a vengeance for the assassination of one of their comrades a few hours earlier) and the government's apathetic search for the culprits spurred the conspirators to carry out their plans within a few days. The *coup d'état* started in Melilla on 17 July 1936 after the government authorities acted pre-emptively and tried to arrest a group of conspirators on the eve of the military uprising. The attempted arrest was a fiasco and, since the secrecy was broken, the rebel officers of the Army of Africa immediately got their units out of the barracks to seize the civil and military authorities loyal to the government. Next day, the conspirators at the home garrisons started joining the rebellion, but they were only half-successful. The Spanish Civil War had begun.

Conclusion

The increasingly strained balance between the military and civil power of the Restoration blew up in 1917. The autonomy given to the officer corps to deal with its own affairs by the dynastic politicians had been useless to solve its main problems and fostered its disaffection from the civil power. When a majority of junior and middle-ranking officers considered their interests harmed in the mid-1910s, it should not come as a surprise to anyone that they defied the government (and the senior generals associated with the political establishment) through the *juntas de defensa*. The dynastic politicians yielded because they needed the army to withstand the increasing social unrest in Spain. However, the *juntas*' success would be marred by the military's internal conflict between Africanist and peninsular officers about promotions and the public's weariness with the penchant of the *juntas* for bullying the government whenever the latter did not comply with their demands. Eventually, the fear of the political consequences of the elucidation of responsibilities for the disaster of Annual led a section of the army headed by General Primo de Rivera to seize the power in order to preserve the position of the military. Ruling as a dictator, Primo de Rivera concluded successfully the Moroccan war, but his military policy did not solve the main problems of the army and caused the disaffection of the learned corps. Politically isolated in the end, Primo de Rivera resigned, but Alfonso XIII's acquiescence to the dictatorship sealed the fate of the monarchy.

The king acknowledged that the game was over in the spring of 1931 and went away. The military, *faute de mieux*, accepted the Second Republic, although few officers sympathized with it. However, the first republican war minister, Azaña, finally implemented many of the reforms the army badly needed, especially the reduction in size of the officer corps. Unfortunately, Azaña's abrasive manner allowed his enemies to present the reforms as an attack against the military, a view accepted by many in the officer corps. And the increasing political polarization aroused the conservative impulses of many officers, who eventually decided that another military intervention was the only way to preserve Spain's integrity and social order. The upshot of these events was the Civil War of 1936–1939. Besides its political consequences, the conflict's aftermath ruined much of the improvement achieved under the Second Republic in allowing the return of an oversized officer corps. About 10,000 of the temporary officers commissioned in Franco's army during the war (see Chap. 6) were incorporated into the active list throughout the next decade; these wartime officers were strongly identified with the values of Franco's regime as a rule. Although the uncertainties of the Second World War (which thwarted a full demobilization and forced Spain to maintain an army larger than usual in peacetime) may have been a temporary rationale for such decision, another was the political purpose of strengthening the loyalty of the officer corps to Franco's regime whatever the price.[38] Alas, the price was the regeneration of the tumour that hindered the full modernization of the Spanish army for over a hundred years.

Notes

1. For a sample of literature on the Spanish army's political and social history during the period 1916–1936, on which this chapter is based, see Mariano Aguilar Olivencia, *El ejército español durante la Segunda*

República (claves de su actuación posterior) (Madrid: Econorte, 1986); Ana Isabel Alonso Ibáñez, *Las Juntas de Defensa Militares (1917–1922)* (Madrid: Ministerio de Defensa, Secretaría General Técnica, 2004); Michael Alpert, *La reforma militar de Azaña (1931–1933)* (Madrid: Siglo XXI, 1982); Carolyn P. Boyd, *Praetorian Politics in Liberal Spain* (Chapel Hill: The University of North Carolina Press, 1979); Carolyn P. Boyd, 'Las reformas militares', in [Luis Suárez Fernández Demetrio Ramos Pérez, José Luis Comellas, and José Andrés-Gallego (eds):] *Historia General de España y América*, 25 vols. (Madrid: Rialp, 1981–1992), XVII. 141–73; Carolyn P. Boyd, *La política pretoriana en el reinado de Alfonso XIII* (Madrid: Alianza, 1990); Gabriel Cardona, *El poder militar en la España contemporánea hasta la guerra civil* (Madrid: Siglo XXI, 1983); Ramón Gómez Martínez, *Constitución y Fuerza Militar (1808–1978)* (Granada: Universidad de Granada, 2010); Carlos Navajas Zubeldía, *Ejército, estado y sociedad en España (1923–1930)* (Logroño: Instituto de Estudios Riojanos, 1991); Stanley G. Payne, *Politics and the Military in Modern Spain* (Stanford: Stanford University Press, 1967); Stanley G. Payne, *Ejército y sociedad en la España liberal (1808–1936)* (Madrid: Akal, 1976); Fernando Puell de la Villa, 'Las fuerzas armadas en la crisis de la Restauración. Las Juntas Militares de Defensa', in Mario Hernández Sánchez-Barba and Miguel Alonso Baquer (eds), *Las fuerzas armadas españolas. Historia institucional y social*, 8 vols. (Madrid: Alhambra, 1986), V. 81–126; Ramón Salas Larrazábal, 'Las reformas de Azaña', in Hernández Sánchez-Barba and Alonso Baquer (eds), *Las fuerzas armadas*, VI. 11–101; Carlos Seco Serrano, *Militarismo y civilismo en la España contemporánea* (Madrid: Instituto de Estudios Económicos, 1984).

2. The dispute about promotion has been tackled recently in Alberto Bru Sánchez-Fortún, 'Los ascensos de guerra (1909–1922). Su repercusión en el nacimiento y desarrollo de las juntas de defensa', *Revista de Historia Militar* (hereafter *RHM*), 119 (2016), pp. 13–65. The Spanish army's official history of the war in Morocco is Servicio Histórico Militar, *Historia de las campañas de Marruecos*, 4 vols. (Madrid: Servicio Histórico Militar, 1947–1981). For shorter accounts, see Carlos Hernández de Herrera and Tomás García Figueras, *Acción de España en Marruecos* (Madrid: Imprenta Municipal, 1929); Federico Villalobos, *El sueño colonial. Las guerras de España en Marruecos* (Barcelona: Ariel, 2004); Roberto Muñoz Bolaños, José Luis De Mesa Gutiérrez, Carlos Lázaro Ávila, and Jesús Narciso Núñez Calvo, *Las campañas de Marruecos (1909–1927)* (Madrid: Almena, 2001). Other works on Spain's involvement in Morocco consulted by the present author are David S. Woolman, *Rebels in the Rif: Abd el Krim and the Rif Rebellion* (Stanford and London: Stanford University Press, 1969); Victor

Morales Lezcano, *El colonialismo hispanofrancés en Marruecos (1898–1927)* (Madrid: Siglo XXI, 1976); Miguel Alonso Baquer, 'El problema de Marruecos', in Hernández Sánchez-Barba and Alonso Baquer (eds), *Las fuerzas armadas*, V. 227–57; Manuel Aragón Reyes, Manuel Gahete Jurado, and Fatiha Benlabbah (eds), *El Protectorado español en Marruecos: la historia trascendida*, 3 vols. (Bilbao: Iberdrola, 2013).

3. Although different in their scope, informative overviews of European armies in colonial warfare can be found in V. G. Kiernan, *European Empires from Conquest to Collapse, 1815–1960* (London: Fontana and Leicester University Press, 1982); Bruce Vandervort, *Wars of Imperial Conquest in Africa, 1830–1914* (London: UCL Press, 1998). On the failure to organize a colonial army in Spain, see Alberto Bru Sánchez-Fortún, 'El debate sobre el ejército colonial en España: 1909–1914', *RHM*, 112 (2012), pp. 65–124. Despite their importance for the Moroccan campaigns and role as breeding ground for Africanist officers, the *regulares* still have not been the subject of scholarly treatments comparable with those devoted to the later Spanish Legion; on the creation and early times of the *regulares*, see José María Jiménez Dominguez, José Luis Pérez Santaella, Juan García del Río Fernández, and Carlos González Rosado, *Fuerzas Regulares Indígenas. De Melilla a Tetuán. 1911–1914. Tiempos de ilusión y de gloria* (Madrid: Almena, 2006). Other types of native forces in Spanish service were the *harkas* and *idalas* (groups of irregular auxiliaries under a Spanish commander and a tribal leader of their own, respectively), the *mehal-las* (regular troops under the nominal authority of the Sultan of Morocco's representative in the Protectorate, trained and partially staffed by Spanish officers), and the paramilitary Indigenous Police (staffed by Spanish army officers, was disbanded and incorporated into the *mehal-las* in 1922). For information on Spain's Moroccan troops, see José Díaz de Villegas, *Lecciones de la experiencia. (Enseñanzas de las campañas de Marruecos)* (Toledo: Colección Bibliográfica Militar, 1930), pp. 98–9; Joaquín de Sotto Montes, 'Notas para la Historia de las fuerzas indígenas del antiguo Protectorado de España en Marruecos', *RHM*, 35 (1973), pp. 117–54; [Arturo Sus Sus (ed.),] *Historia de las fuerzas armadas*, 5 vols. (Saragossa: Palafox, 1983), IV. 171–202; María Rosa de Madariaga, *Los moros que trajo Franco… La intervención de tropas coloniales en la Guerra Civil española* (Barcelona: Martínez Roca, 2002), pp. 75–86, 104–9.
4. Julio Albi de la Cuesta, Leopoldo Stampa Piñero, and Juan Silvela Milans del Bosch, *Un eco de clarines. La Caballería española* ([Madrid:] Tabapress, 1991), p. 307.
5. The term also had a broader meaning which identified both the scholar in African subjects and anyone who advocated the Spanish expansion in

Africa. In fact, only a minority of the Africanist officers was really fluent in any of the local languages or well versed in the native culture. For different treatments of the Africanists, see Andrés Mas Chao, *La formación de la conciencia africanista en el ejército español (1909–1926)* (Madrid: Servicio Geográfico del Ejército, 1988); Sebastian Balfour, *Abrazo mortal. De la guerra colonial a la guerra civil en España y Marruecos (1909–1939)* (Barcelona: Península, 2002), Part 2, Chap. 2 [original title: *Deadly Embrace: Morocco and the Road to the Spanish Civil War*]; Madariaga, *Los moros*, pp. 32–47. It is worth noting that the British officers serving in the Indian Army (that is, the pool of Britain's Indian native regiments) or African native regiments were required to learn the local language used by the soldiers they would lead; however, few of those staffing the latter made any effort to study local cultures. But then they were only seconded for a temporary tour of duty from their parent British regiments, not unlike the Spanish system, whereas Indian Army officers were permanently assigned to native units: Anthony Clayton, *The British Officer: Leading the Army from 1660 to the Present* (Abingdon: Routledge, 2013), pp. 276–7, 280–1.

6. Neither the French nor the Spanish armies had modelled their officers' career pattern on a regimental system; perhaps a case may be made that the unfavourable effect on the officer corps' morale of the colonial soldiers' careerism bears some resemblance to that of the purchase of commissions in the British army before the Cardwell reforms: John Keegan, 'Regimental Ideology', in Geoffrey Best and Andrew Wheatcroft (eds), *War, Economy, and the Military Mind* (London: Croom Helm, 1976), pp. 9–10. See also Douglas Porch, *The March to the Marne: The French Army 1871–1914* (Cambridge: Cambridge University Press, 1981), p. 161; Paul Robinson, *Military Honour and the Conduct of War: From Ancient Greece to Iraq* (Abingdon: Routledge, 2006), pp. 146, 149.
7. Ministerio de la Guerra, *Colección Legislativa del Ejército 1878* (Madrid: n.p., 1878) (hereafter *CLE plus the year of publication*), Item No. 367, Article 30; José Antonio Olmeda Gómez, *Las Fuerzas Armadas en el Estado Franquista. Participación política, influencia presupuestaria y profesionalización, 1939–1975* (Madrid: El Arquero, 1988), p. 98.
8. The figure is taken from Jorge Vigón, 'Breves notas para la historia de las Juntas de Defensa y de la Dictadura', unpublished paper [mimeographed copy filed in Archivo General Militar de Ávila, Fondos del Ejército Popular, Armario 47, Legajo 73, carpeta 8], p. 6. But this and comparable figures circulating in those years might well include campaign medals awarded customarily for service in Morocco, whatever the rank: Julio Albi de la Cuesta, *En torno a Annual* (Madrid: Ministerio de Defensa, 2014), p. 82.

9. Francisco Franco Salgado-Araujo, *Mi vida junto a Franco* (Barcelona: Planeta, 1977), p. 31; Estado Mayor Central del Ejército: *Enseñanzas de la campaña del Rif en 1909* (Madrid: [Depósito de la Guerra], 1911), p. 84.
10. The Africanists initially supported the *juntas* in order to display military corporate solidarity, still unaware of the latter's ultimate intention: Miguel Alonso Baquer, *El ejército en la sociedad española* (Madrid: Ediciones del Movimiento, 1971), p. 258. On the other hand, the boundaries between the Spanish army's peninsular and Africanist factions were not as clear-cut as it could seem at first sight (there were *juntas* in the Protectorate as well) nor every officer was eager to take sides: Balfour, *Abrazo mortal*, pp. 315–17.
11. Fear of revolution was furthered by the Russian Bolsheviks' triumph later in the year. On the 'Red Scare' the latter inspired, see James Matthews, 'Battling Bolshevik Bogeymen: Spain's *Cordon Sanitaire* against Revolution from a European Perspective, 1917–1923', *Journal of Military History*, 80 (2016), pp. 725–55. This fear also led the authorities to suppress fast the enlisted ranks' own associations, set up in imitation of the officers' *juntas de defensa*: Alonso Ibáñez, *Las Juntas de Defensa*, p. 639.
12. Lieutenant Colonel Miguel Núñez de Prado, a commander of *regulares*, articulated this dissatisfaction to the commission of inquiry on the disaster of Annual: [Juan Picasso González,] *Expediente Picasso. Documentos relacionados con la información instruida por el señor general de división D. Juan Picasso sobre las Responsabilidades de la actuación española en Marruecos durante julio de mil novecientos veintiuno*, facsimile edition (Mexico, D.F.: Frente de Afirmación Hispanista, A.C., 1976), pp. 71–2.
13. Data on salary rises taken from Fernando Fernández Bastarreche, 'El Ejército', in [Suárez et al. (eds),] *Historia General*, XVI–1. 665. There were no more significant increases before the Civil War. José Fernando García Moreno stresses that these salary rises only provided a short-lived relief from inflation and the depreciation of the currency: José Fernando García Moreno, 'Servicio militar universal y personal en España. Procedencia social y formación del soldado, clases profesionales y oficial de carrera en el Ejército de Tierra español de 1913 a 1936', 5 vols. (doctoral dissertation, Universidad de Zaragoza, 1985), II. 673–6. But Carlos Navajas concludes that the officers' later complaints about their 'low' salaries are groundless, if the latter are compared with those of workers and civil servants: Navajas, *Ejército*, pp. 194–8.
14. On the staff corps' struggle for survival in the period 1918–1920, see Alonso Ibáñez, *Las Juntas de Defensa*, pp. 534–47.
15. The Legion was founded by Lieutenant Colonel José Millán Astray as a Spanish version of the French Foreign Legion, but most recruits

were actually Spaniards; its first official title was *Tercio de Extranjeros* (Regiment of Foreigners), but Millán Astray always preferred the name *la Legión*, which soon became popular. On the Legion's origins and history during the 1920s, see the first volume of [La Legión,] *La Legión Española. 75 años de Historia (1920–1995)*, 3 vols. (n.p.: [La Legión], 2001–2003) [the first two volumes originally published: [Subinspección de La Legión,] *La Legión Española. (cincuenta años de historia)* (Leganés: Subinspección de La Legión, 1970–1973)]; José E. Álvarez, *The Betrothed of Death: the Spanish Foreign Legion during the Rif Rebellion, 1920–1927* (Westport: Greenwood, 2001); Madariaga, *Los moros*, pp. 86–104; Juan José Amate Blanco, 'La Legión como respuesta a las necesidades militares', in Aragón, Gahete, and Benlabbah (eds), *El Protectorado español*, III. 349–71; Luis Eugenio Togores, *Historia de La Legión. La infantería legendaria. De África a Afganistán* (Madrid: La Esfera de los Libros, 2016), Chaps. 2 and 3.

16. For a recent, detailed account of this debacle, see Albi de la Cuesta, *En torno a Annual.* Another informative account is Juan Pando, *Historia secreta de Annual* (Madrid: Temas de Hoy, 1999).
17. Fernández Silvestre has often been depicted as a 'devil-may-care' cavalryman, but cf. Lucas Canteras Zubieta, 'Manuel Fernández Silvestre: gestación y rehabilitación de un general', *RHM*, 119 (2016), pp. 97–132.
18. In one of the darkest junctures of the Spanish army's history, it is only fair to draw attention to the gallant conduct of the *Cazadores de Alcántara* Cavalry Regiment 14. The unit covered the retreat through three days of combat, losing almost 80 per cent of its strength. For the discredit of those concerned, it took almost a century to award the collective version of the San Fernando Laureate Cross to the regiment in 2012. See Antonio Bellido, *El 'Alcántara' en la retirada de Annual. La Laureada debida* (Madrid: Ministerio de Defensa, Secretaría General Técnica, 2006); Antonio Manzano, *El Regimiento Alcántara, 1911–1921. Disipa, como el Sol, las nubes a su paso* (Madrid: Grupo ATENEA, Seguridad y Defensa, 2012), Chaps. 2 and 3.
19. The strength and casualties of Fernández Silvestre's army are surveyed in detail in Fernando Caballero Poveda, 'Marruecos. La campaña del 21. Cifras reales', *Ejército*, 522 and 523 (July and August 1983), pp. 81–94 and 77–85. But cf. Villalobos, *El sueño colonial*, Annex 3; Manzano, *El Regimiento Alcántara*, pp. 140–6, 190–1.
20. But cf. Miguel Alonso Baquer's view denying a unanimous military agreement: Miguel Alonso Baquer, *El modelo español de pronunciamiento* (Madrid: Rialp, 1983), p. 208.

21. On Primo de Rivera's Moroccan policy, see Shannon E. Fleming, *Primo de Rivera and Abd-el-Krim: The Struggle in Spanish Morocco, 1923–1927* (New York and London: Garland Publishing, 1991); Shannon E. Fleming, 'El problema español de Marruecos y el desembarco en Alhucemas', *RHM*, 35 (1973), pp. 155–72.
22. The figure for European troops is the present author's estimate based on the data in Hernández and García Figueras, *Acción de España en Marruecos*, pp. 411, 683. This source also records 2349 native troops killed or missing for the period 1916–27, but the figure was surely higher.
23. *CLE 1925*, Item No. 427.
24. Nazario Cebreiros, *Las reformas militares. Estudio crítico* (Santander: J. Martínez, 1931), p. 134. Weyler had been dismissed in October.
25. On the republican military conspiracies in the last years of Alfonso XIII's reign, see also Julio Busquets and Juan Carlos Losada, *Ruido de sables. Las conspiraciones militares en la España del siglo XX* (Barcelona: Crítica, 2003), Part 1, Chap. 4.
26. See, for instance, Manuel Azaña, *Estudios de política francesa contemporánea. La política militar* (Madrid: Saturnino Calleja, [1918]).
27. Alpert, *La reforma militar*, pp. 300–8.
28. *Diario Oficial del Ministerio de la Guerra*, 28 April 1931.
29. The regular NCOs disapproved of this step, since they foresaw (quite rightly, as it is shown in Chap. 6) it would restrict considerably their chances of becoming officers: García Moreno, 'Servicio militar', II. 580.
30. Cardona, *El poder militar*, p. 63; García Moreno, 'Servicio militar', II. 484–5, 532–40. On the NCO corps during the Second Republic, see García Moreno, 'Servicio militar', II. 549–63. It is worth noting that, in those years, NCO ranks were given within the regiments in the American army, so a transfer meant demotion to private first class in the new unit; NCOs could not retain their rank when transferred until 1940: L.R. Arms, *A History of the NCO* (n.p.: US Army Museum of the Noncommissioned Officer, 2007), p. 33.
31. *CLE 1931*, Items No. 282, No. 288, No. 337, No. 339, No. 401, No. 444.
32. Miguel Alonso Baquer, 'La selección de la elite militar española', in Hernández Sánchez-Barba and Alonso Baquer (eds), *Las fuerzas armadas*, V. 49–59; Morris Janowitz, *The Professional Soldier: A Social and Political Portrait* (New York: The Free Press, 1971), pp. 74–8.
33. Miguel Alonso Baquer, 'La reforma de la enseñanza militar durante la Segunda República', in [Jean-Pierre Etienvre (ed.),] *Les armées espagnoles et françaises: modernisation et reforme entre les deux guerres mondiales* (Madrid: Casa de Velázquez, 1989), p. 11.
34. Alpert, *La reforma militar*, pp. 21–2, 293–7.

35. Francisco Franco Salgado-Araujo, *Mis conversaciones privadas con Franco* (Barcelona: Planeta, 1976), p. 397.
36. On the political violence during the Second Republic, see also Eduardo González Calleja, 'The symbolism of violence during the Second Republic in Spain, 1931–1936', in Chris Ealham and Michael Richards (eds), *The Splintering of Spain: Cultural History and the Spanish Civil War, 1936–1939* (Cambridge: Cambridge University Press, 2005), pp. 23–44; Eduardo González Calleja, 'Brutalización de la política y banalización de la violencia en la España de entreguerras', in Carlos Navajas Zubeldia and Diego Iturriaga Barco (eds), *Crisis, dictaduras, democracia* (Logroño: Universidad de La Rioja, 2008), pp. 23–38.
37. On the UME's conspiratorial activities, see also Busquets and Losada, *Ruido de sables*, Part 1, Chap. 5. As a reaction, leftist officers created their own underground organization, the Anti-fascist Republican Military Union: see Busquets and Losada, *Ruido de sables*, Part 1, Chap. 6.
38. Fernando Puell de la Villa, *Historia del ejército en España*, 2nd edition (Madrid: Alianza, 2005), pp. 190–1, 204–5; Julio Busquets, *El militar de carrera en España*, revised edition (Barcelona: Ariel, 1984), pp. 112–13; Roberto Muñoz Bolaños, 'La institución militar en la posguerra (1939–1945)', in Fernando Puell de la Villa and Sonia Alda Mejías (eds), *Los ejércitos del franquismo (1939–1975)* ([Madrid]: Instituto Universitario General Gutiérrez Mellado, 2010), pp. 36–8.

CHAPTER 6

Military Education, 1918–1939

This chapter analyses and assesses the reforms of military education under Primo de Rivera's dictatorship and the Second Republic, illustrating the tangle of politics and professional issues which affected Spain's military policy during this period. Its first part concentrates on the purpose behind the creation of the General Military Academy at Saragossa in 1927, the orientation of its training and its teaching methods, and the Africanist influence. The second part discusses the innovative nature of the republican reform of 1931–1932, arguing that its features had already been advocated in the reformist literature of the three previous decades. The chapter also includes a survey of higher military education in the period 1927–1936 and the influence of pre-war models on officer training during the Civil War.

The *Academia General Militar* at Saragossa, 1927–1931

Origins

The idea of training all the army officers in a single academy remained alive in the late 1910s. In 1917 José García Benítez, a graduate of the Toledo general academy and engineer officer, advocated the restoration of a general academy as the best way to solve the problem of the lack of real professional relationships among the corps of the army. Against the argument that a general academy could limit the students' choice of arm, he stated that a common military spirit must be the foundation

J.V. Herrero Pérez, *The Spanish Military and Warfare from 1899 to the Civil War*, DOI 10.1007/978-3-319-54747-3_6

of any career in the army. Thus an engineer officer candidate would have to prove his ability and will to become an officer before thinking of being a military engineer.[1] In 1918 a commission headed by General José Villalba worked to set up a general academy again but, once again, the plan was not implemented.[2] Despite this new miscarried attempt, an article in the cavalry corps journal in 1921 urged the military authorities to set up a general academy by arguing that the First World War had shown the need for cooperation among the fighting arms. Such cooperation would be easier if all the officers knew the essentials of the other arms' tactics through undergoing training in a single academy. The officer corps would also benefit from the comradeship fostered by a general military academy, as had happened with the first one in Toledo.[3] The military authorities had certainly given little, if any regard to practical combined arms training in the academies: indeed, infantry and artillery officer candidates did not carry out joint exercises until May 1923.[4]

General Miguel Primo de Rivera graduated from the Toledo general academy and was convinced of the benefits of common training for officers. He had made no secret of his views: during a celebration of his promotion to general officer rank in 1912, he expressed his purpose of encouraging the restoration of the centre.[5] Primo de Rivera's professional ethos regarded the military as a unitary institution, overriding the narrow corporate attachment to the commissioning branch which had prevailed up to this time, especially among the artillerymen.[6] Once Primo de Rivera became Spain's ruler, it is hardly surprising that he used his position to implement his views. It was perhaps no coincidence that a paper proposing the restoration of a general academy and published in early 1924 (that is, a few months after Primo de Rivera's *coup d'état*) won an essay prize on officer recruitment offered by *Memorial de Caballería*. Its author advanced the idea that the secondary education diploma and the passing of an examination should be the entrance requirements. The examination should not be particularly testing, since the actual selection would be carried out during the two years of training in the general academy. Those officer candidates who passed this stage would choose their branch of the service after being awarded their place in the order of merit. Then they would go to the corps academy with the rank of second lieutenant. After three or four years at the arm school and as officers on probation, they would finally be commissioned as first lieutenants. Thus no new officer would be younger than 21 and he would join his regiment with some practical experience.[7]

The groundwork for the opening of a general academy was done in the period 1924–1925 by a commission under General Antonio Losada, including a working group which designed the syllabus (one of its members was Vicente Rojo). It is odd that Campins, in his essay on the Saragossa academy (finished in 1932), makes no mention at all of this second commission in the pages on the academy's background. Or it may not be, given that the essay leaves the reader with an impression that the entire development of the new academy's innovative teaching orientation and methods took place after the February 1927 decree and the formation of the organizing commission (which included Campins).[8] On the other hand, although the 1924 commission is mentioned in the scholarly literature, as far as the present author is aware its activities have not been surveyed in detail.[9] Unfortunately, the present author came across the thread leading to it too late to undertake further research in order to ascertain to what extent the reformist attempts of the early 1910s influenced the groundwork for Primo de Rivera's academy. In any case, it would appear logical that the planners made use of these previous experiences, which were in line with the orientation Primo de Rivera wanted for the new academy and military education overall.

The new schedule for military education was outlined in the decree, dated 20 February 1927, ordering the creation of the *Academia General Militar* (General Military Academy) in Saragossa.[10] The new academy's primary purpose was clearly expressed at the start of the decree. Prospective officers would learn in the General Military Academy 'the basic [professional] education and, above all, the military spirit that must be common to all the branches of the service'.[11] Thus the Saragossa academy was to perform a role that, in Primo de Rivera's concept of the military profession, the corps academies had failed to do well enough: the vocational testing of the prospective officer. In this sense, the new academy resembled the French military academy of Saint-Cyr, described by an historian as 'a sort of novitiate', where prospective officers tried out their calling and were 'initiated into the rule of the order'.[12] The emphasis on the development of a common military ethos was to be the distinctive trademark of the training in the academy.

The applicants coming straight from civil life had to be at least 17 years old and not older than 21 at the start of the year of admission;

applicants coming from the ranks and the Complement List had to be not older than 25. The junior secondary education (*bachillerato elemental*) diploma was the minimum requirement to take the entrance examination. Training at the academy lasted two years. The choice of arm would be based on the officer candidate's place in the graduation list (if the officer candidate was not satisfied, he could read some subjects for another year to improve his place in the following class's list). Training at the corps academy would last three years; after passing the first year of this stage, the officer candidates would be commissioned as student-second lieutenants. Once their training was finished, they would be given their regular commission as first lieutenants.[13] This last point is noteworthy because after 1893 officer candidates of the learned corps left the academy as first lieutenants whereas those of infantry and cavalry had to soldier two years before getting this rank. According to the decree, this change had the purpose of 'raising, through equalizing, the [officers'] class prestige…'.[14] One cannot help thinking that the infantry and the cavalry had settled a score with the learned corps.

The raised age requirement for officer candidates satisfied an old demand of the military reformists. An article published in 1902 argued that prospective officers should be at least 17 years old before entering a military academy, which would allow their personality to mature at a natural pace, instead of imposing maturity artificially on teenagers during their training; whatever the short-term benefits, demanding excessive responsibility during these early years might be harmful for the officers' personality in the long term.[15] Almost two decades later, the dynastic Liberal politician the Count of Romanones recommended, in an essay on military reform, that the minimum age for officer candidates should be 18.[16] The same year (1920), a contributor to the cavalry's official journal proposed not to commission any officer candidate until he came of age.[17] And in an officer training schedule proposed by Captain José García Miranda and published in the infantry corps journal in 1925, it was proposed that no officer candidate should be taken into an academy before they were 17 years old.[18] However, the age requirement was lowered to 16 years in late 1928, a decision that Campins judged was made for the benefit of the sons of military personnel; he disapproved of this because the academy's course of study was not designed for officer candidates of such a young age, and it also created a potential split between the sons of civilians and the sons of soldiers.[19]

The Commandant and His Staff

Given the rise of the Africanist elite within the Spanish army's pecking order under Primo de Rivera, it should perhaps not be too surprising that the dictator chose one of its members as the first commandant of the Saragossa academy. Millán Astray, the founder of the Spanish Legion, was seemingly the first candidate and actually made a visit to French military schools in 1924. But his passionate, outspoken personality had aroused many enmities within the army and his candidature was eventually discarded. Then Millán Astray apparently nominated his old collaborator in the organization of the Legion, Brigadier General Francisco Franco Bahamonde, for the post.[20] Franco (who, when promoted at 33, became the youngest general officer in Europe) was a rising star on the Spanish military landscape thanks to his achievements on the Moroccan battlefields.[21] He acquired experience in organizing almost from scratch as Millán Astray's first deputy in the Legion, but he had no special qualifications to direct an educational centre. However, this was not an obstacle for the dictator, who wanted battle-hardened instructors to indoctrinate the officer candidates in the tough-minded military spirit which should characterize the officer corps.[22] War merits achieved in Morocco indeed weighed more than anything else in the selection process of the teaching staff.[23] From this point of view, Franco was certainly a good choice as commandant to play an exemplary role in front of the prospective officers. He was a living example of the 'heroic leader' professionalism so cherished by the dictator and practised by the *africanistas*.[24]

The commandant was well-accompanied by officers of the same mould, moreover. Franco was reportedly closely involved in the selection of teachers for the academy and gave preference to those who had at least five years' combat experience in Morocco, especially if they had fought alongside him.[25] Thus the academy staff eventually gathered a select group of Africanists, including several whose careers had been closely associated to Franco's—such as Camilo Alonso Vega, Francisco Franco Salgado-Araujo, and Alvaro Sueiro.[26] And it is certainly notable that only two officers of the teaching staff served in the Republican army during the Civil War.[27] Another remarkable member of the directing staff was the head of studies, Colonel Miguel Campins. He had fought with distinction in Morocco and developed an acquaintanceship with Franco after the late 1910s, although he did not become one of the commandant's closest confidants.[28] But Campins also passed the staff college and displayed a genuine interest in the intellectual development of the officer candidates. This intellectual facet set him apart from the average Africanist.[29]

The new teaching staff was to be a breed apart from the faculties of the former corps academies. The reformist literature had often poured much criticism on the teachers' perfunctory ways and their preference for bookish theory as a sign of their poor quality. Such faults had already been noted in the nineteenth century. General Luis Bermúdez de Castro, who was an infantry officer candidate in the late 1870s, defined the typical academy instructor of the second half of the nineteenth century as unimaginative and easy-going, and this state of affairs seemed to continue after 1900. He compared the easy life of his teachers at the Toledo academy with that of the city cathedral's canons. Bermúdez never saw an instructor on horseback and attributed this fact, as well as the cancellation on rainy days of the single weekly session of practical training, to the poor level of professionalism among the teaching staff.[30] Franco himself agreed that the quality of his instructors at the Toledo academy (in the period 1907–1910) was poor. However, some of them were redeemed in displaying their war wounds and medals for gallantry, which, according to Franco, taught the essentials about the military profession.[31] This last idea was reflected in the Saragossa academy's original faculty, about 90 officers who had collected a dozen Individual Military Medals (the second highest award to military prowess and gallantry) and over 50 promotions through war merit or selection, and a half of them had been wounded in action.[32] The absence of officers with teaching experience in other military academies or academic background has been harshly criticized by some scholars.[33] It must be remembered, however, that in 1924 the Academy of Infantry's Academic Board laid a part of the blame of the failure of the attempted reforms of the early 1910s on the teachers' attachment to old methods.[34] So it made sense to recruit a totally brand new faculty, unpolluted by the practices of the former educational system. Whether or not the officers appointed as teachers were actually proficient enough for their task is a different matter, of course.

According to Stanley G. Payne, Franco was respected as an efficient commandant, although the officer candidates could not help feeling a degree of apprehension due to his disciplinarian ways.[35] This last point is contradictory to the testimony, admittedly partisan, of Franco Salgado-Araujo, who wrote in his memoirs that his cousin imposed arrests only occasionally.[36] There is almost no evidence coming from the officer candidates, but the only memoir published so far states that Franco won high prestige among them. He was remembered as 'a superb commandant' by General Manuel Gutiérrez Mellado—who cannot be accused of diehard Franco-worship. The officer candidates did not attribute the unpleasant aspects of their life in the academy to Franco, but to other members of the faculty.[37] Gutiérrez Mellado's testimony seems to confirm that Franco, given his role as chief organizer and first commandant, took a very personal interest in the functioning of the academy. He had the opportunity to shape the prospective members of the officer corps and this is a privileged position for any officer dedicated to his or her profession. But, following the end of the Moroccan campaigns, Franco's was probably the most attractive assignment for an ambitious officer in the Spanish army as well and a good performance in this post would be of much benefit to Franco's career.[38] As a result, it is understandable that Franco felt such grief when the republican government closed the Saragossa academy in 1931 and unusually lost much of his well-known self-control in his farewell speech, which nearly jeopardized his career. According to Azaña, General Sanjurjo thought that, following the closure, Franco was 'like a kid whose toy is taken away' (*como un chico a quien le quitan un juguete*), while Franco Salgado-Araujo recalled that his cousin was 'deeply depressed, making no effort to conceal his great grief and worry'.[39] It is also plausible that the abrupt closure of the academy, joined to the sudden transition from the monarchy to the republican regime, accentuated Franco's reserved trait.[40] In any event, after being appointed chief of general staff in 1935, Franco lost no time in trying to bring the General Military Academy back to life and urged War Minister Gil Robles to restore it (see below).

A Different Style of Education

The academy's report for the academic year 1929–1930 displays the priorities of the syllabus (which was also a criticism of the old academies' ones). It states that the Saragossa academy avoided 'the long-standing mistake of considering pure mathematics as the sinew of all military education, to the detriment… of the preponderance which must be given to teachings which are essentially tactical and military…'.[41] Moreover, the demanding fitness trials for entrance meant that 'military drill and exercises were not halted by the lack of readiness or stamina in the trainees, as happened in previous times'.[42] This was a significant difference from the previous corps academies (or at least the learned corps' ones), where there was much emphasis on academic achievement. Colonel Campins acknowledged in his essay on the Saragossa academy that the syllabus was more oriented to developing features of character adequate for the military profession than to pursuing knowledge for its own sake, and such character development could not be taught in the classroom or from a book.[43] For example, horsemanship was given great attention because, besides its physical training role, it was reckoned of benefit for the officer candidates' willpower and quick decision-making.[44] Furthermore, any other subject in the syllabus was overridden by the primacy of tactics, which, according to Campins, 'are the fundamental purpose we [the army officers] exist for…'.[45] The Saragossa academy had to be most of all a barracks, since its line was to produce soldiers. In fact, the War Ministry understood it in a very literal way: when Franco requested 12 m^2 per man in the officer candidates' quarters, the ministry only authorized four square metres—the allocation for privates in ordinary barracks. This was too Spartan even for Africanists and after some arguing the ratio allocated was eventually eight square metres (the ratio for sergeants).[46]

Education at the Saragossa academy discarded traditional textbooks and examinations. Instead, theoretical teaching was based on the current service regulations and concise guidelines whose writing is generally attributed to the faculty under the supervision of Campins (see below), and officer candidates were graded through frequent assessment of their coursework (only those who underperformed during the year sat final examinations).[47] Historians usually have credited such methods to Colonel Campins and his interest in the ideas of the educational reformist Fernando Giner de los Ríos (though Campins' knowledge may have been second-hand).[48] Actually the interest of military reformists in Giner's pedagogical ideas dated back to the earliest Restoration at least and some of them influenced teaching at the Toledo general academy (whose tradition the Saragossa academy was supposed to carry on).[49] Although this modernizing trend faded after the 1893 reorganization of military education, it was not forgotten and its gist (or much of it at least) may be detected in the reform of the military academies' syllabi in 1913 and the accompanying teaching instructions, as well as the proposals of the Toledo academy's academic board in the early 1920s and Rojo's May 1926 memorandum (see Chap. 2).[50] It does not seem plausible that the 1924 commission (which included Rojo amongst its members) did not take into account the most recent precedents and the 1927 organizing commission (whose members, it must be remembered, lacked teaching experience) in its turn ignored the groundwork of the former. Moreover, Franco already disliked the liberal ideas that Giner's educational model promoted, so he would not have been very amenable to implement the latter just as it is.[51] But it was a different thing if, perhaps in an unacknowledged way, its essence was integrated into an officially sanctioned, previous project, such as that the 1924 commission had worked on.

The existence of these precedents does not mean that Campins could not have developed in parallel similar educational ideas on his own. And the thorough implementation of a pedagogical system which ran contrary to the mindset of most of the Saragossa teaching staff and Franco himself perhaps would not have been feasible without the presence of Campins. Such achievement would have been creditable and proof of Campins' professional and intellectual prestige.[52] Nonetheless, why Campins did not mention the 1924 commission in his essay is a question still to be explained. He wrote the essay in the months following the closure of the Saragossa academy with the purpose of justifying its *raison d'être* and educational achievements (and, indirectly, the professional ones of its staff). In omitting the groundwork of 1924–1925, was Campins being economical with the truth with regard to the credit for the innovative teaching methods? Or perhaps was he being 'politically correct' by leaving out as many connections of the Saragossa academy with Primo de Rivera's dictatorship as possible? In this regard, presenting the academy in isolation from the political environment certainly helped Campins' case. At any rate, no convincing answer can be given to this little enigma while no new sources on Campins will be available.

Whatever his views about the pedagogical system eventually implemented at the Saragossa academy, Franco did agree with the removal of textbooks, and this was not synonymous of anti-intellectual outlook necessarily.[53] As it was shown in Chap. 2, the reforms of the early 1910s advocated a more practical teaching, less reliant on textbooks (which only should be used for consultation). The distance from that stance to advocating the removal of textbooks is not too long. By the mid-1920s, the Academy of Infantry's Academic Board was proposing to substitute *apuntes* (summaries or guidelines) written by the teaching staff for textbooks.[54] And in his May 1926 memorandum Vicente Rojo (often depicted as the antithesis of the cliché Africanist) reckoned textbooks unnecessary for some subjects, which could be taught with the help of *apuntes* prepared by the teachers themselves.[55] On the other hand, such rejection of textbooks was not unique to the Spanish army (or part of it): the German army officer schools, both in the Imperial and Weimar periods, also used plain guidelines instead of elaborate manuals.[56] In fact, removing textbooks was a rational response to the military academy teachers' deep-rooted, dubious practice of writing new, and often unnecessary, editions of textbooks, which were a source of extra incomes.[57]

Nonetheless, according to the scholar Carlos Blanco Escolá, the Saragossa guidelines were hotchpotches of poor quality, which compiled textbook excerpts and not even identified the sources.[58] In fact, concerning the contents of the syllabus, Franco and his own organizing team did not start from scratch.[59] At least the basic coverage of the subjects was available, since the preparatory working group of 1924–1925 wrote the guidelines for the syllabus' subjects as a part of its tasks. Vicente Rojo himself wrote those for the subjects of firing and telemetry.[60] Was this set of guidelines used by the Saragossa academy's faculty or did the latter write a different set of its own? The answer must be based on a comparison of both sets, if they have survived (that of the Saragossa academy has). Unfortunately, as was mentioned above, this topic came to light too late for the present author being able to undertake further research. On the other hand, Blanco Escolá may be missing the real purpose of the guidelines actually used in the Saragossa academy, for they were learning props intended to help the officer candidate to follow the lessons. They did not include the full content of a subject, so the trainee had to pay attention to the teacher and could even write explanatory notes on them.[61] The excerpts provided the supplementary information to the officer candidate, whose daily schedule hardly left any time for consulting additional bibliography.[62]

The Saragossa academy's approach to officer training actually bore resemblance to that in force in the German army in both the Imperial and Weimar periods. German officer training was divided into two major stages. The officer candidate was initially trained in a regiment as a private and junior NCO. He sat an examination of military proficiency and, if successful, was appointed officer candidate (his superiors also watched him closely all through his service in the enlisted ranks in order to know about his personality). Then he was trained in an officer school. The main difference between the Imperial army and its Weimar successor was that the latter's overall officer training schedule was longer and more demanding academically (applicants had to have the diploma entitling to admission into a university). But even in the Weimar army, although cultural and scientific subjects were not neglected and private study was encouraged after commissioning, 'character' (understood as willpower and a sense of responsibility) rather than intellectual excellence for its own sake was what the selection and training methods looked for.[63]

In fact, while the organization of the Saragossa academy was nearing completion, Franco asked the War Ministry in April 1928 for expense allowances for a trip to the Dresden infantry school. He argued the usefulness of the visit on the basis of this German officer school's resemblance to the projected General Military Academy. Moreover, Franco thought it interesting to look at the way the German army overcame the restraints of the Versailles Treaty in training its officers (this is, by the way, an implicit acknowledgement of the Spanish army's limited availability of modern ordnance).[64] The trip to Germany was approved and carried out, after a delay, in late June.[65] On his return, Franco wrote a report on his visit.[66] Rather than providing new ideas, the visit seemed to confirm that the general orientation of the Saragossa academy's training was right, the development of adequate features of character in the prospective officers taking precedence over the pursuit of intellectual prowess.

Nonetheless, it was a resemblance within limits. In the case of the Saragossa academy, the offspring of military families were apparently reckoned more likely to possess the right values. According to Franco Salgado-Araujo, when the time for selecting officer candidates arrived, the commandant ordered the selection board to assess in a benevolent way the sons of military men killed in action. Though most of them were hardly prepared to satisfy the academic standards for entrance, they supposedly became fine officers due to their moral background: 'Their great spirit made good the [educational] limitations.'[67] By contrast, the builder of the new German army after 1918, General Hans von Seeckt, never permitted his own preference for applicants coming from the traditional military aristocracy to protect officer candidates from the rigorous training schedule, whose educational requirements were demanding.[68]

An Academy for a Dictatorship?

Much has been made of French War Minister André Maginot's visit in October 1930 by laudatory histories of the Saragossa academy (often written by military authors) and Franco's hagiographers. The French politician, whose trip's purpose was just to bestow a French military award on Franco, declared that the academy's organization was 'perfect, and amongst all the military schools in Europe [the Saragossa academy] is doubtless the most modern one'.[69] However, the French historian Bartolomé Bennassar wonders how Maginot reached such conclusion after a visit which was brief (13.45 a.m. to 16.30 p.m.) and superficial (after discounting speeches, other formalities, and the lunch, there only was enough time left for a swift tour of the main building and a parade of the officer candidates). Moreover, Maginot's own military credentials were limited (those of a former reservist sergeant). In short, an underqualified observer saw no more than polished drill and well-kept installations in a rush. Was this enough to assess the real quality of professional training in a European military academy? Bennassar doesn't think so; neither does the present author.

An alternative explanation advanced by Bennassar is that Maginot based his judgement on reports of the French military attaché; unfortunately, it cannot be corroborated because the records of this period hold in the French army's archives are incomplete.[70] Whatever the truth, the noteworthy point is that Bennassar seems to be the first author who applied the historical method to check the accuracy of Maginot's words. Non-judgemental historians of the academy or Franco worshippers accept readily the latter at face value.[71] But even those who adopt a critical stance seem to rate arbitrarily Maginot as a major expert in military professional matters and, as a result of his association with the French debacle of 1940, use his words to question the Saragossa academy syllabus' suitability for modern warfare.[72]

Maginot's visit exemplifies how prejudiced the judgments on the General Military Academy have tended to be as a consequence of the later course of Spanish history and Franco's role in it. Nonetheless, the Saragossa academy had already raised disapproving views in its time, although it was too short-lived to generate too much critical literature from contemporaries. If hostility from the learned corps was to be expected, hostility from within the infantry corps was less so. But Enrique Ruiz-Fornells—an infantryman and a reputable staff officer and military essayist who was appointed undersecretary of the War Ministry in 1930—strongly opposed the academy because it might threaten his arm's relative position within the army (a personal antipathy to Franco himself influenced his opinions as well).[73] Ruiz-Fornells' view is peculiar in the light of the faculty's arm ratios overall: 42% of the teaching staff belonged to the infantry, in contrast with 14% belonging to the artillery.[74]

It is also curious that negative assessments about the academy's syllabus fell into two entirely opposed types. Jesús Pérez Salas, a future senior commander in the Republican army, wrote (after the Civil War) that the teachers of the corps academies were appalled by the ignorance of the officer candidates coming from the Saragossa academy, where scientific and technical matters had been so neglected that their essentials had to be taught again.[75] By contrast, the conservative military essayist Nazario Cebreiros thought that the five years' officer training devised for the 1927 reform was still too long and that the syllabus of the Saragossa academy still included too much academic content—in order to make up for the limited general education of the officer candidates and because this centre allegedly could not help obliging the learned corps' demands.[76] Since Pérez Salas and Cebreiros failed to explain what they meant by 'poor' or 'excessive' scientific content in the education of officers, there is no way to contrast their statements with the academy's syllabus and thus it is difficult to assess the fairness of their judgements. Gutiérrez Mellado (who would be commissioned in the artillery) thought that the syllabus was well balanced in theoretical content and practical training and made no mention of educational shortages when arriving at the corps academy in his recollections.[77]

Nonetheless, it can be argued that Gutiérrez Mellado's testimony is questionable, since he may have been exposed at the Saragossa academy to indoctrination so effective as to prevent him from developing any critical assessment of his own training. Given that the Saragossa academy was created by a dictator and in light of General Franco's future role in the history of Spain, it is almost inevitable to wonder if the syllabus also included in practice an ideological indoctrination oriented to non-professional, political goals.[78] It is again hard to give a straight answer. The Spanish officers' accounts and memoirs are often so biased by ideological partisanship that it is difficult to separate pure professional assessment from political criticism. Franco Salgado-Araujo, a relative and close associate of the commandant, could write nothing but praise of the academy's ethos and achievements.[79] Eduardo Benzo was an infantry officer and a liberal conservative opponent of Primo de Rivera, but he still approved of the Saragossa academy because '[it] provides the future officers with a similar ethos about the problems concerning the virtues that are essentially military by subjecting the former to the same trials in the novitiate, as an affair preceding [specific branch training]'.[80]

Officers who fought in the Republican side during the Civil War expressed more negative judgements of the Saragossa academy. Cordón, a Communist from the early 1930s onwards, acknowledged that a general academy was a good idea, because it strengthened the unity of the officer corps. But he described the Saragossa academy as a reactionary centre which inculcated in the officer candidates a caste spirit and a fascist sense of discipline and loyalty to the leader.[81] Pérez Salas shared this view: the academy produced 'complete horsemen, fine players in several sport games… and, above all, [officers with] a frightening Africanist spirit'. This kind of training, according to Pérez Salas, made the officer candidates very receptive to the ideas of fascist organizations like *Falange* (a political party founded in 1933 by Primo de Rivera's eldest son José Antonio).[82] And in fact Gutiérrez Mellado joined *Falange*.[83] It may not be unreasonable to think that General Franco, as organizer and first commandant, to some degree stamped features of his own *weltanschauung* on the academy's ethos.[84]

The social scientist Julio Busquets concluded in his study of the Saragossa academy classes that no less than 90% of the graduates joined the military rebellion in 1936 (a proportion noticeably higher than in the officer corps as a whole) and they did it as an effect of ideological indoctrination in the academy.[85] Later scholars have generally accepted this figure, although they sometimes disagreed with Busquets' explanation.[86] But Fernando Puell de la Villa has challenged the figure because it is not properly backed by documentary evidence. In his own study of the 701 Saragossa graduates in the service (all of them first lieutenants) when the Civil War broke out, Puell found that only 508 (72%) were promoted to captains in Franco's army with their seniority dated 20 March 1937. Not all of them were actually promoted at the same time; a part was promoted in later dates, whereas others who had to go underground or were imprisoned in the Republican zone did not become captains until the end of the war. Of the remaining 193 (28%), some doubtless died before March 1937, but Puell conjectured that the rest may well have remained loyal to the government and served in the Republican army during the conflict. Therefore the behaviour of the Saragossa graduates in 1936 may have been more similar to their fellow-officers' than it has been thought.[87]

The Corps Academies and the Reform of 1927

At least a brief mention must also be made of the consequences of the 1927 reform for the corps academies, a subject overlooked by most of the academic literature.[88] These academies' new syllabi (whose implementation was to start in 1930) would reproduce the Saragossa model in the distribution of subjects and the practical orientation of teaching. There would be no textbooks and grading would be carried out by assessing frequently of the trainees' coursework. Boarding would be compulsory during the first year; afterwards, once the officer candidates became student-second lieutenants, they could live in an officers' residence.[89] However, the sudden collapse of the monarchy and its replacement with a republican regime dismantled Primo de Rivera's model of military education before the latter had been fully implemented. The corps academies were consolidated into two new ones (see below), which also took in the officer candidates of the closed Saragossa academy's last class.[90] As an interim measure the syllabi would remain in force in the new academies, although in an abridged form in order to reduce to four years overall the training of all the officer candidates who had passed through the Saragossa academy.[91] Therefore, it is still possible to catch a glimpse of how the corps academies coped with the 1927 system.

Puell de la Villa provides in his biography of Gutiérrez Mellado some interesting snippets about the Segovia academy, which also trained engineer officer candidates after 1931. The effect of the 1927 reform on the training of artillery officer candidates was significant, as can be seen in Table 6.1, which compares the allocation of time in the syllabus before the reform and after.

There is a noticeable increase in military and tactical contents to the detriment of the scientific ones. So the Segovia academy syllabus seemed to be in line with the spirit of the 1927 reform, at least on paper. But the old artillery faculty displayed remarkable resilience despite the major cuts in the syllabus after both the 1927 reform and the later republican one: those teachers who could not remain in the academy managed to be posted within the local garrison, marking time until they could return. In such a situation (and helped initially by the presence of the last classes of the 1893 system), the artillery teaching staff could easily warp the ministerial policy by playing with the proportional value of the subjects in the syllabus for the trainee's final grades.[92] And it should not come as a surprise to anyone that prospective artillery officers had to use textbooks

Table 6.1 Allocation of time (in estimated percentage) in the artillery syllabus, Segovia academy, before 1928 and after 1931

Subjects		*Before 1928*	*After 1931*
Theoretical	Science & technology	33	17
	Military subjects	10	20
	Miscellaneous	7[a]	13[b]
Practical	Gunpowder & firing	25	10
	Tactics	12.5	20
	Horsemanship	10	10
	Physical training & others	2.5[c]	10[d]

Notes
[a]French and another foreign language
[b]Economics and a foreign language
[c]Physical training only
[d]Sports (5%) and motor driving (5%)
Source Percentage figures estimated by the present author from information in Fernando Puell de la Villa, *Gutiérrez Mellado. Un militar del siglo XX* (Madrid: Biblioteca Nueva, 1997), pp. 71–72

again after 1931, whose compulsory purchase was justified by arguing that they would be the basis of the officers' private professional libraries.[93] Finding out to what extent there was a backlash against the 1927 reform in the cops academies may well be worthy of further research.

Military Education in the Second Republic, 1931–1936

The Republican Reform of Military Education

Being Primo de Rivera's child sealed the fate of the General Military Academy. The republican government issued a decree dated 30 June 1931 which ordered the closure of the General Military Academy. The arguments for doing so included the illegality of its creation (since Primo de Rivera's dictatorship, in breaking with the constitutional regime, had no legitimacy) and the squandering of personnel and financial resources caused by the existing officer training structure, which was disproportionately large for the real needs of the army.[94] However, it would be naive not to think of additional political and self-serving motivations for this decision. Azaña seemed to be convinced a morally cohesive army threatened civil power more than a split one and saw the indoctrination at the General Military Academy as uncongenial to the new regime.[95]

The republican war minister was not alone. Cordón thought that the Saragossa academy's alleged reactionary indoctrination justified the government's decision, but he deemed that it would have been wiser to substitute a commandant and a faculty more sympathetic to the republican regime for Franco and his men.[96]

In his turn, Cebreiros attributed the decision to the alleged anti-military thinking of Azaña, but also to the pressure of those within the army interested in the disappearance of the Saragossa academy: the interest group of the old academies (formed by teaching staff, textbook writers and preparatory school teachers) and the artillery officers, who wished a return to the system of separate education.[97] To Cebreiros' list Campins added unnamed army 'sages' that loathed the Saragossa academy because they had been left aside during its organization and rejected as teachers.[98] And it should neither be discarded a discontentment in the War Ministry (where Azaña had just confirmed Ruiz-Fornells as undersecretary) with Franco's somewhat self-important way of running the academy.[99]

Military education was restructured by a decree also dated 30 June 1931. Officer candidates would be trained into two new centres: the Academy of Infantry, Cavalry, and Quartermaster (Toledo) and the Academy of Artillery and Engineers (Segovia). Although this was a step backwards in relation to the benefits of common basic training for the officer corps' professionalism, it was not a return to the 1893 system either. Actually the new organization resembled that of France and Britain, which had separate academies for the infantry and the cavalry (Saint-Cyr and Sandhurst) and for the learned corps (*École Polytechnique* and Woolwich).[100] The reorganization cut substantially the number of teachers, from 258 to 85, and the new faculties were mostly formed with the remaining staff of the old ones.[101]

As was mentioned above, the new academies initially trained the classes which had passed through the Saragossa academy. Once these classes had been commissioned, the academies would take in officer candidates who complied with the requirements detailed in the 1932 act on officer recruitment and promotion.[102] Prospective officer candidates were classed into three groups:

a. Any civilian or military man who was 18 years old at least, had the senior secondary education (*bachillerato superior*) diploma and had passed a number of subjects in a university[103];

b. Senior and middle-ranking NCOs who voluntarily took voluntarily an entrance examination;
c. Senior NCOs who took a mandatory entrance examination.

In the annual intakes of the academies, 60% of the vacancies would be reserved for NCOs of groups (b) and (c).[104]

After passing a competitive exam, an applicant of group (a) had to fulfil six months' regimental service in the ranks in his arm of choice; following the satisfactory completion of such service, his commanding officers would issue him with a certificate of military proficiency. Then he would join a military academy and be trained over the course of four semesters in the duties of an officer up to the rank of captain. NCOs of group (b) who passed their examination would join an academy and follow the same syllabus as the officer candidates of group (a); trainees of both groups would be commissioned as first lieutenants at the end of their studies. NCOs of group (c) would take one semester's training whose end would coincide with that of the other groups' syllabus. The three groups would also take together a course of application and practice.[105]

Anticipated Innovation in the Recruitment of Officers

The changes introduced by the 1932 act did not come out of the blue. Evidence from Spanish professional and military-related literature proves that major points of the reform had been put forward during the three previous decades. Therefore, the reform of military education under Azaña could hardly be based on anti-military foundations, since it limited itself to implementing proposals advanced by previous military essayists.

One significant feature of the 1932 act was the rise in the academic standards expected of prospective officers. The older entrance age allowed to demand a full secondary education, as some essayists had requested. Although Romanones did not make any explicit mention of a full secondary education, this requirement is rather obvious when he wrote that applicants should prove an adequate level of general education, in order not to waste time during the first years at the academy by reading non-professional subjects.[106] The same requirement can be deduced in an article published in the cavalry journal which requested that prospective officers had an education similar to that entitling them to university entrance.[107] In fact, the 1932 act went a step beyond the preceding proposals by demanding a certain level of university education, but it gave no detail as to which subjects should be taken. It was not until the release of the first public announcement that such a matter was settled: applicants of group (a) had to provide proof of having passed three first-year subjects (mathematical analysis, geometry and trigonometry, and experimental chemistry) in a *facultad* (university school) of sciences.[108] Romanones had actually suggested that some non-military matters could be read in civil educational institutions and that, therefore, the academies' teaching could be focused on military subjects.[109]

Nevertheless, the 1932 act was timid by comparison with the proposal made by Major Luis Belda in 1924. He suggested that prospective officers should initially take a degree in science. This requirement would provide officer candidates who were more mature physically and psychologically, and whose decision to pursue a military career would be more thoughtful and no longer conditioned by the need to earn a living (since they could look for well-paid civil jobs thanks to their university education). Moreover, sharing higher education would bring the officer corps nearer to the civil society.[110] But if officers like Belda reckoned university education to be compatible with the military spirit, others were hostile to such a combination. Writing in the early 1930s, the Africanist Emilio Mola thought that the requirement of higher education was an anti-military device, an exposure to the indiscipline and the corrosive ideas which, according to Mola, prevailed among the university students. Accordingly, the military values which made up the basis of the officer corps would never take root.[111]

Another significant innovation was the introduction of mandatory service in the ranks before joining the military academy. As early as 1905 Captain Herrera de la Rosa had praised the Japanese practice (taken in turn from the German army) of demanding one year's service in the ranks as a good way to judge whether or not applicants had the endurance and the sense of discipline needed for military life.[112] During the early 1920s, this requirement of preliminary service in the ranks was a common recommendation of the reformist essayists, though they proposed a longer term (one year) than the six months stated in the 1932 act.[113] Moreover, they often demanded that officer candidates achieved previously promotion to NCO rank.[114] Major Belda went further by proposing that applicants should get a commission as reservist officers in the Complement List before sitting the examinations for entrance into a military academy. Since officer candidates would already have been trained as subalterns, the academy syllabi would focus on more advanced subjects needed by a career officer.[115]

Mola saw preliminary training in the ranks as a positive step, in contrast with the studies in the university, where prospective officers would supposedly be exposed to the harmful influence of their civil classmates.[116] But he seems not to have noticed that such training in the ranks could also expose future officers to another disturbing influence. Conscription was still unpopular and the poorly motivated conscripts could be receptive, through the dissemination of anti-militarist propaganda, to leftist doctrines. In fact, the army (or at least the conservative section of the military) seems to have had a relative poor opinion of the political reliability of conscripts. Primo de Rivera's dictatorship undertook a programme of 'moral education' for the enlisted ranks, which was actually a programme of ideological indoctrination for conscripts, especially those from an urban working-class background.[117] Once such indoctrination disappeared under the republican regime, officer candidates would have been potentially exposed to the kind of 'subversive' ideas despised by Mola. Although their influence is impossible to assess for sure and was probably exaggerated by the conservative officers, leftist organizations did certainly attempt to set up underground networks within the barracks and win over the conscripts in the early 1930s.[118] Perhaps the desire to remove prospective officers from this exposure was one of the reasons behind Gil Robles' unsuccessful July 1935 bill to reopen the General Military Academy, which included the removal of the six months' term in the ranks (see below).

It is worth noting that the requirement of service in the ranks was a distinctive feature of the German army and has been reckoned by some historians as one of the reasons behind the German officers' remarkable battlefield performance in the First and Second World Wars.[119] However, this preliminary service in the ranks was also introduced in the French army before the Great War and in this case the result was not so successful.[120] Therefore, any claim that previous contact with the enlisted ranks would have improved the quality of the Spanish junior officers cannot be accepted at face value. In any case the republican officer training system was too short-lived to reach any conclusion on this issue (and, indeed, on the effectiveness of the 1931–1932 reform as a whole).

Reformist essayists had also pointed out before 1932 the need for limiting training in the academies to the essential subjects required by the duties of a junior officer.[121] The republican reform satisfied such a view and even the scathing Cebreiros, one of the harshest critics of Azaña, acknowledged this part of it as a wise move. He wrote that such change was an official acceptance at last that officer candidates had to be trained only as subalterns—not as generals as well—in the military academies.[122] Nonetheless, there has been some questioning the extent to which teaching could be improved, since the faculties and their methods were still those of the old corps academies.[123]

The Democratization of the Officer Corps

Azaña wanted to open the officer corps to candidates from a wider social background in order to get a military in tune with the republican regime's ethos. The 1932 act aimed to achieve such a purpose through the reservation of a large number of academy student place vacancies for NCOs. As was the case with other innovations of the act, there had already been proposals to increase the numbers of officers coming from the ranks. In 1920 an article argued that half of the regular officer corps should be made up of former NCOs.[124] Another one complained of the uselessness in wartime of the reserve list and the dissension this list caused within the officer corps. As a better way to get officers from the enlisted ranks its author suggested reserving up to two-thirds of the student posts in the academies for NCOs with adequate educational qualifications.[125] Cebreiros himself admitted that a larger number of officers coming from the ranks was beneficial, although not because he agreed with Azaña's goals. According to Cebreiros, the officers coming straight from civil life would pursue much better careers than before. Though half of the company grade assignments could be held by former NCOs, the older age of the latter would limit their chances of promotion to senior rank.[126]

However, Azaña's plan had a serious weakness. As was pointed out by the conservative member of parliament Tomás Peire—formerly one of Azaña's military aides in the ministry before leaving the army, the minister's goals were hardly achievable owing to the low educational level of the Spanish NCOs. It seemed disproportionate to reserve 60% of academy posts for NCOs when, on average, they made up only 5% of the academy graduates by 1930.[127] Peire had a point: unless the educational standards for admission were lowered, few NCOs would be able to enter the military academies in the short term under the conditions set by Azaña. The figures of the new military academies' first intake proved Peire right (see Table 6.2).

Table 6.2 Vacancies and intakes of military academies, 1933–1936

1. 5 September 1933 summon		Vacancies	Intake
Applicant category	(a)	112	54
	(b)	84	6
	(c)	84	3
Total		280	63
2. 31 August 1934 summon		Vacancies	Intake
Applicant category	(a)	129	95
	(b)	44	29
	(c)	44	0
Total		217	124
3. 15 March 1935 summon		Vacancies	Intake
Applicant category	(a)	114	151[a]
	(b)	38	42
	(c)	38	30
Total		190	223[a]
4. 1933–1935 summons total		Vacancies	Intake
Applicant category	(a)	355	300[a]
	(b)	166	77
	(c)	166	33
Total		687	410[a]

Notes
[a]Including additional list of 8 February 1936
Applicant categories
(a) Civilians and military men with full academic requirements
(b) Voluntary NCO applicantsMilitary educationNCO applicants
(c) Mandatory senior NCO applicants
Source Diario Oficial del Ministerio de la Guerra, 5 September 1933, 31 July 1934, 31 August 1934, 30 December 1934, 15 March 1935, 21 January 1936, 8 February 1936

Only a dismal 5% of NCO applicants passed the exam in the first summon. The vacancies reserved for NCO applicants in the August 1934 summon were reduced to 40% (at this time a liberal centrist cabinet was in power); even so, only one-third of them was filled and no group (c) NCO made the grade. Actually the poor grounding of NCOs for the examinations had already been recognized, since an army crammer was set up in Avila by War Minister Vicente Iranzo in November 1933.[128] A group of students was schooled in time to sit the exam of the third (and last) summon of the republican military academies. Only 40% of the vacancies were again reserved for NCOs, but the results improved noticeably: 95% of the vacancies were filled. It must be noted that all the successful applicants in group (c) appear as second lieutenants; they were former *subtenientes* (highest enlisted rank) who had been commissioned after the suppression of their rank under Gil Robles (nonetheless, these new second lieutenants would have to pass through a military academy to be promoted to first lieutenants).[129] Yet the figures still fell short of the 1932 act's original ratios and prove beyond doubt the gap between the theory and the reality.

The Avila crammer was a means to bridge such gap, but overly sanguine expectations would be unwise anyway. Two major reasons hindered officer recruitment from the NCO ranks through the military academies. One of them was pure and simple incapacity to cope with the intellectual effort required by the preparatory studies, a consequence of the Spanish lower classes' limited education. The other reason (which affected senior NCOs) was a combination of age, financial interest and professional prospects. Given the time usually required to reach the highest enlisted ranks, the average senior NCO would be commissioned in his mid-forties, whereas retirement age for subalterns was set at 51, so his career as officer was simply too short and not worth the effort invested. On the other hand, reaching retirement age with the top NCO rank meant a better pension than retiring as a subaltern.[130]

The potential of the NCO ranks as a source of officers cannot be underrated, but Azaña's goal of basing officer recruitment on them could not be really achieved until raising their educational level. The 1932 act was removed from Spain's social reality (Alpert suggests that Azaña perhaps was not too involved in the drafting of the bill). Therefore, the republican reform could not attract overnight social groups different from that which traditionally had produced most applicants for a career in the officer corps.[131] In fact, although this is admittedly sheer speculation, promotion from the ranks might have turned into an alternative for traditional background applicants who failed or could not afford to comply with the more demanding educational requirements to enter the military academies. But still better educated than ordinary privates, they might be promoted comparatively fast to middle NCO rank, sat and pass the entrance exam, and impede even more the access of the lower classes to the officer corps. Such had been the actual outcome of a comparable attempt to 'republicanize' the officer corps carried out in the 1900s by the French Radical Party, whose military policy Azaña admired so much.[132]

Only 59% of all the vacancies offered in the period 1933–1935 were filled; 43% were filled by group (a) applicants and 16% by NCOs. Unsurprisingly, the highest ratio is found in the vacancies reserved for group (a) applicants (84%); the overall ratio for NCOs was only 33% (voluntary NCOs filled 46% of their quota, whereas 80% of senior NCOs vacancies were left unoccupied). The reasons for the poor NCO figures have already been discussed above. With regard to the group (a) applicants, the requirement of university studies seems to have deterred or prevented many traditional applicants from sitting the entrance exams (it must be remembered that higher education was very costly for the average army officer's family). However, there was a marked contrast between the first summon and the last one. The first summon was an unmitigated fiasco: only 22% of all vacancies were filled (and only 3% by NCOs). By contrast, there was no vacancies unfilled in the 1935 summon and 95% of the quota for NCOs was filled. These figures indicate better-prepared applicants. But there is a disturbing figure as well: an excess of 17% of admitted applicants, since one out of four civilians admitted was supernumerary. Was this excess an occasional event or the return of an old habit? The outbreak of the Civil War makes impossible to know if, in this issue, the republican reform would have turned into another case of old wine in new skins.

Despite just implementing features which had actually been postulated in the reformist professional literature of the three previous decades, the 1932 act was soon under threat. Following a recommendation of General Franco, then chief of general staff, War Minister Gil Robles presented a bill to parliament in July 1935 in order to restore the General Military Academy. The bill removed the requirements of university studies (which allegedly deterred many applicants) and preliminary regimental service (which was reckoned ineffective) and extended officer training to five years (two in the general academy and three in a corps academy). However, the bill did not obtain enough parliamentary support. Gil Robles was more successful with another bill which extended the military academies' course of study to three years and was approved by the parliament next November.[133]

The Reform of Higher Education, 1927–1932

Primo de Rivera's Reform

The reformist efforts of Primo de Rivera and the Second Republic also applied to the training of staff officers. The same decree which restored the General Military Academy also included the essential features of a new organization of higher military education. The ESG was to be replaced by the *Escuela de Estudios Superiores Militares* (Higher Military Studies College) or EESM. This institution's students would be captains who had at least two years' experience as unit commanders, and field officers.[134] The EESM was made up of two sections: military and industrial.[135] This reorganization meant another triumph of the unitary concept of the military over the more parochial *esprit de corps* of the branches of the service which had prevailed before.

The industrial section would teach three specialities: chemistry and metallurgy, military construction, and electricity and mechanics. Their syllabi would be two years long plus a period of practice with the goal of training specialist engineers in the manufacture of ordnance, and industrial management and mobilization.[136] In other words, this meant that the technical and industrial responsibilities which justified the heavy scientific content in the training of artillery and engineer officers were snatched away from the learned corps. Such a policy had been put forward before. An article had proposed in 1911 the creation of a military industrial corps, which would assume the technical and manufacturing functions of the artillery and engineer corps. Another article, published a decade later, repeated the same proposal.[137] Both essayists were infantry officers and it seems plausible that their views were shared by large sections of the general corps. The Count of Romanones also echoed them in suggesting that a number of artillery and engineer officer candidates should get additional technical training in their academies and be assigned permanently to military factories after being commissioned.[138] Limiting professional training and duties to regimental service meant that all the corps would be on an equal footing, without special responsibilities justifying the different educational levels. Such an egalitarian status could foster a common military spirit within the officer corps and it is no surprise that Primo de Rivera made the idea a part of his policy to promote the corporate unity of the military.

The military section of the EESM would train officers to perform staff duties. After two years' training plus a period of practice, the students would graduate as *diplomados*. Since Article 20 of the decree stated that EESM graduates would never leave their commissioning corps after taking the staff diploma, the 1927 reorganization was also a death sentence for the staff corps, which would disappear through natural wastage.[139] The fact that Primo de Rivera decided to suppress the staff corps at that time seems somewhat paradoxical. According to Mola, senior officers of this corps, far from being absorbed in bureaucratic paperwork, were ready and able to perform as column commanders in the Moroccan campaigns of 1924–1927—in contrast to those of the learned corps. Mola wrote of the latter: 'Not to get out of the headquarters area was more suitable for the colonels and lieutenant colonels of the artillery and engineer corps -and perhaps they reckoned this more scientific.'[140] Actually Primo de Rivera did not harm the staff corps before 1927 and he promoted two of its most outstanding officers, Joaquín Fanjul and Manuel Goded, to general rank in 1926 because of their prowess in the field. These officers' performance in the Moroccan campaigns proved that the preparation and conduct of military operations had replaced the corps' former scientific-oriented functions.[141] It is clear, however, that Primo de Rivera, although acknowledging and rewarding individual merit, was nonetheless determined to erase the sources of discord within the military, including the existence of a separate corps of staff officers.[142]

The Republican Reform

Azaña closed the EESM and restored the *Escuela Superior de Guerra* through a decree dated 21 July 1931.[143] According to Cebreiros, the industrial section was suppressed to satisfy the artillerymen; scientific and technical teaching was thus returned to the Segovia academy, preventing the infantry and the cavalry from having their own ordnance engineering specialists.[144] Azaña's legislation did not mean a full restoration of the old corporatism, however. The 1932 officer recruitment and promotion act ordered that, thenceforward, staff duties would be performed by captains and field officers with the staff diploma, and it declared that the staff corps would be 'wasted away' (*a extinguir*)—that is, no more officers would join the corps, which would disappear through a process of natural wastage.[145] This was the end of a corporate struggle which had lasted from General Cassola's time as war minister.

But this did not necessarily mean a definitive end of the quarrels affecting the staff officers as a whole. Mola wrote about the existence within the military of a faction which rejected the idea that staff college graduates were entitled to priority in assignments near the high command. According to Mola, the destruction of the staff corps was not enough for this faction, which would then turn on the *diplomados*.[146] Mola offered no speculation on this issue. Cebreiros did not hide his hostility to the alleged privileges of the *diplomados*. Following Azaña's appointment as war minister, the staff diploma was required for assignments in the General Staff. However, Cebreiros rejected the idea that staff college graduates deserved to monopolize these assignments; according to him, the former's professional expertise was not (indeed never had been) worthy of such a right.[147] Rejecting an officer of proven ability for a general staff post because he had not taken formal staff training for it would perhaps be unfair. But such possible unfairness is hardly the best argument to lambast—as Cebreiros does—the professional quality of the staff graduates as a whole.

The neglect of professional education after the officers left the academy had already been denounced many years earlier by Captain Gallego.[148] Cordón pondered that as the average officer advanced in his career, his professional knowledge diminished, rather than increasing. He wrote in an outspoken way that 'most of us [the junior officers] reckoned each promotion, from the rank of captain upwards, as a step forward to professional ignorance…'.[149] The Spanish army became aware of this problem and the need of further training to keep the officers' professional expertise up to date. It eventually started amending the situation by introducing courses for captains after the late 1910s.[150] The republican reform went one step further, in a way which also avoided the worst effects of the closed list. The 1932 act required captains to pass a proficiency course in their arm's academy before promotion to the rank of major, and their seniority in the army list was to be reordered according to their marks.[151] This solution had been put forward several years earlier in the pages of *Memorial de Caballería* and in a reformist essay.[152] However, this part of the 1932 act was not put into practice. Even if the proficiency course had been implemented, it would soon have lost much—if not all—of its value as a means of selection for precedence in promotion, since the reordering of seniority after the course was suppressed by Gil Robles.[153]

Promotion to higher command was also influenced by the reforms of military education. At least one instance of demand for a centre which would train the future generals of the Spanish army can be found in a professional journal in 1921.[154] The First World War had shown that the command of major units was increasingly complex and required specialized training. In order to prepare its prospective generals, the Spanish army created a 'Course for Colonels' in 1923.[155] It was initially short, voluntary, and for informational purposes only, in order to prevent it from becoming unpopular, and gradually grew longer and became compulsory. The course, which never had a permanent site and faculty, became a requirement for promotion to general rank in 1927, formally confirmed in 1930. Working on this precedent, the 1932 act ruled that colonels had to pass a course—of one year's duration—in a brand new Centre of Higher Military Studies as a requirement for promotion to general officer rank.[156] However, the centre had no site or faculty of its own before the outbreak of the Civil War and its courses were actually run by the General Staff. And ironically, Azaña, as war minister, would promote some colonels who did not attend these preparatory courses.[157]

The Training of Junior Officers in the Civil War

Antebellum military education left traces in both sides' officer training procedures in the Spanish Civil War. The Spanish army had slightly over 15,000 regular officers (including those of the para-military police forces) in July 1936. But many of these were killed in the initial clashes of the military uprising. Others, who were considered to be out of sympathy with the side which had prevailed in their location, were murdered, executed, sentenced to imprisonment or jailed under suspicion. Finally, there were those who went into hiding because they were in the wrong zone for their political leanings. The Republican side got the worst of the split of the Spanish officer corps into two unequal parts. Including perhaps up to 2000 who came out of retirement (due to age or as a result of Azaña's decree of 1931), the Republican side could resort to a theoretical pool of 5500 professional officers at best.[158] However, many Republican regular officers were 'geographical loyalists' (*leales geográficos*) only and often regarded, rightly or wrongly, as untrustworthy, and were therefore not employed; as yet, definitive data about the actual Republican career officer strength cannot be given, but a figure of about 2000 officers still listed in September 1938 (14% of the 1936 active list) may be reckoned as the minimum number of those who were regarded as reliable.[159] By contrast, 8000 regular officers served dependably in the Nationalist army during the war.[160] The Complement List had 6100 reservist officers in 1936. Perhaps up to 2000 of these joined or were called up by the Nationalists, whereas only 1100 or 1200 served in the Republican army.[161] These numbers of regular and reservist officers were insufficient for the mass armies which the Republicans and the Nationalists built up during the war and both sides had to look for other ways of officering their armies, especially in the junior ranks.[162]

Junior Officer Training in the Nationalist Army

On 4 September 1936, the Nationalist army created the rank of *alférez provisional* (temporary second lieutenant). Following a period of improvisation, temporary officer training was organized in separate arm schools by General Luis Orgaz (head of the Nationalist command for manpower and training) by late April 1937. Although career NCOs could enter these schools, most applicants came from a civilian background. They needed to be at least 18 years old (and no older than 30), to have the senior secondary education diploma and to have at least two months' frontline experience. They would take a platoon leader course for one month (two months after early 1938) and be given a commission with the rank of second lieutenant for the duration. The Nationalist army commissioned 29,000 temporary second lieutenants (with 22,000 in the infantry), although 7000 got their commissions too late in the war to go into action.[163]

The training of the *alféreces provisionales* was shaped after the 'heroic leader' pattern, which was a role model for the Saragossa academy's officer candidates. According to the 'heroic leader' model, intellectual performance was less important than moral fibre. Nevertheless, the Nationalist army—in contrast to the Republican side—was fortunate in having a large enough pool of prospective temporary officers to satisfy both requirements. The *alféreces provisionales* were usually educated youths from a conservative middle- and upper-class background, who identified themselves strongly with the Nationalist cause's basic values (even if they lacked any particular political affiliation).[164] Strong commitment certainly made easier the mission of these officers, who had the primary task of 'pulling' the troops through resolute, exemplary leadership in combat (and accordingly suffered lots of casualties).[165]

Junior Officer Training in the Republican Army

Although in a rather muddled way, in the early weeks of the war the Republican army also started to organize the training of temporary officers through short courses for youths with university or full secondary education (the different political militias also ran their own officer schools in the initial period of the conflict). However, the initial official attempts miscarried because the militiamen did not accept these new officers, who were seen as *señoritos* (rich kids).[166] Eventually, Prime Minister Francisco Largo Caballero and his military adviser, General José Asensio Torrado, managed to bring some order to military education and new separate arm schools were opened in November 1936.[167] The requirements for entrance to these schools, titled *Escuelas Populares de Guerra* (Popular War Schools) from the following December, were roughly similar to those of the Nationalist side—although frontline experience was not required before June 1937—and training lasted three or four months; those who passed the course would be given a temporary commission as *oficiales en campaña* (literally, 'officers in the field'). Military education was reorganized in August 1937: the Popular War Schools were to be gathered in a single institution (although still arranged in arm branches) and junior wartime officers who lacked a formal *en campaña* commission would attend courses ran by divisional schools. Overall, between 10,500 and 11,200 officers graduated in the war schools and another 15,000 passed divisional courses. Additionally, 10,000 men were made officers by the political militias.[168]

The Popular War Schools suffered from several internal problems which reduced their effectiveness in training officer candidates and can be traced back to the flaws of the 1893 system. There was certainly a shortage of adequate teachers (they often were officers reckoned untrustworthy for other assignments), but the syllabi were often also excessively theoretical and the instructors were too fond of perfunctory teaching methods (one can assume that they aped what they had themselves experienced as officer candidates). The Republican war schools also overemphasized—at least until the summer of 1937—the ability to pass entrance examinations to the detriment of previous fighting experience, character and political commitment (features usually found in militia officers).[169] All this reminds one of Spanish military education in the period 1893–1927 and its fondness for syllabi of bookish content and for testing the intellectual powers of officer candidates rather than their potential leadership skills. Insofar as the problems came from institutional inertia in the organization of a new regular army, one cannot help concluding that, in view of the results, the legacy of Spanish military education in the period 1893–1927 was much of an albatross for the Republican war effort.[170]

Republican officer training also seems to have been influenced adversely by another legacy of the pre-war period: corporate factionalism. In his report on the Republican army's performance in the battle of Teruel Lieutenant Colonel Luis Morales (of the artillery corps) complained that the officer training organization lacked enough unity because the branches of the service still considered themselves as watertight compartments. Morales thought it urgent to place all the professional matters of officer training under a single authority in order to impose uniform syllabi for temporary officers.[171] There are reasons to think that his assertions were not totally unfounded if one bears in mind the debate about officer training since the late nineteenth century, compounded by the political militias' factionalism early in the war. In fact, the Republican ministerial authorities had already realized the problem, acknowledging a 'lack of unity, coordination and guidance' in August 1937; thus the Popular War Schools were put under a single command and henceforth all training would be under official control through the ministerial Training Section. [172] But things seem to have seen little improvement, given that the Republican government deemed a new reorganization necessary in October 1938, which set up a new central command for military training, about 18 months later than the Nationalists.[173]

The Republican *oficiales en campaña* did not become an effective counterpart of the Nationalist *alféreces provisionales*, described by the historian Michael Seidman as 'counterrevolutionary middle-class males'.[174] Too many Republican temporary officers were poorly educated, since they often came from a lower-class background, and the best-educated potential officers that truly supported the Republican cause were likely to have joined the militias in the very first days of the war. Thus the later officer training establishment had to deal with trainees of poor educational level or whose commitment was, at best, lukewarm.[175] On the other hand, in an armed conflict between antagonistic ideologies like the Spanish Civil War, the training and performance of military leaders may also be measured according to non-professional standards. This introduces competition between the officers who are ideologically committed and those who are neutral, who may also find more difficulties to inspire their troops.[176] And although not true military leaders, the Republican army's political commissars could also be competitors of sorts through an exemplary behaviour in combat and their responsibilities in maintaining discipline and the functioning of services.[177] One can wonder how many potential officers combining political commitment and adequate education drained away into the commissariat.

Conclusion

The reform of military education was an issue directly affecting corporate interests within the officer corps. Since the defence of these interests was one of the main reasons for the military's involvement in politics, it was inevitable that, as a consequence, political factors weighed on the decisions about military education. Primo de Rivera's reform was essentially sound. As García Moreno illuminates, the Saragossa general academy materialized the reformist aspirations of the early 1910s, which had been particularly prominent in the Academy of Infantry.[178] Moreover, common training in the same academy provided future officers with much of the mutual trust and understanding needed for inter-arm cooperation. On the other hand, since training factory managers and ordnance specialists is not the primary role of a military academy, schooling them in a separate college was a sensible arrangement. Unfortunately, Primo de Rivera´s opponents tended to see only political partisanship behind the 1927 reform (and to the extent that the dictator used military policy as a way to bolster his regime, they had a point). Therefore, when Primo de Rivera fell, his reform was soon killed off, because politics prevailed over professional usefulness. An exception of sorts was the suppression of the staff corps: a step implied in the creation of the EESM and which was formally taken by the republican government.

Nonetheless, the closure of the Saragossa academy before deciding what model of officer training would be implemented by the republican regime seems somewhat ill-judged. In fact, the reform of 1932 was not incompatible with a phase of common training in a general academy. The partial return to a system of separate academies is better understood as another episode in the branches' contest for the promotion of their particular self-interest. On the other hand, Garcia Moreno argues that the republican reform actually limited itself to 'modify' that of Primo de Rivera, since it followed the same orientation. The main modifications were the rise of educational requirements for entering the military academies (which allowed shortening their syllabi) and a greater development of further education through the mandatory courses before promotion to field grade and general rank.[179] As regards training in the military academies proper, this argument is fairly sensible. But the present author thinks that the reform of 1931–1932, especially by requiring a period of regimental service, did in fact mark a significant departure from the Spanish army's traditional model of military education. Initial officer training at last emulated the German model (relatively high academic standards at selection, previous regimental service, junior officer-level practical teaching at the academy), which had been advocated (consciously or unintentionally) by many military reformists during the three previous decades. Nevertheless, assessing the effectiveness of the republican reform will never be possible since no officer candidate who entered the military academies after the implementation of the 1932 act had finished his training when the Civil War broke out.

The soundness of the reforms introduced in the period 1927–1932 is demonstrated by the fact that most of them were adopted in the aftermath of the Civil War and remained in force for the rest of the century.[180] Officer candidates started their training in the restored General Military Academy at Saragossa and completed it in a corps academy; specialist engineers of ordnance and construction (who formed a separate corps thenceforth) were trained in a military technical college (*Escuela Politécnica Superior del Ejército*); the *diplomados* graduated in the Staff College (*Escuela de Estado Mayor*); and passing courses of proficiency for field grade and higher command was necessary for promotion to major and general officer ranks, respectively—the higher command course was taught at a war college (*Escuela Superior del Ejército*). A brand new interservice *Centro Superior de Estudios de la Defensa Nacional* (CESEDEN), which included both the national defence college and the joint staff college, was created in 1964. There was a major reorganization of the armed forces' higher colleges in 1999, when the CESEDEN incorporated all the services' staff and higher command colleges into a single *Escuela Superior de las Fuerzas Armadas*, which was to run joint higher command and staff courses. The Staff College was renamed *Escuela de Guerra del Ejército* in 2000 and assumed responsibility for the field grade course. But it certainly was more momentous that, after 2010, the Saragossa academy, created in 1927 as the nemesis of the 'learned' approach to the initial training of officers, implemented a new four-year course of study which included a university degree in industrial organization engineering. *O tempora o mores!*

Notes

1. Comandante Beta [pseud. of José García Benítez], *Apuntes para historiar tres años de reformas militares (1915–1917)* (Madrid: Imp. de Bernardo Rodríguez, 1917), pp. 136–43.
2. Carlos Blanco Escolá, *La Academia General Militar de Zaragoza (1928–1931)* (Barcelona: Labor, 1989), pp. 50–1.
3. Luis Rivero, 'Unión es fuerza', *Memorial de Caballería* (hereafter *MC*), 6, 1 (1921), pp. 175–6. See also Miguel Campins Aura, 'La Academia General Militar de Zaragoza y sus normas pedagógicas', non-commercial edition (n.p.: [Centro de Ayudas a la Enseñanza de la Armada, 1983]), pp. 6–7, 14–15.
4. 'Prácticas de conjunto de las academias de Infantería y Artillería en el Campamento de Abades (Segovia)', *La Guerra y su preparación*, 8, 2 (1923), p. 14.
5. Blanco Escolá, *La Academia*, p. 60.
6. Carlos Navajas Zubeldía, *Ejército, estado y sociedad en España (1923–1930)* (Logroño: Instituto de Estudios Riojanos, 1991), p. 40. For a description of Primo de Rivera's ideas on military education in his own words, see José Fernando García Moreno, 'Servicio militar universal y personal en España. Procedencia social y formación del soldado, clases profesionales y oficial de carrera en el Ejército de Tierra español de 1913 a 1936', 5 vols. (doctoral dissertation, Universidad de Zaragoza, 1985), II. 753–7.
7. José Durango y Pardini, 'Reclutamiento de la oficialidad del Ejército, en sus diversas escalas, y ascensos en las mismas', *MC*, 9, 1 (1924), pp. 141–3.
8. Campins did mention the 1918 commission and stated that the contents of its final memorandum were greatly taken into account by the Saragossa academy organizing commission, although he did not give details in the rest of the essay: Campins, 'La Academia General Militar', pp. 16–17.
9. See, for instance, Jesús Ignacio Martínez Paricio, Jorge Aspizua Turrión, José Ramón Bernabéu Urbina, and Julio Molina Benayas, *Los papeles del General Vicente Rojo. Un militar de la generación rota* (Madrid: Espasa-Calpe, 1989), p. 35; Manuel Touron Yebra, 'El General Miguel Campins y su época (1880–1936)', 2 vols. (doctoral dissertation, Universidad Complutense de Madrid, 1992), I. 223–4; Blanco Escolá, *La Academia*, p. 59; Navajas, *Ejército*, p. 169.
10. It does not seem accidental that Primo de Rivera made finally the decision after his clash with the artillery corps in 1926: Julio Busquets, *El militar de carrera en España*, revised edition (Barcelona: Ariel, 1984), p. 82.

11. Ministerio de la Guerra, *Colección Legislativa del Ejército 1927* (Madrid: n.p., 1927) (hereafter *CLE plus the year of publication*), Item No. 94, Introduction.
12. John Keegan, 'Regimental Ideology', in Geoffrey Best and Andrew Wheatcroft (eds), *War, Economy, and the Military Mind* (London: Croom Helm, 1976), p. 3.
13. *CLE 1927*, Item No. 94, Article 1, basis 3 to 13.
14. *CLE 1927*, Item No. 94, Introduction.
15. J[uan] Luengo, 'Academias militares', *Revista Científico-Militar* (hereafter *RCM*), 27 (1902), p. 10.
16. [Alvaro de Figueroa,] Conde de Romanones, *El ejército y la política. Apuntes sobre la organización militar y el presupuesto de la guerra* (Madrid: Renacimiento, 1920), pp. 170–1. Romanones was a conspicuous politician of Alfonso XIII's reign and held office several times as minister and prime minister, although he never got the war portfolio.
17. A., 'La oficialidad de nuestros días', *MC*, 5, 2 (1920), p. 9.
18. José García Miranda, 'Proyecto de reformas en el Reclutamiento de la oficialidad de nuestro Ejército', *Memorial de Infantería* (hereafter *MI*), 14, 1 (1925), p. 166.
19. *Diario Oficial del Ministerio del Ejército*, 14 December 1928; Campins, 'La Academia General Militar', pp. 62–3.
20. Blanco Escolá, *La Academia*, pp. 99–100.
21. The biographical literature on Franco ranges from the hagiography to the character assassination. With regard to his military career, two helpful works are Geoffrey Jensen, *Franco: Soldier, Commander, Dictator* (Washington, DC: Potomac Books, 2005); Juan Blázquez Miguel, *Auténtico Franco. Trayectoria militar, 1907–1939* (Madrid: Almena, 2009).
22. Stanley G. Payne, *Politics and the Military in Modern Spain* (Stanford: Stanford University Press, 1967), p. 242.
23. Touron, 'El General Miguel Campins', I. 244–5. When Campins, after being assigned to the Saragossa academy, indicated his own lack of experience as teacher in a military academy to the War Ministry, he was told that he had been chosen precisely because of it, his long regimental service, and volunteering for the Moroccan war: Campins, 'La Academia General Militar', p. 26.
24. On the concept of 'heroic leader' and other military career patterns, see Morris Janowitz, *The Professional Soldier: a Social and Political Portrait* (New York: The Free Press, 1971), pp. 74–8.
25. Fernando Martínez de Baños Carrillo, 'La Academia General Militar (1927–1931). Su historia', offprint of *Cuadernos de Aragon*, 26 ([2001]), pp. 62–3.

26. Alonso Vega, Franco Salgado-Araujo and Sueiro had been officers in the first *banderas* (battalions) of the Spanish Legion; moreover, Franco Salgado-Araujo was the commandant's cousin and Alonso Vega was Franco's friend since boyhood.
27. Blázquez, *Auténtico Franco*, p. 101. For information on the Civil War service of a sample of officers posted to the Saragossa academy, see Blázquez, *Auténtico Franco*, pp. 106–8.
28. Touron, 'El General Miguel Campins', I. 248. Unfortunately for Campins, his acquaintanceship with Franco did not prevent the rebel leader General Gonzalo Queipo de Llano from dismissing him from the command of the Granada garrison because of his hesitation in joining the 1936 military uprising. Campins was court-martialed and shot a few weeks later.
29. Blanco Escolá, *La Academia*, pp. 127–30.
30. Luis Bermúdez de Castro y Tomás, *Mosaico militar. Historias, historietas, anécdotas, episodios, alegrías, tipos y costumbres de la vida militar de antaño* (Madrid: Aldus, 1951), pp. 205–6.
31. Blanco Escolá, *La Academia*, p. 175.
32. Campins, 'La Academia General Militar', p. 121.
33. See, for instance, Busquets, *El militar*, pp. 81–2; Blanco Escolá, *La Academia*, pp. 174–6.
34. García Moreno, 'Servicio militar', II. 723.
35. Payne, *Politics*, pp. 242–3, 243 footnote. On a more positive side, Franco enforced a strict ban of *novatadas* (hazing), helped by the practice of turning each second-year officer candidate into a sort of tutor for one of the following class (his *filio*) and holding the former accountable if the latter was hazed; only *galonistas* (under officers) were exempt from this tutoring role: José Izquierdo, José Ramón Ortiz de Zárate and Ángel Aparicio, *La Academia General Militar. Crisol de la oficialidad española* (Saragossa: Institución 'Fernando el Católico', 2011), pp. 166–8; Martínez de Baños, 'La Academia General Militar', pp. 140–1.
36. Francisco Franco Salgado-Araujo, *Mi vida junto a Franco* (Barcelona: Planeta, 1977), p. 82.
37. Manuel Gutiérrez Mellado, *Un soldado de España: conversaciones con Jesús Picatoste* (Barcelona: Argos Vergara, 1983), p. 52. Gutiérrez Mellado (1912–1995) entered the Saragossa academy in 1929 and was commissioned in the artillery corps (1933); he ran a Nationalist espionage and escape network in the Republican rear during the Civil War; he became lieutenant general and chief of army general staff in 1976; as deputy prime minister for defence affairs (1976–1981), he took charge of making the armed forces accept the democratic regime after Franco's death. For an account of his years in the Saragossa academy,

see Fernando Puell de la Villa, *Gutiérrez Mellado. Un militar del siglo XX (1912–1995)* (Madrid: Biblioteca Nueva, 1997), Chap. 2.

38. If truth be told, Franco even seems to have acted sometimes as if the Saragossa academy were his personal fiefdom, much to the annoyance of the War Ministry staff. For instance, he circumvented the chain of command and used his personal connections at the highest political-military circles to get a sizeable increase in the intake of 1929, in order to admit more orphans of officers who had died in war service; such action thwarted the Ministry's planning to match the numbers of new officers to smaller, more realistic manpower requirements after the end of the war in Morocco: Puell, *Gutiérrez Mellado*, pp. 55–7; Fernando Puell de la Villa, 'Cadetes de la 2ª época, Generales de la Transición', in Congreso de Historia Militar (5°. 2002. Zaragoza), *La enseñanza militar en España: 75 años de la Academia General Militar en Zaragoza* (Madrid: Ministerio de Defensa, Secretaría General Técnica, 2003), pp. 221–3.
39. Manuel Azaña, *Obras Completas*, 7 vols. (Madrid: Ministerio de la Presidencia, Secretaría General Técnica, Centro de Estudios Políticos y Constitucionales, 2007), III. 624 (20 July 1931 diary entry); Franco Salgado-Araujo, *Mi vida*, p. 104.
40. Stanley G. Payne, *Franco. El perfil de la Historia* (Madrid: Espasa-Calpe, 1992), p. 21.
41. Academia General Militar, *Memoria del curso de 1929–30* (Saragossa: n.p., 1931), p. 50.
42. Academia General Militar, *Memoria 1929–30*, p. 51.
43. Campins, 'La Academia General Militar', pp. 37–8. In this regard, it is worth noting that Campins had a approving opinion of the Oxbridge educational system as a reference for the training of officers, because, beyond its academic activities, created a propitious environment for the students' comprehensive development as gentlemen: Campins, 'La Academia General Militar', pp. 31, 38.
44. Julio Ferrer Sequera, *La Academia General Militar. Apuntes para su historia*, 2 vols. (Barcelona: Plaza y Janes, 1985), I. 236; Blanco Escolá, *La Academia*, p. 158.
45. Campins, 'La Academia General Militar', p. 7.
46. Campins, 'La Academia General Militar', p. 42.
47. The officer candidates were initially hard put to adapt to the new methods, used as they were to rote learning just to pass the next exam at schools and crammers. Feeling uneasy about the lack of textbooks, they wrote down almost verbatim the teachers' lectures. They progressively got rid of their 'pass and forget' mindset and learnt to be selective in taking notes and to gain knowledge of the hands-on teaching: Academia General Militar: *Memoria del curso de 1928–29* (Saragossa: n.p., 1930), p. 20.

48. See, for instance, Blanco Escolá, *La Academia*, pp. 145–6; Touron, 'El General Miguel Campins', I. 254–5, 276. However, Blanco Escolá has argued that Campins' reading of Giner and his ideas was so limited and selective that the latter never were genuinely implemented: Blanco Escolá, *La Academia*, pp. 139–48. For Campins' own elucidation of his educational ideas and the academy's teaching methods, see Campins, 'La Academia General Militar', Chaps. 2–5 (especially pp. 77–85, 91–102).
49. On this subject, see, for instance, Miguel Alonso Baquer, *El ejército en la sociedad española* (Madrid: Ediciones del Movimiento, 1971), pp. 199–204; Izquierdo, Ortiz de Zárate, and Aparicio, *La Academia General Militar*, pp. 91–3.
50. For Rojo's ideas on teaching, see Guerrero Martín, Alberto: 'Análisis y trascendencia de la Colección Bibliográfica Militar (1928–1936)' (doctoral dissertation, UNED [Universidad Nacional de Educación a Distancia], 2015), pp. 346–8. It is noteworthy their resemblance to those found in Campins' 1932 essay.
51. Touron, 'El General Miguel Campins', I. 276–7.
52. Touron, 'El General Miguel Campins', I. 277.
53. Jensen, *Franco*, p. 56.
54. García Moreno, 'Servicio militar', II. 726 footnote 43.
55. Guerrero Martín, 'Análisis y trascendencia', p. 342.
56. Luis de la Gándara Marsella, *El oficial alemán de la escala activa y de reserva* (Madrid: n.p., 1916), p. 91; Jesús María Ruiz Vidondo, 'El General Director de la AGM visita la Escuela Militar de Dresde en 1928', *Armas y Cuerpos*, 103 (March 2004), p. 88.
57. Jensen, *Franco*, p. 56.
58. Blanco Escolá, *La Academia*, p. 185.
59. Fernando Martínez de Baños y Carrillo, 'Franco, primer director de la Academia General Militar de Zaragoza', in Congreso de Historia Militar (5°. 2002. Zaragoza), *La enseñanza militar en España: 75 años de la Academia General Militar en Zaragoza* (Madrid: Ministerio de Defensa, Secretaría General Técnica, 2003), p. 183.
60. Martínez Paricio et al., *Los papeles del General Vicente Rojo*, pp. 35–6.
61. García Moreno, 'Servicio militar', II. 785.
62. The present author is grateful to Lieutenant Colonel José Ramón Ortiz de Zárate, a teacher and historian of the Academia General Militar, Saragossa, for his enlightening observations about the function of the guidelines in the training of the officer candidates.
63. For contemporary Spanish surveys of German officer training in both periods, see Gándara, *El oficial alemán*, pp. 69–126; and Luis de la Gándara Marsella, *El moderno ejército alemán* (Madrid: n.p., 1925),

pp. 81–9. On officer training in the Imperial period, see also Steven E. Clemente, *For King and Kaiser! The Making of the Prussian Army Officer, 1860–1914* (Westport: Greenwood Press, 1992), Chaps. 3 and 6. For later overviews and assessments of training in the Weimar period, see Martin van Creveld, *Fighting Power: German and US Army Performance, 1939–1945* (London: Arms and Armour Press, 1983), pp. 133–9; James S. Corum, *The Roots of Blitzkrieg: Hans von Seeckt and German Military Reform* (Lawrence: University Press of Kansas, 1992), pp. 78–84.

64. Archivo General Militar de Segovia, 2ª Sección, División 3ª, Legajo 55 (hereafter AGMS No. Sección/No. División/No. Legajo), Franco to Director General de Instrucción y Administración, 18 April 1928.
65. AGMS 2/3/55, Dirección General de Preparación de Campaña to General Director, Academia General Militar, 11 May 1928; AGMS 2/3/55, Ministerio de la Guerra, Dirección General de Preparación de Campaña, Sección de E.M., 2° Negociado to Teniente Coronel Agregado Militar a la Embajada de España en Berlín, 2 June 1928.
66. AGMS 2/8/122, Academia General Militar, Memoria que emite el General Director de la misma como resultado de la visita efectuada a la Escuela Militar de Dresde (infanterie Schule), 31 July 1928. The content of the report is summarized in Ruiz Vidondo, 'El General Director', pp. 87–8.
67. Franco Salgado-Araujo, *Mi vida*, pp. 80–1.
68. Gordon A. Craig, *The Politics of the Prussian Army 1640–1945*, revised edition (Oxford: Oxford University Press, 1964), p. 394.
69. Quoted in Touron, 'El General Miguel Campins', I. 300.
70. Bartolomé Bennassar, *Franco* (Barcelona: EDAF, 1996), pp. 62–4.
71. See, for instance, Ferrer Sequera, *La Academia*, I. 256–9.
72. See, for instance, Paul Preston, *Franco. Caudillo de España*, updated edition (Barcelona: Debate, 2015), p. 95.
73. Jensen, *Franco*, p. 55.
74. A total of 123 officers were assigned to the Saragossa academy's faculty: 51 were infantrymen, 22 cavalrymen, 17 artillerymen, 13 engineers, and six belonged to the staff corps; the figures are taken from the list in Ferrer Sequera, *La Academia*, I. 275–6. Later scholarship has stated that the Saragossa academy was essentially an infantry establishment: see, for instance, Busquets, *El militar*, p. 82.
75. Jesús Pérez Salas, *Guerra en España (1936 a 1939). Bosquejo del problema militar español; de las causas de la guerra y del desarrollo de la misma* (Mexico, D.F.: n.p., 1947), p. 86. Pérez Salas was an infantry officer who conspired against Primo de Rivera and later became a member of Azaña's military private office at the War Ministry; during the Civil War,

he joined the Republican side, rose to colonel and became divisional and army corps commander; still, his professionalism was hard put to fit in the new model Republican army's environment.

76. Nazario Cebreiros, *Las reformas militares. Estudio crítico* (Santander: J. Martínez, 1931), pp. 217–19. It is worth noting that commissioning branch vacancies were filled by Gutiérrez Mellado's class in the following order (the branches' percentages of the offered vacancies overall are in brackets): engineer corps (7%), artillery (19%), cavalry (12%), quartermaster corps (6%) and infantry (56%); Franco reportedly felt frustration whenever a distinguished infantry officer's son chose a learned corps: Puell, *Gutiérrez Mellado*, p. 69 footnote 40.
77. Gutiérrez Mellado, *Un soldado de España*, p. 52.
78. Such kind of political indoctrination as part of the syllabus is denied in Izquierdo, Ortiz de Zárate and Aparicio, *La Academia General Militar*, p. 74.
79. See Franco Salgado-Araujo, *Mi vida*, pp. 77–104.
80. Quoted in Navajas, *Ejército*, p. 173. Benzo left the army to pursue a political career during the Second Republic and was appointed undersecretary of the Interior by a liberal centrist cabinet in 1934.
81. Antonio Cordón, *Trayectoria (Recuerdos de un artillero)* (París: Librairie du Globe, 1971), pp. 193–4.
82. Pérez Salas, *Guerra en España*, p. 86.
83. Puell, *Gutiérrez Mellado*, p. 99.
84. The alleged indoctrination of the officer candidates in Franco's very conservative ideas is emphasized in Blanco Escolá, *La Academia*, pp. 100–2, 126–7, 181–2.
85. Busquets, *El militar*, pp. 120–1. Depending on the authors, the proportion for the officer corps varies from 60 to 80%.
86. For instance, it has been alleged that the professional prestige and exemplary role of their former instructors led most Saragossa graduates to adopt unconsciously the formers' ideological attitudes: Izquierdo, Ortiz de Zárate, and Aparicio, *La Academia General Militar*, p. 74.
87. Puell, 'Cadetes de la 2ª época', pp. 227–8. Cf. Busquets, *El militar*, pp. 120–7.
88. As far as the present author is aware, the subject only has been addressed in some detail in García Moreno, 'Servicio militar', II. 794–802.
89. *CLE 1927*, Item No. 94, Article 1, basis 13 to 16.
90. On the problems of assigning officer candidates caused by the closure of the Saragossa academy, see Puell, 'Cadetes de la 2ª época', pp. 224–6.
91. *CLE 1931*, Items No. 412, No. 416, No. 417.
92. Puell, *Gutiérrez Mellado*, p. 71.
93. Puell, *Gutiérrez Mellado*, p. 72.

94. *CLE 1931*, Item No. 410.
95. Azaña's political motivation is virtually taken for granted by historians; for a sample, see Ramón Salas Larrazábal, 'Las reformas de Azaña', in Mario Hernández Sánchez-Barba and Miguel Alonso Baquer (eds), *Las fuerzas armadas españolas. Historia institucional y social*, 8 vols. (Madrid: Alhambra, 1986), VI, 64; Michael Alpert, *La reforma militar de Azaña (1931–1933)* (Madrid: Siglo XXI, 1982), pp. 261–2; Boyd, 'Las reformas militares', in [Luis Suárez Fernández, Demetrio Ramos Pérez, José Luis Comellas, and José Andrés-Gallego (eds),] *Historia General de España y América*, 25 vols. (Madrid: Rialp, 1981–1992), XVII. 148; Bennassar, *Franco*, p. 67; García Moreno, 'Servicio militar', II. 981; Touron, 'El General Miguel Campins', I. 295–7, 308.
96. Cordón, *Trayectoria*, pp. 193–4.
97. Cebreiros, *Las reformas militares*, pp. 212–13, 221.
98. Campins, 'La Academia General Militar', p. 115.
99. Puell, *Gutiérrez Mellado*, pp. 66–7.
100. Alpert, *La reforma militar*, p. 255.
101. García Moreno, 'Servicio militar', II. 821.
102. *CLE 1932*, Item No. 506.
103. The public announcements of entrance exams set the upper age limit for civilians at 24 and allowed officers and NCOs of the Complement List taking the examination if they were younger than 30.
104. *CLE 1932*, Item No. 506, Article 3.
105. *CLE 1932*, Item No. 506, Articles 4, 5.
106. Romanones, *El ejército y la política*, pp. 169–70.
107. A., 'La oficialidad', p. 9.
108. *Diario Oficial del Ministerio de la Guerra* (hereafter *DOMG*), 6 September 1933.
109. Romanones, *El ejército y la política*, pp. 172–3.
110. Luis Belda, 'Ideas sobre el reclutamiento, escalas y preparación profesional del Oficial de Infantería', *MI*, 13, 2 (1924), pp. 17–18.
111. Emilio Mola, *Obras completas* (Valladolid: Librería Santaren, 1940), p. 1072.
112. AGMS 2/8/152, Capitán [Eduardo] Herrera de la Rosa, Impresiones recogidas de la campaña ruso-japonesa con el ejército del general barón Nogui, pp. 18–19, 30 November 1905. The Japanese army had been organized on the German model in the late 1870s and the 1880s: Stephen Morillo, Jeremy Black and Paul Lococo, *War in World History: Society, Technology, and War from Ancient Times to the Present*, 2 vols. (New York: McGraw-Hill, 2009), II. 478.
113. A., 'La oficialidad', p. 9; Belda, 'Ideas', p. 18; García Miranda, 'Proyecto', p. 166; Miguel Ponte y Manso de Zuñiga, 'Reclutamiento de la oficialidad del Ejército, en sus diversas escalas, y ascensos en las

mismas', *MC*, 9, 1 (1924), p. 374. Ponte (1882–1952) was commissioned in the cavalry (1897) and reached general officer rank in Morocco; he left the service in 1931 and turned into a monarchist conspirator; during the Civil War, Ponte held senior commands on the Nationalist Aragonese and central fronts; he signed the letter sent in 1943 by a group of generals asking Franco for the prompt restoration of the monarchy.

114. García Miranda, 'Proyecto', p. 16; Ponte, 'Reclutamiento', pp. 375–6; Romanones, *El ejército y la política*, p. 169.
115. Belda, 'Ideas', p. 18.
116. Mola, *Obras completas*, pp. 1069–70.
117. Navajas, *Ejército*, pp. 242–59.
118. Payne, *Politics*, pp. 304–5.
119. Corelli Barnett, 'The Education of Military Elites', *Journal of Contemporary History*, 2, 3 (1967), pp. 25–6; Creveld, *Fighting Power*, p. 129.
120. Barnett, 'The Education of Military Elites', p. 26.
121. Rufino Ginés, 'Reclutamiento de oficiales', *MI*, 10, 1 (1921), p. 315; A.F.B. y P. de A., *Nuestro ejército. Lo que es y lo que puede ser* (n.p.: S. Fernando, 1923), p. 55.
122. Cebreiros, *Las reformas militares*, pp. 224–5, 228–9.
123. Gabriel Cardona, 'La reforma de la enseñanza militar en la II República (1931–1932)', in Julio Busquets and Valentina Fernández Vargas (eds), *La enseñanza militar en España: un análisis sociológico* (Madrid: Consejo Superior de Investigaciones Científicas, 1986), p. 79.
124. A., 'La oficialidad', p. 97.
125. Ginés, 'Reclutamiento de oficiales', pp. 313–15.
126. Cebreiros, *Las reformas militares*, pp. 269–70.
127. Tomás Peire, *Una política militar expuesta ante las Cortes Constituyentes* (Madrid: Vallinas, 1933), p. 23. For other critical views during the parliamentary discussion, see Alpert, *La reforma militar*, p. 249.
128. On the Avila preparatory school, see García Moreno, 'Servicio militar', II. 622–32. Amongst the successful NCO applicants, 35 in group (b) and 29 in group (c) were Avila alumni: García Moreno, 'Servicio militar', II. 630.
129. *DOMG*, 11 December 1935; Salas Larrazábal, 'Reformas de Azaña', VI. 80, 330 note 239; Alpert, *La reforma militar*, p. 251.
130. García Moreno, 'Servicio militar', II. 628–9, 635.
131. Alpert, *La reforma militar*, p. 250. On the other hand, Azaña's reform had retained the financial benefits for sons of military men and the precedence of the latter and military men over ordinary civilians in case of similar grades in the entrance exam: García Moreno, 'Servicio militar', II. 822, 850.

132. Douglas Porch, *The March to the Marne. The French Army 1871–1914* (Cambridge: Cambridge University Press, 1981), pp. 81–4.
133. *DOMG*, 6 July 1935, 3 November 1935, 26 November 1935; José María Gil Robles, *No fue posible la paz* (Barcelona: Planeta, 1978), p. 247; Ferrer Sequera, *La Academia*, I. 295; Salas Larrazábal, 'Reformas de Azaña', VI. 75, 329 note 215.
134. *CLE 1927*, Item No. 94, Article 1, basis 17.
135. *CLE 1927*, Article 1, basis 18.
136. *CLE 1927*, Article 1, basis 18, 20.
137. Miguel A. Moreno y Álvarez, 'Instrucción y reclutamiento de la oficialidad', *RCM*, 36 (1911), p. 261; M[anuel] de Pazos y Zamora, 'Las especialidades', *MI*, 10, 1 (1921), pp. 317–18.
138. Romanones, *El ejército y la política*, pp. 171–2.
139. *CLE 1927*, Item No. 94, Article 1, basis 18, 20.
140. Mola, *Obras completas*, pp. 1025–6.
141. Alonso Baquer, *El ejército*, pp. 287–8.
142. On Primo de Rivera's reform of higher military education, see also García Moreno, 'Servicio militar', II. 802–9.
143. *CLE 1931*, Item No. 513.
144. Cebreiros, *Las reformas militares*, pp. 225–6.
145. *CLE 1932*, Item No. 506, Article 6, transitory arrangement No. 3.
146. Mola, *Obras completas*, pp. 971–2.
147. Cebreiros, *Las reformas militares*, pp. 150–2.
148. Eduardo Gallego Ramos, *Proyecto de reorganización y mejora del ejército de tierra* (Guadalajara: n.p., 1905), p. 18.
149. Cordón, *Trayectoria*, p. 71.
150. García Moreno, 'Servicio militar', II. 809–12; Miguel Alonso Baquer, 'El Ejército y la formación de oficiales en el reinado de Alfonso XIII', in Congreso de Historia Militar (5°. 2002. Zaragoza), *La enseñanza militar en España: 75 años de la Academia General Militar en Zaragoza* (Madrid: Ministerio de Defensa, Secretaría General Técnica, 2003), pp. 70–1, 74–5.
151. *CLE 1932*, Item No. 506, Article 14.
152. A., 'La oficialidad', pp. 97–8; A.F.B. y P. de A., *Nuestro ejército*, p. 55.
153. Salas Larrazábal, 'Reformas de Azaña', VI. 66, 76, 329 note 217.
154. A., 'El problema del generalato', *MC*, 6, 1 (1921), pp. 228–30.
155. More information on this course in García Moreno, 'Servicio militar', II. 812–14; Jesús María Ruiz Vidondo, 'Los cursos de coroneles, génesis y primer desarrollo (1923-1930)', *Revista de Historia Militar* (hereafter *RHM*), 99 (2006), pp. 22, 30–1, 35–53.
156. *CLE 1932*, Item No. 506, Article 16.

157. More information on the centre and its courses in Jesús María Ruiz Vidondo, 'La enseñanza militar en el alto mando: historia, organización y metodología', *Educación XX1*, 9 (2006), pp. 206–8.
158. These figures are taken from Ramón Salas Larrazábal, *Historia del Ejército Popular de la República*, 4 vols. (Madrid: Editora Nacional, 1973), I. 186–7; see also Michael Alpert, *The Republican Army in the Spanish Civil War, 1936–1939* (Cambridge: Cambridge University Press, 2013), pp. 85–91.
159. Alpert, *Republican Army*, pp. 91, 107–9.
160. The figure is taken from Jesús María Salas Larrazábal and Ramón Salas Larrazábal, 'La Guerra', in [Suárez et al. (eds),] *Historia General*, XVII. 316, 318; it may include 1500 to 2000 retired officers: José María Gárate Córdoba, *Alféreces provisionales. La improvisación de oficiales en la guerra del 36* (Madrid: San Martín, 1976), pp. 28, 31.
161. Gárate, *Alféreces provisionales*, pp. 31–3; see also Alpert, *Republican Army*, p. 127.
162. The fullest studies on the recruitment and training of wartime officers, which this section has drawn mostly on, are Gárate, *Alféreces provisionales*; and José María Gárate Córdoba, *Tenientes en campaña. La improvisación de oficiales en la guerra del 36* (Madrid: San Martín, 1976). See also Busquets, *El militar*, pp. 106–13; Alpert, *Republican Army*, pp. 126–8, 144–54.
163. An abridged version of Gárate Córdoba's previous work on the Nationalist temporary officers can be found in José María Gárate Córdoba, 'Improvisación de mandos en 1936. I. Alféreces Provisionales', *RHM*, Special Issue (2010), pp. 53–114.
164. Busquets, *El militar*, pp. 109–11.
165. On the motivational effect of this sort of leadership, see Anthony Kellett, 'The Soldier in Battle: Motivational and Behavioral Aspects of the Combat Experience', in Betty Glad (ed.), *Psychological Dimensions of War* (Newbury Park: Sage, 1990), p. 224.
166. Alpert, *Republican Army*, p. 126.
167. The isolated Republican territory in the North also had several officer schools.
168. An abridged version of Gárate Córdoba's previous work on the Republican temporary officers can be found in José María Gárate Córdoba, 'Improvisación de mandos en 1936. II. Tenientes en Campaña y III. La enseñanza militar en guerra: estudio comparativo', *RHM*, Special Issue (2010), pp. 115–63.
169. Alpert, *Republican Army*, pp. 149–52, 154.
170. But such state of affairs was not preordained, since some instructors implemented effective training methods and an infantry school in

Asturias was allegedly able to train officers as fast and successfully as the Nationalist ones: Gárate, *Tenientes en campaña*, pp. 59, 165–79; Alpert, *Republican Army*, p. 152.

171. Archivo General Militar de Ávila, Fondos del Ejército Popular, Armario 65, Legajo 787, carpeta 6 (hereafter AGMA-EP No. Armario/No. Legajo/No. carpeta), Las operaciones de Teruel, 25 February 1938.
172. *Diario Oficial del Ministerio de Defensa Nacional*, 11 August 1937.
173. Alpert, *Republican Army*, pp. 149, 153.
174. Michael Seidman, *A ras de suelo. Historia social de la República en la Guerra Civil* (Madrid: Alianza, 2003), p. 249 [original title: *Republic of Egos: A Social History of the Spanish Civil War*]. After the war, a majority of them held diehard stances within Franco's regime: Busquets, *El militar*, pp. 112–13.
175. Alpert, *Republican Army*, p. 150; Salas Larrazábal, *Historia del Ejército Popular*, II. 1509–10; Gárate, *Tenientes en campaña*, pp. 234–5; Seidman, *A ras de suelo*, pp. 248–50. It should come as no surprise to anyone that many Republican temporary officers were hard put to apply professional skills requiring more than a very elementary literacy, such as map-reading: after one year of war, training instructions for the crack manoeuvre army corps issued in October 1937 emphasized map-reading as a skill which officers had to learn: AGMA-EP 53/463/1, Ministerio de Defensa Nacional, Estado Mayor Central, Plan para los cuerpos de ejército de maniobra, 17 October 1937.
176. Gerke Teitler, *The Genesis of the Professional Officers' Corps* (Beverly Hills and London: Sage, 1977), p. 19. Teitler is referring to professional officers, but the present author thinks that this argument can be applied to the politically nonpartisan *en campaña* officer as well.
177. Gárate, 'Improvisación de mandos en 1936. II', p. 148. On the political commissars, see Alpert, *Republican Army*, Chap. 8.
178. García Moreno, 'Servicio militar', II. 872–3.
179. García Moreno, 'Servicio militar', II. 873–4, 983.
180. The information in this paragraph comes from [Arturo Sus Sus (ed.),] *Historia de las fuerzas armadas*, 5 vols. (Saragossa: Palafox, 1983), IV. 257–8, 260–5, 276–9, V. 61–2; Eladio Baldovín Ruiz, *Historia del Cuerpo y Servicio de Estado Mayor* (Madrid: Ministerio de Defensa, Secretaría General Técnica, 2001), Chap. 12; Izquierdo, Ortiz de Zárate, and Aparicio, *La Academia General Militar*, Chap. 7.

CHAPTER 7

Firepower and Battlefield Doctrine, 1914–1939

On the subject of the Spanish military's battlefield doctrine in the Civil War, it appears as if the interest of academic historians has been satisfied by arguing tentatively, as does Michael Alpert, that 'although officers on both sides had imbibed the same military ideas... French defensive concepts, highly respected in the [pre-war] Spanish military academy', seemed to predominate on the loyalist or Republican professional officer corps; by contrast, the *africanista* background of many field commanders on the rebel or Nationalist side would foster 'the more aggressive concepts of colonial warfare'.[1] However, there was no such clash of different styles of warfare in the Spanish civil conflict. Here there was a single style, developed from the early 1920s, which each army applied with different success on the battlefield, and in which firepower played a very important role.

This chapter will provide (as far as the available sources allow) a fuller and better documented account of the evolution of the Spanish army doctrine in relation to the importance of firepower in the battlefield. First it will consider the Spanish military's reactions to the changes in warfare brought about by the First World War and to its own experience in Morocco.[2] It will then survey the development of a new battlefield doctrine (embodied in the 1925 regulations for the employment of major units) after 1918 and how effectively was implemented by the armies on both sides in the Civil War. Although this chapter is not focused solely on the artillery, it is this arm that will naturally receive much more attention, and the last part will survey how each side

J.V. Herrero Pérez, *The Spanish Military and Warfare from 1899 to the Civil War*, DOI 10.1007/978-3-319-54747-3_7

adjusted the use of its artillery to the pre-war doctrine under the conditions found in the Civil War.[3]

The Lessons of the First World War

The New Face of Battle and Bayonet Diehards

The Spanish army kept an eye on the evolution of warfare in the First World War from the vantage point of neutrality.[4] The War Ministry made an effort to gather information on the professional aspects of the war through the Spanish military attachés in the belligerent countries and commissions of observers which visited the fighting armies at the invitation of their respective governments. A permanent commission was set up in France after November 1914 and three commissions visited the British army (one in 1916 and two in 1917). Two commissions visited the German army in 1917 and another one was sent to the Austro-Hungarian army the same year. Some renowned essayists (for instance, General Ricardo Burguete, Captain Luis de la Gándara) were among these observers. The military attachés' reports and, to a lesser extent, digests of memoranda written by the commissioners were published in the General Staff journal *La Guerra y su preparación*, created in 1916.[5]

Contrary to pre-war expectations, ascertaining when fire superiority had been achieved was more problematic than expected and, combined with inexperience and the commanders' eagerness to reach a quick decision on the battlefield, this led too often to premature attacks that turned into bloodbaths. Moreover, all the combat arms had skipped over training for a real close cooperation with the others, an oversight that produced devastating consequences in 1914.[6] A section of the Spanish military refused to accept the transformations that technology (that is, the material, non-human factor) had caused in the face of battle and still interpreted the latter according to the accepted terms of the pre-war paradigm. Luis Fernández España, an infantry colonel and editor of the journal *Estudios Militares*, recognized in early 1915 the increases in the use of field fortifications and in infantry and artillery firepower as outstanding features of the conflict. However, he saw the stagnation of operations on the Western Front as a temporary pause, mainly due to the exhaustion of the armies. The absence of mobile warfare was a result of the belligerents' desire to rest, not of an inability to undertake offensives.[7] Another essayist, who was studying at the staff college, recognized by 1916 that firepower, in forcing armies to dig successive lines of trenches, had brought about a halt to manoeuvre warfare. But he also maintained that men remained the basic factor due to the increase of the armies' numerical strength, which was necessary to break through the enemy front.[8] Even the countermeasures forced by the increased firepower were explained as the triumph of the human factor: '[E]xperience has told us that man has beaten materiel…and he has beaten it because, in putting the means of his intelligence and will to work in the service of his self-preservation, he has deduced the absolute need of underground warfare…'[9]

For those fond of the pre-war morale-based values the essence of infantry fighting remained allegedly the same: a vigorous bayonet attack. Although the bayonet proved not to be the tactical solution against firepower, it does not mean that this weapon could not produce some psychological effect on the battlefields of the First World War and training and encouraging troops to use this weapon was regarded as a morale booster.[10] The cult of the bayonet and its rhetoric were echoed, for instance, by the Spanish military attaché in Vienna, Major José Fernández de Villa-Abrille, in his report of April 1916 on the Austro-Hungarian infantry assault tactics: '[Q]uick and resolute leaps, small volume of fire and a lot of heart, the ultimate argument is the bayonet; the latter, as usual, is what writes the last word on the breast of the conquered!'[11] Even an essayist advocating higher numbers of machine guns in the infantry units concluded that infantry had 'proven that, despite the progress of firearms, the bayonet… still is what settles the fiercest fighting and preserves unharmed [the infantry's] renown as *queen of battles*'.[12] The problem was that such attitude could (and did) foster the idea that technology had a negative impact on the fighting man's morale. It should come as no surprise to anyone that this criticism of the increasing role of ordnance in infantry tactics found a voice in Ricardo Burguete (now a brigadier general), who wrote in 1917 that 'the main weapon is the heart and [the latter] is proven by marching forward; instinct must drag [the soldier] to thrust the bayonet into the enemy's breast'.[13] Since the machine could not beat the spirit, there was no reason for the introduction of new weaponry. Quite the contrary:

> [P]oisoned in spirit and stunned in mind with the wrong idea that 'the machine is all and the man nothing', [some people] look to the machine for the remedy and turn the infantry into a collection of coolie companies

> carrying... a thousand devices, of occasional use [whose only effect is] the disappearance of infantry, creating another [false infantry] which only keeps the name of this arm...[14]

Burguete's view was not unusual. As late as 1918, another officer could still lament that the infantryman had to carry equipment such as gas masks, hand grenades, and so on, because 'victory lies in the point of the bayonet'.[15]

Coping with a New Way of Warfare

If part of the Spanish military disliked the new features of combat, other officers understood soon the changed nature of the battlefield, so different from the pre-war expectations. Major General José Marvá, who had advocated the technological modernization of the army for years, made lucid remarks about the increased dehumanization and material nature of war by early 1916:

> [The current war is] on the technical side the rule of the machine; on the side of art, the drowning of inspiration; on the tactical side, the diminished importance of manoeuvre; on the strategic side, the enlargement of the stage to encompass the entire planet.
>
> The drama is gaining in magnitude what it loses in gallantry; the sense of destruction grows sharper and victory comes through the mechanical accumulation of battalions and devices. Success does not surrender to the bayonet but to the missile; not to the most skilful deployment but to the highest explosive; rather than man, it is due to industry and money.[16]

Captain Epifanio Gascueña had elaborated on the willingness to accept great losses in the offensive by the time the war broke out.[17] But he had already admitted, in the spring of 1915, that firepower had reduced the fighting on the battlefield to the two arms (infantry and artillery) whose tactics depended on it; even more, victory could be achieved with less bloodshed by relying on artillery fire. Thus the battle was to depend on the close liaison and cooperation of both arms.[18] Gascueña also warned about the need to adapt tactics to the increased firepower even in a colonial war against an irregular enemy. He deplored the careless methods used by the Spanish forces in the Moroccan campaigns during the 1910s. Commanders were too eager to go into action and move forward. This meant sometimes that the selection of emplacements for the artillery was neglected and this arm was not given enough time to provide effective fire support. Gascueña urged his colleagues, either fighting in Morocco or training at home, to take more account of the time needed by the artillery to go into action and provide effective fire support. And the officers at the headquarters should also restrain their impulses to ride to the firing line to the detriment of their main duties in the coordination of operations.[19] Artillery fire certainly lost much of their effectiveness in Morocco because it was not coordinated by a single command during an operation. The artillery units were divided into penny packets and scattered among the field columns.[20] General Dámaso Berenguer, Spain's ablest colonial commander during this period, later admitted that the tactical doctrines of the Spanish army were outdated by 1919. There were no official regulations on using automatic firearms in the front line, large masses of artillery under a single command, or aircraft for offensive actions and intelligence gathering.[21]

Discerning Lessons After 1918: From the Western Front to Morocco

After the guns fell silent in Europe, the French army became a model to be imitated due to its final victory in 1918. But some Spanish essayists regarded it as a questionable model. For the French army was to base its post-war doctrine on the predominance of firepower over manoeuvre.[22] A consequence, which was not unnoticed in Spain, was that the French tactical regulations were virtually to take for granted that a future war would see fortified fronts again.[23] In 1921 Captain Enrique Maquieira, who was against the slavish imitation of the French doctrine, deplored such thinking. It fostered a defensive mentality (such as that which got hold of the French by 1917, according to Maquieira) that could not be dispelled by planning limited counteroffensives. Moreover, carrying out these counteroffensives required substantial superiority of military means.[24] Maquieira seems to suggest in this last point that the Spanish army's lasting shortage of material resources was another reason not to follow the French doctrine too closely.

Though the French army doctrine was the main guideline, as will be shown below, some professional articles leave no doubt that ideas on the predominance of morale-related factors in warfare still had force in sections of the Spanish military, though they were not so aired as before. In the period 1919–1920 they were echoed by several articles in the cavalry corps journal. These articles asserted the value of the 'psychological' principles of warfare, unaffected by technological change, and argued that tactical doctrine could not depend on the number and quality of new weapons.[25] They considered machine guns, mortars, and hand grenades as weapons of occasional use and deplored the creation of subunits of specialist troops in the fighting arms.[26]

Such thinking met a forceful reply in an article by Major Luis de la Gándara, a distinguished military essayist, in 1921. He criticized the attitudes of the section of the military that deplored the adoption of new weaponry for the infantry by alleging that such a step would restrain mobility and lead to the death of manoeuvre warfare. De la Gándara thought that these views were retrogressive and based on armchair strategist thinking, which reasoned as if the clock should be put back half a century. On the contrary, the empirical experience of warfare had forced the armies to adopt new weapons in order to cope with innovations unknown 25 years before, such as the automatic machine gun. If the progress of technology and industry had brought obstacles to mobility in warfare, this same progress was providing the means to overcome the obstacles through motor transport and, perhaps, the aircraft in the not too distant future. De la Gándara admitted that these retrograde essayists had a point in arguing that a repetition of the Western Front conditions would be hardly possible. But even if future conflicts did not create long fortified fronts, the weaker armies would resort to the advantages of the ground and the tactical use of field fortifications to compensate for their numerical or material inferiority.[27]

The present author's survey of the Spanish professional literature on tactics has found no significant evidence of anti-material thinking after the early 1920s. Perhaps the fact that a French-inspired doctrine was officially adopted by the Spanish army inhibited the essayists from expressing their views openly, since the latter were not totally in tune, so to speak, with official thinking. Moreover, there was the undeniable fact that the French army had won the war. But the penchant for making sense of war as a psychological contest remained alive. Colonel Juan García Gómez-Caminero published in 1925 an essay which was reminiscent of Burguete's rhetoric and arguments depicting war as a biological need, an indispensable trial for getting nations tougher, and was awarded an army prize.[28] Another essayist emphasized as late as 1933 the importance of morale in terms echoing the literature before 1914: 'The combat is nothing but a struggle between two wills. Above… all the armaments, always shines this intense, purifying flame, where selfishness and ambitions are melted down, which is called military morale and decides victory.'[29]

Meanwhile, the Spanish army resumed operations in Morocco after 1918 against natives that, during the early 1920s, learned to use field fortifications and availed themselves of the artillery captured in their victory of Annual.[30] This development and the events of the Great War led a staff officer with Moroccan experience to frown on 'the mad, unrestrained [*desenfrenada*] offensive. Graves are filled with fanatics and airheads [*alucinados*], sons of the doctrinal extremism previous to the [1914–1918] war'.[31] The Spanish artillery had to develop a new doctrine for the use of their firepower in this theatre of operations. The Spanish artillerymen were now required to destroy enemy field fortifications and deliver counter-battery fire, in addition to the previous missions of support to the columns and punitive shelling. This enlarged role led to the creation of the *masa de artillería* (artillery mass). It grouped under a single commander all the artillery that was not attached to the field columns during an operation. The *masa* carried out general support fire plans and deliver additional artillery fire to individual columns when needed.[32] This was one of the few practical teachings the Spanish army learned from its Moroccan campaigns, which had not provided many useful lessons, according to the General Staff's new guidelines on tactical doctrine (*Doctrina para el empleo táctico de las Armas y los Servicios*), issued in 1924.[33]

Actually the forces in Morocco might well have applied the military doctrine developed north of the Gibraltar Strait, since they adopted a new operational approach for the decisive French–Spanish offensive against Abd el-Krim's rebellion in the spring of 1926, that is, *after* the issue of the 1925 field service regulations (which will be discussed below).[34] Instead of setting up small outposts piecemeal through short advances of columns that had returned to the base by sunset, the Spanish army undertook a relentless offensive on a wide front with strong task forces moving forward along several axis of advance, heavy artillery support, improved logistics, and no halt until subduing the enemy.[35] It is surely not accidental that this approach was similar to the French one.[36] Under the command of Marshal Philippe Pétain, the French army also assembled a big force with strong artillery support. This force undertook a methodical advance, taking advantage of its overwhelming superiority in manpower and firepower to reduce casualties.[37]

And these methods worked. The Spanish and French armies attacked in early May 1926 and overwhelmed the Rif tribesmen within two months (Abd el-Krim himself surrendered to the French on 27 May).[38] But it was not a *promenade militaire* from the start and in order to win 'none of the built-up means was superfluous'.[39] The Spanish break-out from the Alhucemas beachhead in the sector of the Lomas de los Morabos (8–10 May) was described as 'a modern-style battle to break through a defensive front organized according to a rational fire plan'.[40] So there were grounds to assert that the last Rif campaigns had demonstrated that the methods of European regular warfare could be used successfully in limited wars, with some adaptation due to terrain and the nature of the enemy.[41] This was indeed the conclusion of Manuel Goded, a field commander and, during the campaigns of 1926–1927, chief of staff of the Spanish army in Morocco. Morocco was not a 'special war' and only small unit (battalion and below) tactics required some adaptation to the terrain and the enemy's methods of fighting.[42] Incidentally, in relation to an alleged 'special' nature of war in Morocco, Enrique Ruíz-Fornells maintained that stressing it was no more than an excuse to conceal that the Spanish army's problems derived from neglecting a thorough, intellectually demanding study of warfare.[43]

Towards a New Spanish Battlefield Doctrine

The Making of the 1925 Regulations for Major Units

The Spanish military did not wait to the climax of the Moroccan war to take the French army as their major source of battlefield doctrine in the 1920s.[44] The present author's research has only found one piece of official documentary evidence on this issue, but it goes a long way to proving that the French post-war doctrine weighed heavily on the Spanish one. This piece of evidence is a report on a paper written by Colonel Enrique Ruíz-Fornells, head of the General Staff's Sixth Section (military doctrine), for the working group on the reform of the field service regulations.[45] The report is undated, but it was written after February 1923, when the General Staff was reorganized and the 6th Section took charge of doctrinal affairs.[46]

The General Staff had set up a working group (*ponencia*) to survey the main features of modern warfare, and the role, equipment and organization of the fighting arms. The group, in its turn, asked Colonel

Ruíz-Fornells for guidelines on which it could base its work. Actually the guidelines submitted by Ruíz-Fornells were apparently an outline of new field service regulations. The report on Ruíz-Fornells' paper praised the soundness of its doctrinal foundations and pointed out that this soundness was backed by its coincidence with the doctrine of the French army, the highest contemporary authority in warfare matters, according to the working group. Such a doctrine had been set forth, among other French regulations, in the 1921 Provisional Instructions for the Tactical Employment of Major Units.[47] Nonetheless, the working group had ideas of its own. Ruíz-Fornells' paper had proposed organic infantry guns and mortars for the infantry battalion and creating regimental platoons of assault pioneers (equipped with flame-throwers, demolition charges and smoke grenades). The working group disagreed. It reckoned that all this ordnance would limit the mobility of the infantry in so hilly a country as Spain and that infantry guns could not stay for long in very advanced emplacements without being exposed to enemy fire. Moreover, all that infantry support weaponry would be operated better by artillery and engineer troops.[48]

The Spanish battlefield doctrine finally took shape in the *Reglamento para el empleo táctico de las Grandes Unidades* (Regulations for the Tactical Employment of Major Units) of 1925 (RGU 1925). Since Ruíz-Fornells was one of the writers, it should come as no surprise to anyone that the regulations' contents were influenced by those of his 1923 paper. The Spanish officers seem to have had small doubt about the Gallic influence. Major Román López Muñiz went so far as to state that the Spanish regulations were essentially a copy of the French ones of 1921.[49] Major José Ungría, another writer of RGU 1925, felt no shame about copying foreign models: '[The regulations] are copies because they must be a copy; are we falling into the naive aim of inventing tactics of our own alone?' Ungría based his assertion on Spain's dependence on foreign military technology and the fact that the Spanish army's professional experience during the last hundred years had been limited to colonial and civil wars.[50]

Firepower and Manoeuvre in RGU 1925

RGU 1925 stated that modern weapons had radically increased the importance of firepower: 'Fire plays a prevailing role in combat... Fire superiority is the best protection against the enemy's elements of destruction.'[51] Although tactics were based on the combination of fire and movement, movement seemed implicitly subordinated to fire. RGU 1925 stated that the effect of fire made movement easier and the latter should be carried out by echelons; every echelon would move forward covered by the fire of the others until reaching the objective.[52] The aim of manoeuvre would be to achieve a material advantage over the enemy forces by inflicting physical damage in order to weaken their morale until they gave up. Moral superiority was a consequence of the material superiority, and it must not be reckoned as a direct effect of a manoeuvre that had just started.[53] As a result, the superiority of the offensive was no longer asserted as an axiom and, according to a contemporary essayist, the Spanish doctrine was rather impartial on this point.[54] The emphasis was put now on the goal of the battle, and the ways to achieve it would depend on the circumstances. The purpose of battle, according to RGU 1925, was 'the moral and physical destruction of the enemy; if [the battle] is offensive, by driving him out of his positions, by breaking his lines, and by pursuing him…; if it is defensive, by driving his attacks back and by thwarting his advance'.[55]

The prevalence of fire was echoed in infantry tactics. Thus this arm's new tactical regulations of 1926 stated that the attack arrangement of infantry units had to be 'essentially a fire arrangement as thorough and powerful as possible, in order to achieve from the start the fullness of firepower, since it is difficult to know beforehand the number and quality of the enemy defences…'.[56] López Muñiz pointed out how the firing regulations for small arms gave a pre-eminent role in combat to firepower. In order to neutralize the enemy fire during the attack the advancing infantry had to achieve fire superiority. No attack would be successful without being prepared and accompanied by fire supremacy over the enemy. This key idea of fire supremacy had to be applied in the offensive (firepower that moves forward) and in the defence (firepower that holds back).[57] Since fire supremacy should always accompany the infantry's advance, it was logical that the infantry units had increased after the First World War their scale of organic means to deliver firepower, from heavy machine guns to mortars and rifle grenades.[58]

Infantry–Artillery Cooperation in RGU 1925

RGU 1925 also put the Spanish army in contact with the French concept of the 'methodical battle' (*bataille conduit*), which was the foundation of

Gallic military doctrine after 1918.[59] The conduct of the methodical battle was characterized by the overall subordination of operations to the action of the artillery. This, in its turn, required a specific style of command and control of the fighting arms.[60] RGU 1925 exalted the status of the infantry as the queen of the battlefield, 'the arm whose success all the other ones must work for'.[61] However, the regulations later made a more sober assessment of the capacities of this arm. Its capacity to hold ground unsupported was recognized, but 'its limited offensive capacity requires the support… of the artillery, with which it has to keep so close a liaison that both actually form a whole'.[62] In practice, the division's infantry commanders had no authority over the artillery, which was controlled by the divisional commander. Centralization was the rule for the tactical employment of the artillery.[63] Such centralization was a logical development of RGU 1925, which asserted that the artillery 'will have all the more facility to concentrate their fire the more centralized is their command'.[64] Only in exceptional cases would artillery be attached temporarily to an infantry unit and subordinated to its commander.[65]

Ruíz-Fornells (now a retired general officer) justified the need for such centralization in 1935. If the divisional artillery was divided among the infantry units, the latter would be embarrassed in their movements. There would be no unity in the division's manoeuvre because every subordinate commander would carry out the plan according to his personal judgement. Thus targets of tactical importance for the division as a whole could be left unshelled and the divisional commander would be unable to shift firepower all around the division's operational area.[66] On the contrary, Major López Muñiz reckoned that the result of this procedure was inflexibility in the divisional tactics, hardly suitable for manoeuvre warfare. Moreover, centralization of tactical command and artillery support at the divisional headquarters made it difficult to organize and use combined arms task forces at brigade and regimental level.[67] He summed up the official doctrine and his own judgement on it this way: 'Centralization, artillery support not under the infantry's command, narrow combat fronts, systematic deployment of units, stereotyped procedures, absence of manoeuvre, overall rigidity, and too detailed orders are the features [of] our current tactical rules.'[68]

A centralized procedure of command such as that adopted by the French army after 1918 also put a heavy burden on the staff officers at headquarters.[69] The reluctance to delegate authority to lower command levels meant long, excessively detailed processes for the preparation of operations orders. RGU 1925 suffered from this malady, as was candidly revealed by General Ángel Rodríguez del Barrio. He was the director of the army corps-level manoeuvres carried out in the Pisuerga River area in the autumn of 1932. Rodríguez del Barrio wrote in his report that the strict application of RGU 1925 caused insoluble situations. As he pointed out, this was unsurprising because the regulations were based on the French ones and these had been born from a kind of warfare where time and material means were plentiful.[70] Rodríguez del Barrio described the preparation of orders for offensive operations at army corps level according to Article 164 of RGU 1925, which had to include a detailed study of the terrain (incorporating maps). The experience of the manoeuvres had shown that writing the orders required eight hours at least; the copying process required a roughly similar time, and the delivery added several hours more. The total time consumed in the procedure was a full day. Since the corps commander could not make decisions before knowing all the events of the day and being informed by his intelligence officer, the work could not start until late afternoon. This meant that the orders could not be issued to the divisions and the brigades early next morning (and after being issued, the lower headquarters had to prepare their own orders). Therefore, Rodríguez del Barrio wrote that new procedures (faster and more suited to manoeuvre warfare) should replace those in force in the regulations.[71]

The spirit of RGU 1925 also clashed with the interest in a new model of major unit that the Spanish army developed during the 1920s and early 1930s (see Chap. 8). Spain's rugged ground (which made the creation of a duplicate of the Western Front unworkable) and dissatisfaction with the organization of the existing infantry division fostered the interest in the combined arms mixed brigade (*brigada mixta*). Such interest dissented from RGU 1925, which did not contemplate combined arms tactics below the divisional level. The organization of mixed brigades also meant decentralizing the command of divisional artillery in order to allow these formations to operate with a degree of independence. This was at odds with the RGU 1925, which stated that the divisional artillery commander was also the commanding officer of the division's artillery units.[72] Nonetheless, the Spanish army started to adopt this new type of major unit under the form of separate and divisional mountain brigades during the Second Republic.

Ordeal: Spanish Battlefield Doctrine at War, 1936–1939

Firepower and Manoeuvre in the Spanish Civil War

How far did the Civil War match the pre-war expectations?[73] As many essayists anticipated, the Civil War was not the Western Front redux. Even when both armies reached their maximum strength, the force–space ratio still was too low to prepare deep defensive lines (the density of troops per kilometre of front has been estimated one sixth of that on the Western Front, whereas the Republican general staff reckoned that halting an enemy attack required at least one battalion per kilometre).[74] This gave opportunities for manoeuvre warfare, since it was easier to break through the enemy front. But as De la Gándara warned in his 1921 article, a defending modern army did not need a continuous front to thwart an offensive. And when the attacker failed to exploit the initial success and the defender could bring reserves to the threatened sector in time, the fighting became more static and attritional in character, as the Republicans were to find out at their cost in several of their major offensives of 1937–1938, the crucial years of the war. The Nationalist army's offensive operations of the same period hinged on the exploitation of heavy artillery bombardments (often in combination with air bombing) and achieved a high rate of success. The *Ejército Popular de la República* (EPR)—the brand new Republican regular army—never matched the Nationalist overall effectiveness in the offensive use of artillery and paid dearly for it.[75]

Shortage of heavy artillery for front breakthroughs was a major handicap for the Republicans. In this context it is worth noting that in the summer of 1937 the artillery command of the EPR's Army of the Centre actually advocated seeking the offensive decision through the manoeuvre on weakly fortified fronts, rather than attempting breakthroughs of stronger fronts without adequate artillery support.[76] Indeed, it does not seem casual that after a series of failed attacks with artillery preparation in the spring of 1937 the Republican command decided to start four of its next major offensives (Brunete, Saragossa, Teruel and the Ebro River) by using surprise and night infiltration. Even so, only one (Teruel) was fully successful and took place against a very vulnerable and weakly defended salient.[77] The other offensives lost momentum and stalled very soon because the Republicans were held up by isolated Nationalist strongpoints that withstood against the odds.

The blame is usually put on the Republican field commanders, whose lack of experience and initiative prevented their troops from bypassing enemy positions of lesser importance. But the Nationalist forces did not attempt to bypass enemy positions either. A Nationalist junior officer who took part in the most successful exploitation of the war (the initial phase of the offensive of Aragon in March 1938) recalled that whenever his column met resistance the infantry deployed in fighting order behind a line of tanks extended abreast. If the enemy position was strong, a short preparatory shelling preceded the assault.[78] This happened although the Nationalist command actually had issued instructions before the offensive that prescribed bypassing any resistance too stiff.[79] Morale and discipline played certainly a role: the Nationalist troops were readier to stand fast in exposed or surrounded positions than the Republican ones, which were very fearful of getting encircled.[80] But it may also be arguable that a poor use of artillery prevented the Republicans from taking more advantage of their initial success because their firepower was insufficient to crush enemy resistances fast enough.

By contrast, the Nationalists' record for the period 1937–1938 was considerably more successful. After the failures at the Jarama River and Guadalajara (February–March 1937), almost all their offensives reached its main objectives; only the campaign of the Levant (late April–July 1938) did not achieve its aim of seizing Valencia quickly and was totally called off after the Republican attack across the Ebro River. Colonel (later Brigadier General) Juan Vigón, the Army of the North's chief of staff (and perhaps the best military strategist in the Nationalist army), planned deliberately these offensives to be carried out in stages, with pauses to assess the results and introduce modifications if necessary.[81] Probably the most adequate term for their tactical nature might be 'a form of mobile attrition', whose rate of advance depended on circumstances such as terrain, the defenders' strength and morale, weather or the availability of motor transport.[82] But they were not always crude 'big pushes'. As far as circumstances allowed it, Vigón sought to manoeuvre at the operational level (Santander and Aragon are the finest examples). Only when Franco accepted a Republican challenge (or took the bait, according to taste), as in Brunete, Teruel, and the Ebro, the Nationalist troops became engaged in a kind of fighting that could remind of Verdun or the Somme.[83]

Race for Victory: Implementing RGU 1925

Anyway, all the Nationalist offensives in the northern fronts shared a common feature: massive bombardments to prepare breakthroughs of enemy defensive lines. This was in line with RGU 1925's emphasis on the concept of concentrated use of artillery under centralized command. This concept also affected the divisional organization. When the Nationalist army was reorganized in late 1937 the standard infantry division was given a small organic field artillery (two light battalions). All non-divisional units were grouped in the *Reserva General de Artillería*, which would attach them temporarily to army corps and field army echelons as needed.[84] Nonetheless, Carlos Martínez de Campos (who, as artillery commander of the Navarrese Brigades and later the Army of the North, was responsible for the major bombardments that opened the way for the Nationalist troops after March 1937) mused after the war that giving such reduced means to the new divisional artillery commanders perhaps was overcautious.[85]

One issue beyond the scope of this chapter but which the reader must bear in mind as well is that these Nationalist major artillery bombardments were carried out in combination with air attacks on the Republican positions, since there was not enough cannon for the artillery to fulfil all the tasks that were its responsibility.[86] This air bombing was indeed the Nationalists' 'flying heavy artillery', which made up the shortage of the conventional type and gave Franco's army the edge in the firepower contest.[87] Air bombing also strengthened the attacking troops' morale so much in the first days of the Biscay offensive (from 31 March 1937 onwards) that air support became indispensable for major attacks.[88] By the spring of 1938 the combined artillery-aviation bombardment opened with a period of artillery preparatory fire; during its last stage, a bomber force dropped its load on the enemy positions. Then the assault forces advanced under the cover of direct support artillery and ground attack aircraft (*aviación de cooperación*).[89] This combined bombardment required artillery officers well-grounded in its techniques. Artillery battalion and battery commanders were reminded in late March 1938 to take care of their fire when Nationalist ground attack aircraft operated on the target area: two aircraft had been shot down in the preceding days by direct hits of Nationalist field artillery shells.[90]

RGU 1925 was also the official doctrine in the Republican army. However, its implementation left much to be desired and this prevented the EPR from making the best use of the cannon available. For instance, more centralized use of artillery was missed until well advanced in the war. Reports about failed attacks in the spring of 1937 criticized the dispersal of batteries amongst the mixed brigades and assault task forces (*agrupaciones*), thus leaving the higher command echelon without enough artillery resources for counter-battery fire and additional support fire in the decisive moments of the operations.[91] About nine months later a report on the Republican debacle south of the Ebro in March 1938 concluded that a reason for the complete failure of the artillery (which never delivered fire with groupings stronger than a battalion) was the absence of a centralized command for the arm.[92]

Such underperformance should not come as a surprise to anyone given the severe handicap that afflicted the EPR and especially its artillery, which required a high level of technical expertise: the shortage of competent officers, which wartime efforts could not amend.[93] Moreover, both temporary officers trained in hurry and professional ones used to an undemanding peacetime routine did not to reach an adequate level of proficiency in wartime overnight. The Nationalist army also had to cope with this problem.[94] But the Republican officers had to undergo a longer learning process than their Nationalist counterparts.

It did not help either that the EPR adopted the mixed brigade as the building block for its force structure (see Chap. 8). Although it has been stated that the mixed brigades lacked in reality any organic field artillery, a number of them did have, at least during the first year of war.[95] Creating a brigade with organic artillery was a sound idea in theory, but in the circumstances of the Republican army it became a self-defeating one.[96] After-action reports concluded that the mixed brigades should not have organic artillery, because its existence fragmented the arm and went against the guiding principle of concentrated use (the only way to achieve effective outcomes).[97] Although the brigade artillery was not officially disbanded, it seems that the Republican army eventually centralized the command of artillery at the divisional level for offensive operations and at army corps level for the defence in the later stages of the war.[98]

The EPR did not use the air force to enhance its firepower on the battlefield as successfully as the Nationalists. The Republican army high command seems to have neglected to attend to a close liaison with the air force in its operational planning. Matters were made worse by the fact that the Republican air force was under the actual control of the Soviet advisors, who had concerns of their own, for most of the war. The result was that the army often did not get the air support required for its operations (although, according to Soviet observers, 80% of the Republican air effort was devoted to the support of ground forces by June 1938). This happened even if, for at least the first year, the Soviet airmen seemed very interested in carrying out battlefield support missions, whose effectiveness was proved in the battle of Guadalajara (March 1937). The Soviet Union also was less steady than Germany and Italy in providing adequate amounts of aircraft and pilots and rotated the latter faster, which made more difficult to develop techniques similar to the Nationalist ones.[99]

The Republican learning process toward a more centralized command of artillery was not exempt from *faux pas*. For instance, a report written after the battle of Brunete concluded that assigning the control of all the direct support missions to the army corps, instead of the division, was a mistake. It forced the liaison between infantry and artillery to pass through more command echelons, which delayed and distorted the support calls.[100] But as their officers underwent their learning process, the Republican artillery did improve their performance. Reports of both sides on the operations in the Levant and Catalonian fronts in the spring of 1938 noted a more concentrated use of the Republican artillery, which enhanced its effectiveness.[101] And a Nationalist report of June 1938 pointed out that the Republicans in the Levant front had figured out how to emplace artillery far in the rear, but still close enough to the frontline to deliver a heavy defensive barrage in front of their positions. The result was that they could often shell the Nationalist infantry without exposure to counter-battery fire, since the range of the Nationalist cannon was too short.[102]

Infantry–artillery cooperation was a harder craft to master, however, and the EPR never did it enough. Both arms had their share of blame. Some reasons of the artillery's poor performance have been mentioned above. As regards the infantry's faults, one of them was the misunderstanding of the different modes of artillery fire. During the first year of war, at least, direct support fire to cover the assault was mistaken for preparatory fire and therefore the infantry did not advance when the artillery opened up the former. Nor the assault units moved forward fast enough to exploit the artillery's stunning effect, so the defenders had time to man their positions again.[103] Another fault noted during all the war was a reluctance to advance without continuous artillery support.

The infantry seemed to forget that the artillery could not destroy, or even neutralize totally, all enemy resistance, and in the battle of Brunete division and brigade commanders even resorted to coercion when asking for artillery support. The result were long, ineffective bombardments that wasted ammunition and wore down cannon because the infantry asked for 'five minutes more' without end.[104]

The Nationalist army did not avoid totally the problems of inter-arm cooperation either. General Alfredo Kindelán, commander-in-chief of the Nationalist air force, criticized the artillery during the battles of Teruel and the Ebro because of its poor support to the infantry.[105] Martínez de Campos answered to criticisms like these by reminding that cannon fire was not a tactical cure-all. For instance, he felt that most of the responsibility for the artillery's poor performance in the Teruel front from 29 December 1937 to 7 January 1938 fell on the army corps and divisional commanders, who made the final decisions about artillery fire but lacked understanding of the principles and capabilities of the arm. Artillery units were ordered to fire too soon or too far and could not give effective support to infantry units that did not use signal flags and panels because they were afraid to reveal their positions to the enemy. Other infantry units expected that artillery fire would deliver them from fighting at close quarters in the last stage of the assault and did not even come near to the Republican positions when it did not.[106] If true, incidentally, this last contention seems to demand more nuance in Michael Seidman's assertion that the Republican army's option of surprise infiltration was due to the inferior combativeness of its troops, besides its inferiority in materiel.[107]

During the battle of the Ebro, Martínez de Campos acknowledged that poor liaison means and a shortage of good observation posts limited the effectiveness of the divisional artillery. This led sometimes to use artillery attached at army corps or army level for direct support missions. However, these echelons could never be as effective in such missions as the division.[108] But according to Martínez de Campos, the main reason for the limited success of the Nationalist attacks was the infantry commanders' poor tactical skill. Insufficient reconnaissance of the terrain limited the divisions' attacks to one or two axis of advance only. And regimental and battalion commanders' poor use of machine guns and mortars prevented their units from suppressing the strongpoints that came through the shelling and were the real obstacles.[109]

One final point to consider is the relevance of the Spanish military's experience in the Civil War within the context of the evolution of land warfare in the interwar period. A British military observer noted in the spring of 1938 that tactics were 'largely based on Great War principles' and the Nationalists' recent successes were due to 'their ability to concentrate in secrecy a large preponderance of field artillery in the sector selected for the break-through'.[110] This should not come as a surprise to anyone since RGU 1925 was the Spanish adaptation of the effective Allied offensive methods eventually developed in the First World War, whose French version was the *bataille conduit*. Was this a sign of professional backwardness? Some authors seem to think so.[111] However, it has to be borne in mind that the utter debacle of the French army facing the German blitzkrieg in 1940 has fostered a perception of any fighting doctrine or technique related to 1918 as outmoded.[112]

In fact, most of the fighting methods at the end of the First World War may well be described 'prototypes' for those of the Second World War.[113] Advances in technology (especially in the fields of communications and armoured vehicles) basically added flexibility and speed to the methods proven in 1918, allowing to react to battlefield events and to exploit them at a faster tempo, but they did not remove attrition. A different matter is how successfully individual armies adapted their doctrines to these advances. As regards the Spanish case, such advances were not mature enough when the war came in 1936 and field commanders still fought with many of the shackles worn by their Great War predecessors. So the Spanish armies were not wrong to fight after RGU 1925.[114] Nevertheless, it is worth noting that the Nationalists readily incorporated tactical air support (a tool of attrition, by the way) to the pre-war doctrine and both sides were prepared to try blitzkrieg-style operations when the conditions were appropriate, as in the Saragossa (August 1937) and Aragon (March 1938) offensives (see Chap. 10).

Conclusion

In the decade after 1914 the Spanish army battlefield doctrine left aside the paradigm of 'the psychological battlefield'—where the non-material, morale-related factors predominated—as a consequence of the realities (trench warfare and preponderance of firepower) seen on the battlefields of the First World War and the experience in the Moroccan campaigns. Nonetheless, a section of the military was still attached to the former concept and fended for it in the professional literature of the late 1910s and early 1920s. But this can be described as a rearguard action, and the increased tactical importance of firepower was not called into question thereafter.[115] A new, French-inspired doctrine that stressed the methodical conduct of offensive operations based on centralized use of artillery firepower was introduced and materialized in RGU 1925 (and probably provided the intellectual background for the successful offensives of 1926 that broke the back of native resistance in the Rif).

Nonetheless, some officers expressed dissatisfaction about the 1925 doctrine's implications at the operational level. They thought that any future war in Spain would be a less static affair than the Western Front and therefore the Spanish army did not need to centralize command and artillery firepower so much. They urged the army to develop a military doctrine and organization of its own, better adapted to the features of the Spanish territory, in contrast with the penchant for copying French models, which seems to have weighed much on official decision-making circles. This gap between the official doctrine and a current of professional thinking which envisaged Spain's future conflicts in terms of more dynamic manoeuvre warfare shows that the Spanish army of this period was not the intellectual wilderness that it is usually supposed to be.

The Civil War proved right the critics of the official doctrine as far it was not a conflict of static fortified fronts. The force–space ratio was so low that it was hard to set up solid defences, especially defences in depth. Thus there were chances for the armies to break through a front and make substantial gains. However, once the fronts became stabilized by the end of 1936 and each side started deploying a mass army, the war also proved the validity of the pre-war doctrine as regards the key role of firepower on the contemporary battlefield. The side which applied sooner and better the essentials of RGU 1925 was also the more successful. The Nationalist army managed to be ready to start applying its own version of the methodical battle in the northern fronts by the early spring of 1937, whereas the Republican army underwent a longer learning process because of its smaller professional cadre. Progress in the use of artillery seems not to have been noticeable in the EPR until the spring of 1938, too late to alter scales that were already tipped in favour of the Nationalists. Above all, the Republicans never were able to match their opponents' effectiveness in infantry–artillery cooperation. And mastering the latter was critical in a conflict that proved again that, in modern conventional warfare, operational schemes, however brilliant, came to naught when the attacker lacked effective firepower to suppress in time the defences of key tactical positions.

In sum, the Nationalist victory on the battlefield was not delivered by an alleged manoeuvrist *elán* of the *africanista* field commanders, who mostly served in Franco's army. The Republican commanders resorted to manoeuvre as well, and even in more ambitious ways. The fact is that both sides' professional officers went into war sharing the same firepower-oriented battlefield doctrine and the Republican departure from it was rather an attempt to make a virtue of necessity by circumventing the EPR's handicaps. For relying on overwhelming firepower had the great advantage that it enabled troops to achieve their objectives without intolerable casualties.[116] It could also lead to a slower rate of advance, but this proved to be a price that the Nationalist side (or at least its *caudillo* Franco) was prepared to pay.

Notes

1. Michael Alpert, 'The Popular Army of the Spanish Republic, 1936–39', in Wayne H. Bowen and José E. Álvarez (eds), *A Military History of Modern Spain: From the Napoleonic Era to the International War on Terror* (Westport: Praeger, 2007), p. 108.
2. For a concise overview of the basic tactical problem of the First World War, see David T. Zabecki, *The German 1918 Offensives: A Case Study in the Operational Level of War* (Abingdon and New York: Routledge, 2006), Chap. 3. For a sample of literature on military developments after the First World War, see Allan R. Millett and Williamson Murray (eds), *Military Effectiveness*, new edition, 3 vols. (New York: Cambridge University Press, 2010), Volume 2; Roger Chickering and Stig Förster (eds), *The Shadows of Total War: Europe, East Asia, and the United States, 1919–1939* (New York: Cambridge University Press, 2003). On the experience of another neutral country's military, see M.M. Olsansky, 'The Development of the Swiss Army's Combat Methods after the First World War', in Herman Amersfoort and Wim Klinkert (eds), *Small Powers in the Age of Total War, 1900–1940* (Leiden: Brill, 2011), pp. 307–20.
3. On the evolution of artillery during the First World War, the following works are enlightening: Paul Strong and Sanders Marble, *Artillery in the Great War* (Barnsley: Pen and Sword Military, 2009); Dale Clarke, *World War I Battlefield Artillery Tactics* (Oxford: Osprey, 2014); Sanders Marble (ed.), *King of Battle: Artillery in World War I* (Leiden: Brill, 2015).

4. Although Spain was a neutral country, most officers were *germanófilos* (Germanophile) because they admired the German military prowess and sympathised with the authoritarian tinges of Wilhelmine Germany: Carolyn P. Boyd, *Praetorian Politics in Liberal Spain* (Chapel Hill: The University of North Carolina Press, 1979), pp. 45–6.
5. Fernando Redondo Díaz, 'Los observadores militares españoles en la Primera Guerra Mundial', *Revista de Historia Militar*, 59 (1985), pp. 197–208.
6. Jonathan M. House, *Toward Combined Arms Warfare: A Survey of 20th Century Tactics, Doctrine, and Organization* (Fort Leavenworth: United States Army Command and General Staff College, Combat Studies Institute, 1984), pp. 17–18; Hew Strachan, *The First World War. Volume I: To Arms* (Oxford: Oxford University Press, 2001), pp. 227–8, 237–9; Geoffrey Wawro, *A Mad Catastrophe: The Outbreak of World War I and the Collapse of the Habsburg Empire* (New York: Basic Books, 2014), pp. 155–6, 195–7, 201, 220–4, 227.
7. Luis Fernández España, 'Orientaciones alrededor de la guerra de hoy', *Estudios Militares* (hereafter *EM*), 34, 1 (1915), pp. 70–1, 131–3.
8. Francisco Anaya Ruíz, 'De la guerra mundial. Impresiones hispanófilas', *Revista Técnica de Infantería y Caballería*, 27, 1 (1916), pp. 412–14, 457.
9. Enrique Albert, 'La Infantería en la Gran Guerra', *Memorial de Infantería* (hereafter *MI*), 7, 1 (1918), p. 288.
10. Rob Engen, 'Steel against Fire: the Bayonet in the First World War', *Journal of Military and Strategic Studies*, 8, 3 (2006), pp. 13–21.
11. Archivo General Militar de Madrid, Fondos del Ministerio de la Guerra, Legajo 153, Respuestas a las 14 preguntas de la circular del E.M.C. de 19 febrero último, 18 April 1916. Fernández de Villa-Abrile (1878–1946) was the commander of 2nd Division (Seville) in July 1936; arrested by the rebel officers because of his loyalty to the government, he was discharged from the service in late 1936; a court martial sentenced him later to six years' imprisonment.
12. Juan Génova, 'Las armas automáticas modernas', *MI*, 5, Special Issue (1916), p. 350. Italics in the original.
13. Ricardo Burguete, *La ciencia militar ante la guerra europea. Su evolución y transformación hasta el presente y el porvenir* (Barcelona: Maucci, 1917), p. 503.
14. Burguete, *La ciencia militar*, pp. 473–4.
15. Antonio Vera Salas, 'Impresiones de un curso de tiro en Valdemoro', *MI*, 7, 2 (1918), p. 442.
16. José Marvá y Mayer, 'Algo sobre la guerra actual', *MI*, 5, Special Issue (1916), pp. 457–8.
17. Epifanio Gascueña, 'El fuego y la maniobra como medios de combate', *MI*, 3, 2 (1914), p. 321.

18. Epifanio Gascueña, 'La Artillería es hoy el Arma hermana de la Infantería', *MI*, 4, 2 (1915), pp. 424–8.
19. Epifanio Gascueña, 'El enlace de las armas con aplicación al estudio táctico de una campaña moderna', *EM*, 35, 2 (1916), p. 119. For a relation of the operations of the Spanish artillery in Morocco, see Guillermo Frontelas Cabrera, 'Campañas de Africa del siglo XX (1909–1927)', in Aurelio Valdés Sánchez (ed.), *Al pie de los cañones: la Artillería española* ([Madrid:] Tabapress, 1994), pp. 437–81.
20. Jorge Vigón, *Historia de la Artillería española*, 3 vols. (Madrid: Consejo Superior de Investigaciones Científicas, 1947), III. 183.
21. Dámaso Berenguer, *Campañas en el Rif y Yebala*, 2 vols. (Madrid: Servicio Histórico Militar, 1948), I. 250–1.
22. Michael Alpert, *La reforma militar de Azaña (1931–1933)* (Madrid: Siglo XXI, 1982), pp. 74–5; Robert A. Doughty, *The Seeds of Disaster: The Development of French Army Doctrine, 1919–1939* (Mechanicsburg: Stackpole Books, 2014), Kindle edition, Chaps. 4 and 5 [originally published: Hamden: Archon Books, 1986].
23. A., 'La doctrina francesa acerca de la Caballería', *Memorial de Caballería* (hereafter *MC*), 5, 2 (1920), p. 175.
24. Enrique Maquieira, 'Ideas francesas sobre la ofensiva', *MI*, 10, 1 (1921), pp. 402–6.
25. A., '¿Potencia de fuego o potencia de maniobra?', *MC*, 5, 1 (1920), p. 421.
26. A., 'Lo ocasional y lo permanente. La aviación y la disimulación', *MC*, 4, 1 (1919), p. 102; A., 'Contra las tendencias modernistas', *MC*, 4, 1 (1919), p. 363.
27. Luis de la Gándara, 'El material y el armamento de la infantería en la guerra mundial', *MI*, 10, 2 (1921), pp. 106–9, 177–8.
28. [Juan] García [Gómez-] Caminero, *De la guerra* (Cadiz: Revista Médica, [1925]).
29. [Francisco] Rodríguez Urbano, *Polémica sobre el combate (ensayo de una organización militar)* (Barcelona: Araluce, 1933), p. 156. It is worth noting that this kind of thinking still influenced attitudes in the interwar American army and shaped Italian army doctrine in the 1930s: Brian McAllister Linn, *The Echo of Battle. The Army's Way of War* (Cambridge, MA: Harvard University Press, 2007), pp. 129–32, 139–40; McGregor Knox, *Hitler's Italian Allies: Royal Armed Forces, Fascist Regime, and the War of 1940–43* (Cambridge: Cambridge University Press, 2000), pp. 51–2; Amedeo Osti Guerrazzi, *The Italian Army in Slovenia: Strategies of Antipartisan Repression, 1941–1943* (New York: Palgrave Macmillan, 2013), pp. 9–10.
30. Abd el-Krim tried to organize a core of regular troops for his army, helped by some European mercenary advisors: [Manuel] Goded, *Marruecos. Las etapas de la pacificación* (Madrid: Compañía Ibero-Americana de

Publicaciones, 1932), pp. 93–103; David S. Woolman, *Abd el-Krim y la guerra del Rif* (Barcelona: Oikos-Tau, 1971), pp. 165–71 [original title: *Rebels in the Rif: Abd el Krim and the Rif Rebellion*].

31. José Díaz de Villegas, *Lecciones de la experiencia. (Enseñanzas de las campañas de Marruecos)* (Toledo: Colección Bibliográfica Militar, 1930), p. 53. Díaz de Villegas was a staff officer in the headquarters of the Army of Operations in Morocco during the campaigns of 1926–27.
32. Goded, *Marruecos*, pp. 66–8; Dámaso Berenguer, *Campañas en el Rif y Yebala*, I. 251–2; Vigón, *Artillería española*, III. 188.
33. Estado Mayor Central del Ejército, *Doctrina para el empleo táctico de las Armas y los Servicios* (Madrid: n.p., 1924), p. 3.
34. The employment of the artillery in the campaigns of 1925–1926 was essentially in accordance with the 1925 field service regulations: see Díaz de Villegas, *Lecciones de la experiencia*, pp. 79–89.
35. Goded, *Marruecos*, pp. 41–72.
36. For the French army in the Rif War, see Auguste Laure, *La victoire franco-espagnole dans le Rif* (Paris: Plon, 1927). Short overviews in English in Doughty, *Seeds of Disaster*, Chap. 4; Anthony Clayton, *France, Soldiers and Africa* (Oxford: Brassey's, 1988), pp. 106–11.
37. Nonetheless, some French officers expressed discontent about '*une théorie aussi materialiste*' that could be detrimental of tactical prowess: Clayton, *France, Soldiers and Africa*, p. 111. On the limitations of Petain's approach, see also Jan Pascal, 'L'Armée française face à Abdelkrim ou la tentation de mener une guerre conventionnelle dans une guerre irrégulière 1924–1927', *Stratégique*, 93, 94, 95, and 96 (2009), pp. 319–38; Frédéric Danigo, 'France and the Rif War: lessons from a forgotten counterinsurgency war. (Northern Morocco—April 1925–May 1927)' (student research paper, United States Marine Corps Command and Staff College, 2010).
38. Accounts of this campaign in Goded, *Marruecos*, Chaps. 4 and 5; [Andrés] Sánchez Pérez, *La acción decisiva contra Abd el-Krim. Operaciones en el Rif central en colaboración con el Ejército francés* (Toledo: Colección Bibliográfica Militar, 1930); Servicio Histórico Militar, *Historia de las campañas de Marruecos*, 4 vols. (Madrid: Servicio Histórico Militar, 1947–1981), IV. 112–39.
39. Sánchez Pérez, *La acción decisiva*, p. 146.
40. Sánchez Pérez, *La acción decisiva*, p. 146.
41. Doughty, *Seeds of Disaster*, Chap. 4. See, for instance, Lieutenant Colonel Gemeau, 'L'emploi des feux dans la guerre du Rif', *Revue militaire française*, 95, 67 (1927), pp. 116–17.
42. Goded, *Marruecos*, pp. 45–7.

43. Geoffrey Jensen, *Cultura militar española. Modernistas, tradicionalistas y liberales* (Madrid: Biblioteca Nueva, 2014), p. 197.
44. Alpert has brought out an alleged interest in German army organization and doctrine in the period 1928–1931, but his argument is based on the contributions to professional journals of Lieutenant Colonel Juan Beigbeder, the military attaché in Berlin: Alpert, *Reforma militar*, pp. 69–71. Certainly Beigbeder took a keen interest in increasing the contacts with the German army and urged the Spanish military authorities to send hand-picked officers to Germany. They would get first-hand knowledge and experience of German doctrine in artillery, infantry, and gas warfare. Beigbeder argued that the German military had to work under conditions of under-equipment (imposed by the Versailles Treaty), which were overcome through efforts of imagination, hard work, and study. Thus the Spanish army could learn from the German ways in order to improve its professional proficiency despite its traditional shortage of resources. But, apart from the letters sent by Beigbeder to the War Ministry, there is no further evidence available pointing to a widespread desire for closer cooperation with the German military. Beigbeder's letters are filed in Archivo General Militar de Segovia, 2ª Sección, División 3ª, Legajo 55 (hereafter AGMS No. Sección/No. División/No. Legajo).
45. AGMS 2/8/535, Doctrina Táctica. Informe del Cor de EM Jefe de la 2ª [sic] Sección que llevará a la Ponencia designada por la Junta de Jefes del Centro, n.d.
46. The General Staff was reorganized on 21 February 1923: Ministerio de la Guerra, *Colección Legislativa del Ejército 1923* (Madrid: n.p., 1923), Item No. 67.
47. AGMS 2/8/535, Doctrina Táctica. Informe del Cor de EM Jefe de la 2ª [sic] Sección que llevará a la Ponencia designada por la Junta de Jefes del Centro, pp. i–ii, n.d. Curiously, the artillery corps was displeased by the French regulations of 1921. An editorial review in the artillery journal pointed out that, although giving great preponderance to the artillery, they stressed too much the importance of the aircraft, infantry heavy weapons, and the tank: 'L'Instruction provisoire sur l'emploie tactique des grandes unités', *Memorial de Artillería*, 79, 2 (1922), p. 511.
48. AGMS 2/8/535, Doctrina Táctica. Informe del Cor de EM Jefe de la 2ª [sic] Sección que llevará a la Ponencia designada por la Junta de Jefes del Centro, pp. xiv, xviii–xix, n.d.
49. R[omán] López Muñiz, *Los procedimientos tácticos vigentes en la actualidad (ensayo de táctica comparada)* (Toledo: Colección Bibliográfica Militar, 1929), p. 6. This would not have been unique: the American army's 1923 Provisional Manual of Tactics for Large Units was a direct translation of the French 1921 Provisional Instructions: House, *Toward Combined Arms Warfare*, p. 71.

50. José Ungría, 'Los estudios tácticos en la Escuela de Guerra de París', *La Guerra y su preparación*, 10, 1 (1925), p. 518.
51. Estado Mayor Central del Ejército, *Reglamento para el empleo táctico de las Grandes Unidades* (Madrid: n.p., 1925) (hereafter RGU 1925), Article 111.
52. RGU 1925, Article 111.
53. RGU 1925, Article 136.
54. Emilio Pardo, 'La doctrina militar española después de la guerra de 1914–1918', *MI*, 25, 1 (1936), p. 96.
55. RGU 1925, Article 106.
56. Dirección General de Preparación de Campaña, *Reglamento Táctico de Infantería*, 2 vols. (Madrid: n.p., 1926), II. Article 431.
57. López Muñiz, *Los procedimientos tácticos*, pp. 11–12.
58. Pardo, 'La doctrina militar española', p. 26.
59. The idea that modern combat after the First World War was to turn into a slow and methodical affair, due to the extended use of strong field fortifications and massive firepower (especially artillery fire), was not totally unknown in the Spanish professional literature: see Carlos García Nieto, 'Comentarios tácticos', *MI*, 10, 1 (1921), pp. 146–7.
60. Doughty, *Seeds of Disaster*, Chaps. 4 and 5.
61. RGU 1925, Article 29.
62. RGU 1925, Article 32.
63. López Muñiz, *Los procedimientos tácticos*, pp. 53–8.
64. RGU 1925, Article 48.
65. RGU 1925, Article 48.
66. Enrique Ruíz-Fornells, 'Datos para la decisión', *MI*, 24, 1 (1935), p. 180.
67. R[omán] López Muñiz, *Comentarios sobre doctrina, organización y procedimientos tácticos. (Notas orientadas al estudio del problema en nuestro Ejército)* (Toledo: Colección Bibliográfica Militar, 1934), pp. 140–2, 155–6.
68. López Muñiz, *Comentarios sobre doctrina*, p. 146.
69. Doughty, *Seeds of Disaster*, Chap. 5.
70. [Ángel Rodríguez del Barrio,] *Grandes Maniobras del Pisuerga. Otoño de 1932. Memoria* (Madrid: n.p., 1932), p. 8.
71. [Rodríguez del Barrio,] *Grandes Maniobras del Pisuerga*, pp. 106–7.
72. RGU 1925, Article 25.
73. Even the mere listing of bibliography on the Civil War would take a whole book, so the present author will limit himself to citing a few works which may be useful for those readers who may not be too well-versed in the military events of the conflict. Antony Beevor, *The Battle for Spain: The Spanish Civil War 1936–1939* (London: Weidenfeld & Nicolson, 2006) has become the current standard history of the war

in English. Jorge Martínez Reverte, *El arte de matar. Cómo se hizo la Guerra Civil española* (Barcelona: RBA, 2009), and Jesús María Salas Larrazábal and Ramón Salas Larrazábal, *Historia general de la guerra de España* (Valladolid: AF Editores, 2006) are two informative, single-volume military accounts in Spanish; the second one is actually a book version of Jesús María Salas Larrazábal and Ramón Salas Larrazábal, 'La Guerra', in [Luis Suárez Fernández, Demetrio Ramos Pérez, José Luis Comellas, and José Andrés-Gallego (eds),] *Historia General de España y América*, 25 vols. (Madrid: Rialp, 1981–1992), XVII. 263–680. Further detail can be found in the volumes of the series written by José Manuel Martínez Bande, some of which are cited in other chapter notes. Some of the controversies of the war are concisely presented in Kenneth W. Estes and Daniel Kowalsky (eds), *History in Dispute. Volume 18: The Spanish Civil War* (Detroit: St. James Press, 2005).

74. G[regorio] López Muñiz (ed.), *Diccionario enciclopédico de la guerra*, 13 vols. (Madrid: Gesta, 1954–1958), *sv* 'Fortificación'; Martínez Reverte, *El arte de matar*, p. 301; Michael Seidman, *The Victorious Counterrevolution: the Nationalist Effort in the Spanish Civil War* (Madison: University of Wisconsin Press, 2011), p. 65.
75. For an overview of the Spanish artillery during the Civil War, see Carlos Medina Ávila, 'La Guerra Civil', in Valdés (ed.). *Al pie de los cañones*, pp. 483–519.
76. Archivo General Militar de Ávila, Fondos del Ejército Popular, Armario 55, Legajo 545, carpeta 8 (hereafter AGMA-EP No. Armario/No. Legajo/No. carpeta), Observaciones sobre las operaciones de Julio, 10 August 1937.
77. The Republican offensive manoeuvre in Teruel had achieved its main tactical goals by 18 December 1937 and seizing the city itself was a mopping-up operation (although it lasted longer than expected because of stout Nationalist resistance).
78. Peter Kemp, *Legionario en España* (Barcelona: Luis de Caralt, 1975), p. 207 [original title: *Mine Were of Trouble*].
79. José Manuel Martínez Bande, *La llegada al mar* (Madrid: San Martín, 1975), p. 61.
80. Michael Seidman, *A ras de suelo. Historia social de la República en la Guerra Civil* (Madrid: Alianza, 2003), p. 352 [original title: *Republic of Egos: A Social History of the Spanish Civil War*]; Seidman, *The Victorious Counterrevolution*, p. 64; Stanley G. Payne, *The Spanish Civil War* (Cambridge: Cambridge University Press, 2012), p. 86; Carlos Engel, *Estrategia y táctica de la guerra de España, 1936–1939* (Madrid: Almena, 2008), p. 393.

81. Engel, *Estrategia y táctica*, p. 395. Juan Vigón (1880–1955) was commissioned in the engineer corps, passed the staff college and served in Morocco; a monarchist, he left the army in 1933 and joined the Nationalist side in 1936; after the war he was air minister and chief of the inter-service general staff (actually an advisory body).
82. The term is taken from Jonathan Boff, *Winning and losing on the Western Front: the British Third Army and the Defeat of Germany in 1918* (Cambridge: Cambridge University Press, 2014), p. 37.
83. Even so, during the battle of Teruel the Nationalists eliminated the Republican salient of Sierra Palomera through a double envelopment in the so-called 'battle of the Alfambra River' (5–7 February 1938).
84. López Muñiz (ed.), *Diccionario enciclopédico*, *sv* 'División'; [Arturo Sus Sus (ed.),] *Historia de las fuerzas armadas*, 5 vols. (Saragossa: Palafox, 1983), II. 125.
85. Carlos Martínez de Campos, *Apuntes sobre empleo de Artillería* (Madrid: Ediciones Ejército, 1942), p. 129. Martínez de Campos' contribution, like Vigon's, to the Nationalist military victory has remained rather overlooked.
86. Klaus A. Maier, 'The Condor Legion: An Instrument of Total War?', in Chickering and Förster (eds), *Shadows of Total War*, p. 287; Rainer Waelde, 'The Experience of the Japanese–Chinese War and of the Spanish Civil War for the Development of the German "Blitzkrieg Doctrine" and its Lessons for the Transformation Process' (School of Advanced Military Studies monograph, United States Army Command and General Staff College, 2003), p. 30. On the development of air support in the Nationalist side, see Richard P. Hallion, *Strike from the Sky: the History of Battlefield Air Attack, 1911–1945* (Washington, DC: Smithsonian Institution Press, 1989), pp. 102–8; Michael Alpert, 'The Clash of Spanish Armies: Contrasting Ways of War in Spain, 1936–1939', *War in History*, 6 (1999), pp. 341–3.
87. For a small sample of literature on the interwar development of air support, see Hallion, *Strike from the Sky*, Chaps. 5–9; Richard R. Muller, 'Close Air Support', in Allan R. Millett and Williamson Murray (eds), *Military Innovation in the Interwar Period* (Cambridge: Cambridge University Press, 1996), pp. 144–90; James S. Corum, 'The Luftwaffe's Army Support Doctrine, 1918–1941', *Journal of Military History*, 59 (1995), pp. 53–76.
88. Carlos Martínez de Campos y Serrano, *Ayer*, 2 vols. (Madrid: Instituto de Estudios Políticos, 1946–1970), II. 78–9. For an account of the effects of air attacks on the enemy in these early days, see Herbert Molloy Mason, Jnr., *The Rise of the Luftwaffe: Forging the Secret German Air Weapon 1918–1940* (New York: Ballantine Books, 1975), pp. 230–1.

89. Archivo General Militar de Ávila, Fondos del Ejército Nacional, Armario 15, Legajo 2, carpeta 4 (hereafter AGMA-EN No. Armario/No. Legajo/No. carpeta), Ejército del Norte, Comandancia de Artillería, Adición a la instrucción N° 6, 20 March 1938. The ground attack aircraft usually employed a circular pattern of low-level bombing and strafing upon a target dubbed *cadena* ('chain'), which the Spanish airmen had tried out in a more rudimentary form in Morocco: López Muñiz (ed.), *Diccionario enciclopédico*, *sv* 'Cadena'; Hallion, *Strike from the Sky*, pp. 105–6.
90. AGMA-EN 15/2/4, Ejército del Norte, Comandancia General de Artillería, Instrucción n. 9, 25 March 1938.
91. AGMA-EP 55/545/8, Observaciones al empleo de la artillería en el ataque a Huesca del 12 de Junio 1937, 13 May [sic] 1937; AGMA-EP 55/545/7, Ejército del Centro, Comandancia General de Artillería, Deducciones y enseñanzas de la actuación artillera en las operaciones realizadas por el I Cuerpo de Ejército en los días 30 y 31 de Mayo y la de 5 de Junio en Guadarrama y Navacerrada, 9 June 1937.
92. AGMA-EP 64/796/1, De cómo ganó Yagüe la batalla del Sur Ebro, para ganar después los franquistas la del frente del Este, n.d.
93. On the Republican army's officer corps, see Michael Alpert, *The Republican Army in the Spanish Civil War, 1936–1939* (Cambridge: Cambridge University Press, 2013), Chaps. 5 and 6; José María Gárate Córdoba, *Tenientes en campaña. La improvisación de oficiales en la guerra del 36* (Madrid: San Martín, 1976).
94. Martínez de Campos, *Ayer*, II. 70–1.
95. See, for instance, Manuel Tagüeña Lacorte, *Testimonio de dos guerras* (Barcelona: Planeta, 1978), p. 106.
96. See also Alpert, 'The Popular Army', pp. 100–2.
97. AGMA-EP 55/545/8, Observaciones al empleo de la artillería en el ataque a Huesca del día 16 de Junio de 1937, 17 June 1937; AGMA-EP 65/787/6, Las operaciones de Teruel, 25 February 1938.
98. *Enciclopedia universal ilustrada europeo-americana. Suplemento anual 1936-1939*, 2 vols. (Madrid: Espasa-Calpe, 1944), *sv* 'España' [II. 1477].
99. Alpert, *Republican Army*, pp. 235–7, 248–51, 309–10; Hallion, *Strike from the Sky*, pp. 95–103, 114.
100. AGMA-EP 55/545/8, Observaciones sobre las operaciones de Julio, 10 August 1937.
101. AGMA-EP 55/545/7, Reserva General de Artillería, Informe relativo a la actuación de esta Unidad en las operaciones de los días 22 al 26 de Mayo de 1938, 31 May 1938; Archivo General Militar de Ávila, Fondos del Cuartel General del Generalísimo (N), Armario 7, Legajo 375, carpeta 103 (hereafter AGMA-CGG No. Armario/No. Legajo/No.

carpeta), Resumen de las observaciones hechas por el Jefe que suscribe en relación con las operaciones realizadas por el C. de E. de Castilla durante los días 22 a 28 de Mayo, ambos inclusive, 30 May 1938; AGMA-EN 15/2/8, Ejército del Norte, Estado Mayor, 3ª Sección, Directiva general N° 11 bis, 18 June 1938.

102. AGMA-CGG 7/376/7, Informe del enlace con el Cuerpo de Ejército de Castilla desde el 11 al 18 de Junio de 1938, 28 June 1938.
103. AGMA-EP 55/545/8, Observaciones al empleo de la artillería en el ataque a Huesca del día 16 de Junio de 1937, 17 June 1937; AGMA-EP 55/545/8, Observaciones sobre las operaciones de Julio, 10 August 1937.
104. AGMA-EP 64/778/22, Ejército de Maniobra, Estado Mayor, 3ª Sección, Instrucción reservada num. 27, Observaciones deducidas en el periodo de operaciones ofensivas del 6 al 16 del corriente mes, 19 July 1937; AGMA-EP 55/545/7, Ejército del Centro, Comandancia General de Artillería, Informe que lleva el Jefe que suscribe al Inspector General del Arma en relación a los extremos que abarca la Directiva formulada por la citada Inspección en 29 de Junio del corriente año con motivo de la ofensiva efectuada en el mes de Julio por el Ejército de Maniobra, 5 August 1937; AGMA-EP 55/545/8, Actuación de la artillería sobre la cabeza de puente enemiga de Serós, n.d.
105. Kindelán to Franco, 6 January 1938, in Fundación Nacional Francisco Franco, *Documentos inéditos para la historia del Generalísimo Franco*, 4 vols. (Madrid: Fundación Nacional Francisco Franco, 1992–1994), I. 175; Kindelán to Franco, 16 August 1938, in Fundación Franco, *Documentos inéditos*, I. 194.
106. AGMA-EN 15/2/1, Ejército del Norte, Comandancia General de Artillería to Comandante General de Artillería de los Ejércitos Nacionales, 8 January 1938. The Nationalist infantry's overreliance on the artillery fire had been noticed by the Republican side as well: AGMA-EP 64/778/20, Ejército de Maniobra, Estado Mayor, Sección III, Instrucciones sobre conducta a seguir por los mandos y tropas en caso de fuerte ataque enemigo, 5 January 1938.
107. Seidman, *A ras de suelo*, pp. 285–6.
108. AGMA-EN 15/2/10, Ejército del Norte, Comandancia de Artillería, Instrucción n. 16, Sobre la necesidad de asegurar la precisión del propio tiro, antes de que sea reforzado por el fuego del escalón siguiente, 30 August 1938.
109. AGMA-EN 15/2/12, Ejército del Norte, Comandancia de Artillería to General Jefe del Ejército del Norte, 14 September 1938.
110. Quoted in Jeremy Black, *Avoiding Armageddon. From the Great War to the Fall of France, 1918–40* (London: Bloomsbury, 2012), p. 225.

111. See, for instance, Antony Beevor, *La Guerra Civil Española* (Barcelona: Crítica, 2005), pp. 285, 677 [original title: *The Battle for Spain: The Spanish Civil War 1936–1939*]; Juan Carlos Losada, *Los mitos militares en España. La historia al servicio del poder* (Madrid: Biblioteca Nueva, 2005), p. 195.
112. Tim Gale, *The French Army's Tank Force and Armoured Warfare in the Great War. The Artillerie Spéciale* (Farnham: Ashgate, 2013), Kindle edition, Conclusion.
113. Gary Sheffield, *Forgotten Victory. The First World War: Myths and Realities* (n.p.: Endeavour Press, 2014), Kindle edition, Chap. 10 [originally published: London: Headline, 2001]. This argument is developed in a fuller way in G[ary] D. Sheffield, '*Blitzkrieg* and Attrition: Land Operations in Europe 1919–45', in Colin McInnes and G.D. Sheffield (eds), *Warfare in the Twentieth Century. Theory and Practice* (London: Unwin Hyman, 1988), pp. 51–79.
114. A historian who pulls no punches in exposing the deficiencies of Italy's armed forces accepts that the Italian artillery's centralized direction was more effective in cooperating with other arms than the decentralization of the German gunners in El Alamein, a set-piece 1918-style battle: Knox, *Hitler's Italian Allies*, pp. 154–5.
115. Although it is a subject beyond this book's scope, one finds tempting and almost inevitable to link this kind of thinking to fascism, or at least to non-democratic conservative political ideologies. Major Francisco Rodríguez Urbano complained that, notwithstanding the value of the spirit, 'the power of numbers triumphs in the elections, and this same power prevails, or tries to prevail, in the social struggles which turn more virulent every day'. In his eyes almost every social class seemed to be dominated by a materialistic attitude to life: Rodríguez Urbano, *Polémica sobre el combate*, p. 83. However, Burguete and García Gómez-Caminero had shifted to leftist political stances by the 1930s, so the link may be less inherent than it is usually thought: Jensen, *Cultura militar*, pp. 250–1. And fascism was not an anti-machine movement either: Azar Gat, *A History of Military Thought: From the Enlightenment to the Cold War* (Oxford: Oxford University Press, 2001), pp. 521–4.
116. A foreign observer, recalling the massive losses of the attacking troops in the First World War, noted that the Nationalist infantry 'has seldom experienced the terrible depression resulting from heavy loss in an attack bravely carried forward…': quoted in Alpert, 'The Clash of Spanish Armies', p. 341.

CHAPTER 8

From the Organic Division to the Mixed Brigade

The organization of military units and its evolution is an underdeveloped subject in military history. It is often studied in combination with change imposed by technological innovation, but technology is not the only factor influencing the way armies organize their fighting units, large and small. Their unit organizations may also reflect their conception of warfare (usually articulated in their military doctrine) or strategic considerations. This chapter will survey the evolution of the infantry (or 'organic') division in the Spanish army from 1914 to 1931, the alternative proposals (of which the mixed brigade was the most advocated one in the professional literature) and the development of the basic major units in each side during the Civil War. Much attention will be paid to the mixed brigade because the Republican army based its force structure on it. So far the present author is aware there is no published study on this major unit spanning both its development before 1936—within the context of contemporary Spanish military thinking—and the consequences of its adoption during the Civil War. Filling such gap is a purpose of this chapter.

Towards a New Model of Major Unit

The Infantry Division in the Spanish Army, 1914–1931

Spanish military doctrine in the mid-1920s defined a 'major unit' (*gran unidad*) as 'the assembling, under a single command, of troops of all arms together with their services'.[1] The Spanish army's major unit *par*

J.V. Herrero Pérez, *The Spanish Military and Warfare from 1899 to the Civil War*, DOI 10.1007/978-3-319-54747-3_8

excellence at the start of the twentieth century was the infantry division. By 1914 its theoretical organization was roughly similar in combat troops to that of the French infantry division. Both had four infantry regiments (of three battalions each), grouped into two brigades, as well as one field artillery regiment, one or two engineer companies, and service support units. The main differences were that the Spanish division had one organic cavalry regiment (the French division had only one squadron) and also was stronger in engineer and service support units. This was because the army corps was the main operational unit in France, whereas the division was considered in the Spanish army as a self-contained formation with a capacity to operate on its own.[2] The experience of the First World War (and manpower shortages) led most belligerent armies to reduce the number of divisional infantry units, which increased the proportion of artillery to infantry: by 1918 a typical division had nine infantry battalions, divided into three regiments or brigades.[3] This triangular organization, known as *ternaria* in the Spanish military vocabulary, was to be the norm after the end of the conflict.

The 1918 army reform act introduced a new organization of the Spanish infantry division. This major unit—the main one in peacetime and whose organization would remain unchanged in case of war—was called *división orgánica* (organic or established division). Its organization included two infantry brigades (two regiments each), one field artillery brigade (one light and one heavy regiments), one sapper battalion, one signals (telegraph) company, one lighting (searchlight) platoon, and service troops; when necessary for training or war operations, the organic division would be joined by cavalry troops from the cavalry divisions. There would be 16 organic divisions in peacetime.[4] It is clear that the Spanish army followed the foreign experience to only a limited extent. The strength of the divisional artillery was increased hugely, but there was no cut in infantry troops. Moreover, the infantry's organization at the divisional and brigade levels remained 'binary' (*binaria*)—that is, based on the pairing of two manoeuvre units.[5] This kind of organization has the disadvantage of limiting the commander's capacity to intervene in the conduct of an operation through the employment of a third (or reserve) manoeuvre element. If both organic manoeuvre elements are deployed on the first line, the commander lacks a reserve or must improvise one with troops detached from the organic elements (thus weakening the parent units).[6]

An issue discussed in the professional literature was whether the 1918 divisional organization was too cumbersome and inflexible for contemporary warfare in Spanish territory. Considering the experience of the First World War, an article in the cavalry journal argued in 1919 that the weaker army in a conflict had to combine the use of fortified zones fixing the enemy with the manoeuvre of the field forces in order not to be enticed into an early decisive battle. Given her limited military power, Spain had to set up fortified zones along her borders (after the French model) to channel enemy offensives and organize her army in small, very mobile, strategic formations (not larger than a division).[7] This line of reasoning was in line with the views set forth a few years later by the working group on the reform of the field service regulations. The group reckoned that Spain, unless she went to war beside powerful allies, could only fight to defend her own territory. Such a defence would be based on Spain's mountainous land border. Given the features of the ground, which made the maintenance of extended fortified fronts impossible, a repetition of the trenches of the Western Front was impractical. On the contrary, there was a need for a mobile army, capable of driving back the enemy advance, which would proceed from a limited number of mountain passes. Within this framework, the working group's report noted that the establishment of animal transport for the 1918 division was set at more than 6000 horses and mules; according to the working group, such a figure was revealing of the large amounts of divisional equipment and the difficulties that the latter presented to the division's mobility.[8]

Despite this view, the official policy did not introduce major changes in the organization of the infantry division during the 1920s. RGU 1925 defined the organic division as the major tactical unit and the basis of the organization of the field forces. It was the only major unit organized in peacetime, since formations above the divisional level would be organized, in general, at the start of the conflict.[9] RGU 1925 also followed in essence the divisional model of 1918. The organization of the 1925 *división orgánica* included two infantry brigades (two three-battalion regiments each), one cavalry squadron (detached from an army corps cavalry regiment in wartime), one light field artillery brigade (one gun and one howitzer three-battalion regiments), one sapper battalion, one signals group (including a wireless platoon), one aerostatic platoon (one balloon), and one air observation flight.[10]

The French influence certainly weighed to some extent on the divisional organization. But the French infantry division, although strong in artillery, was a leaner unit due to the triangular organization of its infantry. Nonetheless, the French non-motorized infantry division of 1929–1930 was neither very supple nor very mobile.[11] Therefore, the Spanish 1925 division, a heavier unit of 12 infantry battalions and six artillery *grupos* (battalions), was even less adequate for manoeuvre warfare. Thus there was a paradoxical situation. Although enthusiasm for doctrines envisaging continuous fronts to be overcome through firepower-based procedures does not seem to have been widespread, the Spanish army kept a divisional organization which fitted better with that kind of warfare. The present author's research has found no evidence regarding the reasons for keeping the binary organization, but there are records suggesting that those responsible for developing new military doctrine disliked the idea of cutting the infantry strength within the division.[12] Perhaps another reason could be a desire to provide the division with strong reserves of its own.[13]

A decree dated 8 September 1926 disbanded the divisional artillery brigades and put all the artillery regiments under the command of the military regions where they were stationed; the generals commanding the regions would assign artillery units to divisions and brigades on a temporary or permanent basis at their discretion. The decree argued that the artillery brigade was too cumbersome.[14] It was an odd explanation, since RGU 1925 saw no problem with the brigade the previous year. Moreover, the decision came few days after the artillery officers had been suspended without salary due to their opposition to Primo de Rivera's promotion policy.[15] And given that another decree reduced the field artillery to 16 light and eight 'foot' (heavy) regiments in November, there are grounds to conclude that this reorganization of the divisional artillery had little to do with military doctrine and was much of a punishment of the artillery officers' disaffection from the regime.[16] Another cut in the strength of the artillery corps (the light field regiments were reduced to eight) after the attempted coup d'état in January 1929 provides further evidence.[17] These events are another example of how the politics of the Spanish army could eventually harm its own professional effectiveness.

The reforms of Azaña restored the divisional model of 1925. The 25 May 1931 decree on the peacetime organization of the army acknowledged that the triangular division prevailed by the end of the Great War, but it later argued that the triangular model's drawbacks were prompting abroad proposals to bring back the older organization. Given that the issue was still undecided, the four-regiment organization, which provided the division with a greater operational capability, had been retained.[18] The 1931 organic division included two infantry brigades (two infantry regiments each), one cavalry squadron plus one machine gun and one cyclist infantry platoons (all detached from army corps units), one light artillery brigade (one gun and one howitzer regiments), one sapper-miner battalion, one signals group (four platoon-strong), one lighting (searchlight) platoon, one aviation flight and one balloon unit.[19]

But the lack of major changes in 1931 does not mean that there was uncritical acceptance of the current model of division. The organization of the infantry division after the 1925 regulations was reckoned unsuitable for the Spanish theatres of operations by a number of essayists, who advocated more flexible models of major unit. One line of thinking argued for the introduction of the triangular division. Major Martínez de Campos proposed divisions of nine infantry battalions and a proportionate scale of artillery; these smaller divisions would give the army corps more operational flexibility.[20] Captain Manuel Vicario considered the infantry brigade echelon redundant in the contemporary infantry division.[21] But the triangular division had other advantages, besides the operational ones. Cebreiros expressed his amazement because Azaña, who liked imitating French military models, had not substituted the triangular division for the Spanish binary organization. He argued that a triangular organization of the infantry, based on smaller peacetime battalions (three active companies), would allow Spain to keep 16 standing divisions (432 companies or 144 battalions) with only a slight increase on the actual strength of the eight divisions and the mountain units (360 companies making 64 line and 8 mountain battalions) envisaged by Azaña's reform. Thus it would not have been necessary to bring about the premature retirement of many infantry officers. For Cebreiros, all this was explained by Azaña's desire to expel politically untrustworthy personnel.[22]

The Mixed Brigade

Most reformist essayists, however, advocated the so-called *brigada mixta* (mixed brigade) as the basic major unit. In point of fact, from the beginning of the twentieth century the Spanish army's force structure included formations which matched the concept. The 1902 army yearbook enters a *cazadores* half-brigade, stationed in southern Andalusia, with three light and one mountain infantry battalions, plus one artillery battery and an element of cavalry (both detached from regiments).[23] By early 1904 this unit had been upgraded to brigade, with two half-brigades (two *cazadores* and one mountain infantry battalions each), one mountain artillery battalion, and one cavalry squadron (detached).[24] Following the reorganization of the army later in this year, the brigade (now titled the 2nd *Cazadores* Brigade) also included one engineer company, one signals and one quartermaster platoons, and one ambulance unit (and its infantry were all *cazadores* battalions).[25] However, the cavalry, engineer, signals and quartermaster elements were sub-units from other commands, which cast some doubt about their full organic character as brigade troops.

The three *cazadores* brigades were sent to Melilla for the 1909 campaign and, through attachments, they all became combined arms formations with two half-brigade headquarters, six infantry battalions, two infantry machine gun platoons, one mountain artillery battalion, one cavalry squadron, one engineer group (one sapper and one signals companies), one quartermaster company, and one ambulance unit.[26] This organization did not last long: the campaigns in Morocco fostered the use of ad hoc columns grouping temporarily the elements needed for an operation and the reshuffle of units. The *cazadores* brigades (two still in Morocco and one back in the peninsula) were no longer all-arms formations by 1916 and were disbanded after the 1918 reform.[27]

Nonetheless, disbanding these brigades did not have any serious consequence. As Martínez de Campos remarked in a lecture at the staff college in May 1933, the mixed brigade was no more than the formal organization of the traditional combined arms column. The latter had become the main operational unit of the Spanish army during its recent campaigns, which saw at their start the dismemberment of larger and heavier formations, such as the division.[28] This procedure bore some resemblance to the German army's contemporary concept of *kampfgruppe* (battle group). A *kampfgruppe* was a combined arms task force organized within the division in order to carry out a specific mission. It was the organizational expression of the German doctrine, which imbued commanders at all levels with the need for decentralization and initiative on the battlefield. This also meant that the manoeuvre elements should be accompanied by their own support artillery, which would be able to attend immediately to their calls.[29] The Spanish mixed brigade provided full combat support to a group of infantry battalions (the manoeuvre element) as well; the difference lay in the fact that manoeuvre and support elements were grouped in a standing formation, whereas the *kampfgruppe* was formed ad hoc with divisional resources.

In the same year that RGU 1925 were issued Lieutenant Colonel Vicente Rodríguez Carril, of the artillery corps, delivered a lecture in which he described the disadvantages of keeping single-arm brigades within the division. The infantry brigades introduced a rather unnecessary partition of the divisional front, which complicated the liaison between artillery and infantry, while the artillery brigade fostered in this arm a sense of separation to the detriment of the infantry's needs. The solution was to create self-sufficient mixed brigades as organic combined arms units within the division.[30] A staff corps officer, Lieutenant Colonel Carlos Noreña, pointed out in 1933 that the mixed brigade gave a greater flexibility to the binary division. This was very useful in a country like Spain, whose rugged territory was unfavourable for the movement of large major units. Therefore, the smaller ones should keep a certain capacity to operate with autonomy. The mixed brigade also made training in liaison and cooperation between arms easier. Finally, the mixed brigade allowed the division to assume the functions of the army corps as an operational manoeuvre unit (capable of undertaking actions of some amplitude in time and space with homogeneous units), an arrangement which fitted better with the conditions of the Spanish army.[31]

The reorganization of 1931 introduced the mixed brigade in the Spanish army again (in the form of a specialist unit). The 25 May decree included the organization of two brand new mixed mountain infantry brigades. Each brigade was made up of two half-brigades (two mountain infantry battalions each), one mountain artillery regiment (two howitzer battalions), one sapper company, one signals group and service units.[32] The idea of the mixed brigade being most suitable for many theatres of operations in Spanish territory had gained strength by the mid-1930s and was reflected in the expansion of formations for mountain warfare during Gil Robles' term of office as war minister.[33] A new mixed mountain brigade was created in September 1935; the brigade included two infantry half-brigades (two battalions each), at least one mountain artillery group, and one mixed sapper and signals company.[34] At the same time, two organic divisions were to be fully transformed into mountain formations (so it seems logical that their brigades became mixed brigades); another two divisions would turn one of their infantry brigades, one howitzer battalion, and a corresponding slice of service troops into mountain units.[35] According to Ramón Salas Larrazábal (who unfortunately does not mention his source), one of the army's general inspectorates proposed the theoretical organization for a mixed brigade in 1934. It had two infantry regiments, one artillery regiment, one group of cavalry squadrons, one engineer group, one signals company and service units; the brigade's strength was about 12,500 troops.[36] However, if compared with the contemporary models outside Spain, this unit can be classified rather as a light division.[37]

It is clear, after the above evidence, that by the mid-1930s the Spanish army was shaping a force structure in which new self-contained major units smaller than the *divisiones orgánicas* loomed large. If the reorganizations of 1935 had been implemented, almost a half of the standing infantry brigades would have been based on the mixed model by the late 1930s. Such a development went against the 1925 regulations, which did not contemplate combined arms tactics below the divisional level. Actually, they stated that the divisional artillery commander was also the commanding officer of the division's artillery units.[38] This was at odds with the organization of mixed brigades, which required decentralizing the command of divisional artillery in order to allow the brigades to operate with a degree of independence.

Major Units in the Civil War

The Republican New Model Army and the Mixed Brigade

The militia columns that made up the bulk of the government forces after the breaking up of many loyal army units suffered many defeats in the summer of 1936. Their failure convinced the Republican cabinet to organize a new regular army: the *Ejército Popular de la República* (EPR). This army was to be innovative in its organization by choosing the mixed brigade as the basic major unit. Since it was a model of unit discussed by Spanish military essayists and partly adopted for the army force structure before July 1936, the Republican officers in charge of the organization of the new army needed little, if any, encouragement and guidance from foreign advisers to choose this major unit model.[39] Moreover, it is plausible that the Republican command thought that combined arms brigades could be ready for combat sooner than conventional divisions, at a time when the Nationalist columns of the Army of Africa were heading fast towards Madrid. In any case the mixed brigade became the basis of the Republican army's force structure.[40]

The decision to create mixed brigades was seemingly made in a cabinet meeting on 27 September 1936 and formalized in a restricted decree dated next day. Until 18 October, there is no mention of the mixed brigades in military official records.[41] The Republican mixed brigade had originally four infantry battalions, one cavalry squadron, one field artillery battalion, one mixed (sappers and signals) engineer company and service units; its original strength was about 3800 troops. This establishment underwent several changes during the war. The most significant one came in June 1937, when there was a cut in the strength of the brigade artillery (reduced to one single three-piece battery) and service troops.[42] The cause of this cut lay in the insufficient capacity of the Republican side to satisfy the equipment needs of its fast-grown army. Other changes were the split of the engineer element into one sapper and one signals companies, the substitution of armoured cars for horses in one platoon of the cavalry squadron, and the downgrading of the medical *grupo* (small battalion) to company. On the other hand, the strength of the infantry battalion increased steadily (from 633 to 828 officers and men). In its final form (October 1938) the mixed brigade had one accompanying battery and a new depot company (training and holding unit for replacements); its cavalry was reduced to one platoon or replaced by one armoured car platoon (it was envisaged that one brigade per division would have armoured cars). The establishment of the mixed brigade during the war fluctuated between 3500 and 4200 troops.

The course of the conflict made it necessary to expand the Republican army force structure to include divisions, army corps, field armies and army groups. The Republican division was born on 31 December 1936 and, besides the brigades (usually three), would eventually include one artillery battalion, one engineer company (later suppressed), a few service units and (later in the war) one machine gun battalion; a three-brigade division became officially over 14,000-strong. The structure of army corps and field armies was implemented during the spring of 1937 and the Republican side had (on paper) four armies, one army-level territorial command, 17 army corps, 63 divisions and 205 mixed brigades by the early summer of 1937. Over a year and quite a few defeats later, in October 1938 the Republican order of battle still included two army groups, six armies, 23 army corps, 70 divisions and 200 brigades. Such an organizational effort was noteworthy. In fact, turning the hotchpotch of political militias and the rump pre-war military forces into a regular army capable of operating with some professional efficiency says a lot about the career officers who led such process. Nonetheless, the adoption of the mixed brigade did not provide the EPR with any advantage over its opponent. Why did not the performance of the mixed brigade come up to the pre-war expectations?

The material resources at the disposal of the Republican army are part of the answer. This study does not intend to join the controversies about how much ordnance was supplied to the Republican side during the war.[43] But whatever the answer to such a question, the sources consulted by the present author show that the actual strength of the mixed brigades was usually below their table of organization and equipment and, in particular, that their firepower decrease gradually during the war. This conclusion partly disagrees with Ramón Salas Larrazábal, who argues that the EPR forces were usually much stronger than the Republicans' memoirs and accounts, and many later histories of the war, admit.[44] His point is not groundless, but he sometimes goes too far and overestimates the actual fighting strength of the EPR. For instance, he resorts to the official establishment of the Republican division and points out that this was stronger in manpower than the Nationalist one after late 1937. Thus, whenever both sides deployed the same number of divisions on the eve of a campaign or battle, the Republicans should in theory have had an advantage in strength. And Salas Larrazábal also seems to assume that the Republicans were always able to field fully equipped units.[45] However, this argument is contradicted by his own criticism (also shared by Michael Alpert) of the EPR high command's penchant for organizing large numbers of units in a hurry, in which he states that the Nationalist army did not send new units to the front until they had a proper establishment of men and weapons.[46]

The shortages in weaponry were perhaps more significant than those in manpower. The original mixed brigade had a theoretical scale of 108 light and 36 heavy machine guns. In late 1937 the scale of automatic weapons was reduced to 98 light and 32 heavy machine guns. And the figures became just 24 light and 24 heavy machine guns by the summer of 1938.[47] This last scale seems like a way of making a virtue of necessity, because it comes near some brigades' actual overall figures. For instance, on the eve of the battle of the Ebro, after a period of refitting, the brigades of the 35th Division (a seasoned unit which was to spearhead the Republican offensive) had less than 50 light machine guns and a dozen heavy machine

Table 8.1 Manpower, rifles and machine guns available in the mixed brigades of the Republican 35th Division, July 1938

Brigade	*Strength*	*Rifles*	*Light machine guns*	*Heavy machine guns*
11th	3373	1764	45	12
13th	3046	1822	49	4
15th	3233	1782	49	7

The division included a machine gun battalion with 14 light and 34 heavy machine guns
Source José Manuel Martínez Bande, ***La batalla del Ebro*** (Madrid: San Martín, 1978), p. 84 footnote 103

guns at best (see the exact figures in Table 8.1). In short, the mixed brigades lost much machine gun firepower (if they ever had the original scales) as the war went on. Something rather similar can be said about rifles. The original rifle scale for the Republican mixed brigade was 2897 rifles; the figure was later cut to 2413 (June 1937) and then increased to 2969 (late 1937). Finally, the scale of the summer of 1938 was 1860 rifles.[48] The increase in late 1937 may reflect an increase in the Republican rifle stocks, but this situation must obviously have been temporary, given the sharp fall in the 1938 scale (whose numbers come interestingly near those of the 35th Division's brigades on the eve of the battle of the Ebro—see the exact figures in Table 8.1).

A post-war source (which is not partial to the Republican cause) states that the theoretical scale of rifles for the mixed brigade was 2200 rifles (the figure seems to refer to those allocated to the infantry battalions), but in practice the usual numbers available in many Republican brigades were from 1600 to 1800.[49] This would mean that the average Republican brigade often could not field more than four–fifths of its theoretical scale of rifles. Therefore, it seems plausible to presume that the EPR mixed brigades (probably excepting those which were reckoned as crack units) would often be under strength in terms of firepower. Nonetheless, there could be significant differences between fronts.[50] The evidence the present author has come across suggests that the subject of the relationship between available weaponry and battlefield effectiveness surely needs further research and more refined analysis.

The limitations of the Republican officer corps also affected the performance of the mixed brigade. This was based on a sound, self-contained organization, but it also required a large establishment of trained staff personnel for its headquarters, which should take charge of issues previously managed at the divisional level. As Alpert points out, the mistake of the Republican organizers was that they had created sophisticated units without realizing that the EPR lacked enough officers able to command them effectively.[51] This flaw did not go unnoticed, nor was it the only one. The mixed brigade seemed to have outsized service support elements as well.[52] And, as was shown in Chap. 7, the existence of organic field artillery in the brigades contributed to the ineffectiveness of the arm. A report of June 1937 concluded that the brigade should not have organic batteries with standard field cannon and that grouping the artillery in divisional battalions was a better arrangement.[53] Lieutenant Colonel Morales' report after the battle of Teruel reckoned the performance of the mixed brigade unsatisfactory and proposed a return to the conventional three-battalion infantry regiment, without other arms' support units.[54] Nonetheless, in order to put the criticisms about the brigade artillery and services in perspective, the reader must bear in mind that many brigades—probably most—lacked much of their service elements and never had any field cannon at all.

The introduction of the mixed brigade also had effects on the doctrine for the tactical use of the division. As was shown above, Lieutenant Colonel Noreña had pointed out before the war that the mixed brigade meant that the division commander would conduct the operations of self-sufficient subordinate formations, instead of combining the efforts of separate arm units. The EPR indeed followed this guiding principle. Its instructions (undated but surely issued after early 1937) on the organization and combat of a division defined the latter as a major unit which grouped several brigades and coordinated their combat. In order to focus the divisional command on this operational role the division was devoid of logistical units of its own (excepting the medical services), since motor transport made it possible to supply the brigades from army corps or army depots. Therefore the Republican new model division only had the headquarters, two or three brigades, the medical echelon, and attached artillery.[55]

This organization turned the Republican divisions into highly flexible formations, which could exchange, detach or be given mixed brigades according to the tactical situation without overloading the divisional command with logistical issues. To illustrate this flexibility, Table 8.2 offers a sample of four Republican divisions and their orders of battle in different moments of the war, starting from their date of creation

Table 8.2 Orders of battle of four Republican divisions during the Civil War

Division	*Date*	*Mixed Brigades*
6th	31 December 1936	4th, 42nd, 43rd
	May 1937	4th, 42nd, 43rd, 75th
	December 1937	43rd, 75th, 149th
	13 April 1938	107th, 129th, 209th
	11 August 1938	107th, 209th, 220th
22nd	3 April 1937	51st, 77th, 78th
	May 1937	51st, 78th, 93rd
	30 April 1938	79th, 211st, 220th
	12 May 1938	6th, 129th, 220th
	November 1938	51st, 78th
	December 1938	51st, 55th
39th	13 April 1937	22nd, 64th, 81st
	December 1937	22nd, 64th, 96th
	August 1938	64th, 96th, 129th
65th	22 June 1937	42nd, 43rd
	12 July 1937	7th, 21st, 42nd
	August 1937	42nd, 152nd
	December 1937	36th, 42nd
	30 April 1938	36th, 42nd, 75th
	24 May 1938	43rd, 61st
	November 1938	18th, 112th, 200th

Source Carlos Engel, *Historia de las brigadas mixtas del Ejército Popular de la República, 1936–1939* (Madrid: Almena, 1999), Appendix 4

and irrespective of any other possible changes between two dates. Such flexibility was used much during the defensive campaign of the Levant (April–July 1938), when the divisions became mere headquarters units in practice.[56] Nonetheless, given the Republican problems in getting able senior commanders and staffs (surveyed below), it is unsurprising that this capacity to modify the divisional orders of battle took a toll on the overall effectiveness. The historian José Manuel Martínez Bande, after noting that quite a lot of mixed brigades were transferred from one division to another, were fielded separately within the Republican orders of battle or simply disappeared during the same campaign of the Levant, concludes that such organizational flexibility brought confusion and disorganization as well.[57]

Organizational Pragmatism in the Nationalist Army

By contrast with the rational-minded model developed by the Republican army, the evolution of the Nationalist force structure was a rough-and-ready process.[58] Since describing the many changes in the organization of the Nationalist army until late 1937 would be a long-winded task, suffice to say that the Nationalist order of battle in the first half of the war was broadly an assortment of territorial and field (or manoeuvre) commands. Their strength depended on the local situation or their operational responsibilities. The territorial armies were the highest operational commands, each one embracing several fronts. In their turn, the fronts covered by territorial army corps, divisions and brigades resembled more those of conventional armies, army corps and divisions respectively, whereas several field major units were stronger than their name suggested. For instance, the area of responsibility of the 6th Organic Division (under the Army of the North) spanned the fronts of Biscay, Burgos and Palencia by late 1936. Its forces were divided into two sectors or *agrupaciones*. The 1st *Agrupación* covered the Biscay front and included the first four of the so-called 'Navarrese Brigades', which would operate as field major units and bore no resemblance to the pre-war brigades, since they were born as all-arm columns. Each brigade had two half-brigades and a reserve by March 1937, but it still had not a fixed number of battalions and artillery batteries (one was 9900-strong, doubling the strength of the others); they grouped overall 32 infantry battalions. The 6th Division became the VI Army Corps and its *agrupaciones* were renamed divisions in April 1937. The 1st *Agrupación*, augmented by two new Navarrese brigades and a strong Spanish–Italian mixed brigade during the Biscay offensive, would be eventually titled 61st Division next May. And two Navarrese brigades had reached divisional size by early July 1937.[59] According to one of their former commanders, the Navarrese brigades were 'small divisions, very supple, manoeuvring, and could be adapted easily to any sort of terrain, including high mountain'.[60]

By early summer of 1937 the Nationalist order of battle included three armies, six army corps, 12 divisions and five separate brigades as territorial commands grouping the bulk of its fighting units; in addition, there were six reserve infantry divisions, one separate brigade and the *Corpo Truppe Volontarie* (three Italian divisions and two Spanish–Italian brigades).[61] The Nationalist army was finally organized in field army corps and divisions in late 1937. By December 1938, it fielded four armies, 12 army corps (plus two army corps-level commands), and 56 infantry and two cavalry divisions.[62] The new Nationalist infantry division was initially based on the pre-war model, but it was weaker in support troops. According to the establishment of late 1937, the Nationalist division had 12 infantry battalions, grouped into four regiment-sized commands (they had no relation with the pre-war regiments) which in turn formed two brigades, and two (sometimes three) field artillery battalions plus small numbers of engineer and service troops. Its strength was between 10,500 and 12,000 troops.[63] This division became the basic major unit in the Nationalist army; brigades and regiments operated only as tactical headquarters for the infantry battalions. Later in 1938, the Nationalist divisions started organizing their infantry into three regiments or regimental groups (*agrupaciones*) of four battalions each.[64] The author has not found in the sources consulted any explanation for this last change in the Nationalist divisional organization, which suppressed the brigade echelon and one regimental headquarters. Perhaps operational experience convinced the Nationalist army to adopt a leaner and more flexible divisional command structure by simply increasing the infantry regiment's strength by one battalion.

Command and Performance of Major Units

Theoretically, the Republican organization, based on the mixed brigade (a model of unit already present in the Spanish military organization before the war), was a sound structure allegedly more suitable for manoeuvre warfare on extended fronts like those of the Spanish Civil War. On the contrary, the Nationalist army operated with ad hoc formations for the first half of the war, and, after setting up a standard force structure, initially followed the 1925 divisional organization, which was more suitable for methodical fighting on fortified fronts. Nonetheless, the Nationalists carried out manoeuvre warfare operations more successfully than the Republicans. An explanation for this paradox is found in the quality of the command: no organizational arrangement could make up for the professional limitations of the Republican officer corps.[65] The Republican army lacked trained cadres to staff the huge number of senior headquarters required by its expansion.

Moreover, the 1925 regulations were still the official doctrine (as seen at the end of the Republican instructions on the organization and combat of a division).[66] Their guidelines, with their methodical approach, matched badly with the would-be flexibility allowed by the mixed brigade and the actual conditions of the Civil War (with extended fronts often allowing manoeuvre warfare). Well-trained, experienced officers perhaps would have been capable of overcoming the frictions between the official doctrine and the reality. But the EPR had to resort to officers who lacked either the proficiency or an appropriate environment (or both) to tackle such frictions.

The Nationalist army, in its turn, certainly did not display innovative organizational thinking with its attachment to the division as its basic major unit, but on this occasion conservatism paid off. After all, Spanish officers were already familiar with the 1925 regulations and had only to put into practice their contents as far as the situation made feasible. Although centralization of command at the divisional headquarters meant a loss of tactical flexibility, it did not scatter professional officers amongst lots of lower-level headquarters and allowed the Nationalists a more effective use of the limited amount of artillery available, since the divisional commander could concentrate all the artillery resources of his formation for supporting any part of it.[67] Centralization of staff work at higher headquarters also prevented the dispersal of the limited numbers of trained staff officers serving on the Nationalist side (106 out of 245 captains and field officers of the pre-war staff corps, plus 31 who came back from the reserve or retirement, were serving on the Nationalist side by April 1937).[68]

Performance was also influenced by the professional background of the senior officers. Alpert's analysis of the background of the field commanders of both armies about mid-war (by the time of the battle of Teruel) shows that, when the war broke out in 1936, the Republican regular officers commanding major units had, on average, a lower rank (and therefore were theoretically less trained for senior command) than their counterparts in the Nationalist army. Surveying the Nationalist side's major unit commanders in January 1938, Alpert points out that the three army commanders were already general officers when the war broke out in 1936; the army corps commanders were colonels at least, and most infantry division commanders had reached the rank of lieutenant colonel.[69]

On the other side of the hill, the EPR's field army commanders in December 1937 were (in terms of their ranks in 1936) two brigadier generals, an artillery lieutenant colonel, an infantry major who had been seconded to the police before the war, and a retired infantry major. Professional officers commanded most army corps, but only one was a colonel and the rest had reached the rank of lieutenant colonel at best. A similar situation existed among the division commanders; even so, 17 out of 59 divisional commander positions (28%) had been given to militia officers. There were only 49 regular commanders or chiefs of staff in the existing 188 brigades and most of them had been captains with little, if any, operational experience. Moreover, throughout the conflict, many regular officers held assignments that did not fit their professional training and pre-war experience: for instance, artillery officers were appointed as commanders of mixed brigades or infantry divisions, or subalterns as chiefs of staff. Headquarters down to army corps level had enough trained staff officers, but these were very scarce in the divisions and virtually unknown in the brigades.[70]

A legacy of the Spanish military academies of the early twentieth century can be seen in the performance of Nationalist major unit commanders. They had passed through the military academies in the 1900s and 1910s, and it was pointed out in Chap. 2 that the teaching methods of these establishments did not encourage the development of individual initiative. On the contrary, officer candidates got used to unthinking obedience. Such behaviour was described by Kindelán in a letter to Franco during the battle of Teruel.[71] Kindelán attributed the field commanders' poor performance to the lack of willing obedience: if the orders they were given could not be carried out easily at the first attempt, the commanders did not make any further effort to overcome the difficulties—and did not even look for excuses. No initiative on the battlefield could be expected from such commanders. Kindelán reported how on 5 January 1938 Nationalist aircraft had bombed and strafed the enemy positions for half an hour in front of a Nationalist division, but instead of taking advantage of the air attack (in which one aircraft was shot down) the Nationalist troops limited themselves to watching the action and eating their midday meal. When the divisional commander was asked about this inaction, he answered that he had only been ordered to protect the advance of the division operating on his right flank.

In view of cases like the one described above, it is not surprising that senior commanders became annoyed. Kindelán pointed out that General Fidel Dávila (the commander of the Army of the North) reckoned his divisional commanders as 'bad' and Vigón (Dávila's chief of staff) had the same opinion about most divisional commanders operating in Teruel.

But the latter was even more dissatisfied with the army corps commanders, who did not know 'the matter they have in hand' (*lo que traen entre manos*). These command problems in the Nationalist army probably stemmed to a large extent from limited communications networks together with poor command and control arrangements, which could not assure contact between field commanders and their staffs. Kindelán pointed out in the same letter that there were always staff officers in Dávila's headquarters, but if asked about the situation at the front, they did not know what was going on. The situation at the army corps headquarters was worse: during the day, they had only a subaltern, who often was ignorant even of the whereabouts of the army corps commander and his staff.

The limitations of the Nationalist command at divisional and brigade level can also be seen in evidence some months later during the campaign of the Levant. On 24 May 1938, General Antonio Aranda (the commander of Galicia Army Corps) reported to Dávila that the commander of the 4th Division (one of the best Nationalist divisions) was running into command difficulties because he had no able brigade commanders; this forced him to intervene personally in the operations of smaller units and meant that he could not delegate the division's command to any subordinate. If this was the situation in an elite division, one cannot help wondering what happened in the others.[72] And about the same time, the Castile Army Corps progressed too slowly because of bad weather and terrain, but also because the divisional commanders neglected to fulfil keenly the army corps' operations plan. They were so concerned with their own commands' individual endeavours that they were unable to carry out concerted actions in an effective way.[73]

Conclusion

The study of sources concerning the organization of major units reveals that there was a gap between the official doctrine and a strong current of professional thinking which envisaged Spain's future conflicts in terms of mobile warfare. The former was more suitable for methodical operations along fortified fronts and for armies in which the division was the smallest, permanent combined arms unit. The latter advocated the organization of self-contained formations below the divisional level because they fitted better with the Spanish army's operational experience and the theatres of operations in Spanish territory. Such gap is another example of an uneasy relationship during the 1920s and early 1930s between the tradition of imitation of the French model, which seems to have weighed much on official decision-making circles, and the thinking of a section of the military which advocated the development of distinctively Spanish solutions to Spain's military problems. This last section urged the army to develop a military doctrine and organization of its own, better adapted to the features of the Spanish territory, rather than yield to the penchant for regulations and organizations copied from the French experiences on the Western Front. Summing up this section's stance, Captain Vicente Rojo wrote in late 1929: 'Thinking in Spanish and doing in Spanish; there it is the real knot of our intellectual independence and our military effectiveness.'[74] As regards the major units, the organization of a growing number of mixed brigades during the period 1931–1936 may be seen as a sign of rising influence of this 'Spanish approach' to military problems.

However, the conditions under which the Republican army had to fight in the Civil War prevented it from organizing its mixed brigades properly. Shortages of weapons and other equipment meant that they resembled large four-battalion infantry regiments, instead of being real all-arms formations. And its performance suffered from the EPR's general malady: the shortage of trained command cadres. Therefore, the mixed brigade could not be really put to the test during the Civil War and its operational effectiveness cannot be judged fairly. Nonetheless, it is worth reminding that the Nationalist army eventually organized its divisions' infantry into four-battalion groups as well, which meant the demise of the traditional divisional brigade. And the advocates of the mixed brigade may have felt vindicated by the evolution of major unit organization in other armies after 1945, which tended toward the creation of all-arms brigade-sized formations.[75] The Spanish army, which was built upon the triangular infantry division after the war, followed suit and adopted the combined arms brigade as its basic major unit in 1965 (the regiment remained as an administrative unit).[76] And one cannot help bringing to mind Lieutenant Colonel Noreña's article and the flexible orders of battle of the Republican divisions when reading that by the mid-2010s the American divisions would operate 'as plug-and-play headquarters commands (similar to corps) instead of fixed formations with permanently assigned units'.[77] Maybe part of the Republican organizers' fault only lay in having confidence in an idea whose time had not come yet.

Notes

1. Estado Mayor Central del Ejército, *Reglamento para el empleo táctico de las Grandes Unidades* (Madrid: n.p., 1925) (hereafter RGU 1925), Article 92.
2. Emilio Pardo, 'La doctrina militar española después de la guerra de 1914–1918', *Memorial de Infantería* (hereafter *MI*), 25, 1 (1936), p. 58.
3. F[ernando] Ahumada, *La Infantería en la Gran Guerra. Su evolución táctica*, 2 vols. (Toledo: Colección Bibliográfica Militar, 1928), II. 26–7; Pardo, 'La doctrina militar española', pp. 58–9; G[regorio] López Muñiz (ed.), *Diccionario enciclopédico de la guerra*, 13 vols. (Madrid: Gesta, 1954–1958), *sv* 'División'; Jonathan M. House, *Toward Combined Arms Warfare: A Survey of 20th Century Tactics, Doctrine, and Organization* (Fort Leavenworth: United States Army Command and General Staff College, Combat Studies Institute, 1984), pp. 40–2; Shelford Bidwell and Dominick Graham: *Fire Power. British Army Weapons and Theories of War, 1904–1945* (London: Allen & Unwin, 1985), pp. 139, 303 note 134.
4. Ministerio de la Guerra, *Colección Legislativa del Ejército 1918* (Madrid: n.p., 1918) (hereafter *CLE plus the year of publication*), Item No. 169, Addendum No. 1, 3rd basis, sections (a), (b).
5. The two-brigade, four-regiment division is called 'square division' as well.
6. López Muñiz (ed.), *Diccionario*, *sv* 'Binaria, Organización'; House, *Toward Combined Arms Warfare*, p. 40.
7. A., 'La posición del Ejército frente a las enseñanzas de la guerra mundial', *Memorial de Caballería*, 4, 1 (1919), pp. 50–1.
8. Archivo General Militar de Segovia, 2ª Sección, División 8ª, Legajo 535 (hereafter AGMS No. Sección/No. División/No. Legajo), Doctrina Táctica. Informe del Cor de EM Jefe de la 2ª [sic] Sección que llevará a la Ponencia designada por la Junta de Jefes del Centro, n.d., pp. iv–vii.
9. RGU 1925, Article 94.
10. RGU 1925, Article 95.
11. Robert A. Doughty, *The Seeds of Disaster. The Development of French Army Doctrine, 1919–1939* (Mechanicsburg: Stackpole Books, 2014), Kindle edition, Chap. 5 [originally published: Hamden: Archon Books, 1986].
12. AGMS 2/8/535, Doctrina Táctica. Informe del Cor de EM Jefe de la 2ª [sic] Sección que llevará a la Ponencia designada por la Junta de Jefes del Centro, n.d., p. xxiii.
13. Pardo, 'La doctrina militar española', p. 59.
14. *CLE 1926*, Item No. 318.
15. *CLE 1926*, Item No. 312.
16. *CLE 1926*, Item No. 394. The light regiments were organized into two gun and two howitzer battalions, and the foot ones into one gun and one howitzer battalions.

17. *CLE 1929*, Item No. 197. On the reorganizations of the artillery corps in 1926 and 1929, see also Carlos Navajas Zubeldía, *Ejército, estado y sociedad en España (1923–1930)* (Logroño: Instituto de Estudios Riojanos, 1991), pp. 165–6.
18. *CLE 1931*, No. 282. See also Michael Alpert, *La reforma militar de Azaña (1931–1933)* (Madrid: Siglo XXI, 1982), pp. 300–8.
19. *CLE 1931*, Item No. 282, Articles 2, 3. It was planned to turn later the gun regiment into a mountain howitzer one in four divisions.
20. Carlos Martínez de Campos, 'Reflexiones practicables sobre asuntos artilleros', *Memorial de Artillería* (hereafter *MA*), 86, 2 (1929), pp. 223–4.
21. Manuel Vicario, 'Las grandes maniobras del Pisuerga', *MI*, 22, 1 (1933), pp. 520–1.
22. Nazario Cebreiros, *Las reformas militares. Estudio crítico* (Santander: J. Martínez, 1931), pp. 43–7. The Spanish infantry battalion had four rifle and one machine gun companies, so either Cebreiros' peacetime battalion would only have two active rifle companies or he was taking for granted the machine gun companies in the additional battalions.
23. Ministerio de la Guerra, *Anuario Militar de España 1902* (Madrid: n.p., 1902) (hereafter *AME plus the year of publication*). A half-brigade (*media brigada*) was usually a group of two or three self-administering, separate infantry battalions under the command of a regimental level headquarters.
24. *AME 1904*.
25. *CLE 1904*, Item No. 205; *AME 1905*. The other *cazadores* brigades (1st and 3rd) only had two half-brigades plus the signals and quartermaster platoons and the ambulance unit.
26. Carlos Hernández de Herrera and Tomás García Figueras *Acción de España en Marruecos* (Madrid: Imprenta Municipal, 1929), p. 66.
27. *AME 1916*; Jesús Martínez de Merlo, 'De la Brigada de Cazadores a la Brigada Polivalente', *Ejército*, 903 (June 2016), p. 39.
28. Carlos Martínez de Campos, *El problema orgánico de las grandes unidades desde el punto de vista táctico-terrestre* (Madrid: n.p., 1933), p. 10.
29. R[omán] López Muñiz, *Los procedimientos tácticos vigentes en la actualidad (ensayo de táctica comparada)* (Toledo: Colección Bibliográfica Militar, 1929), p. 58; Doughty, *Seeds of Disaster*, Chap. 5.
30. Vicente Rodríguez Carril, 'Ejercicios en cuadro de la Brigada de Artillería de la 5.ª División realizadas en el año actual', *MA*, 82, 2 (1925), pp. 255–6.
31. Carlos Noreña, 'Nuestra División como unidad de maniobra', *Revista de Estudios Militares* (hereafter *REM*), 2, 1 (1933), pp. 331–4. Noreña (1889–1936) had studied in the French war college; unassigned when the war broke out, the government ordered him to rejoin the service; Noreña refused to obey and professed his moral support for the uprising, though he had not been involved in the conspiracy; he was court

martialled and shot despite the French government's pleas for mercy; after the war, his name would be permanently at the top of the list of staff corps lieutenant colonels. Other articles making the case for the mixed brigade: Secundino Serrano, 'La Brigada, unidad de combinación de acciones de Infantería y Artillería', *REM*, 2, 2 (1933), pp. 97–109; José Luis Soraluce, 'La maniobra en las distintas unidades de la División', *REM*, 2, 2 (1933), pp. 191–6; José García Colomo, 'Las Brigadas con capacidad de combate autónomo', *REM*, 2, 2 (1933), pp. 259–69.

32. *CLE 1931*, Item No. 282, Article 3.
33. The historian Puell de la Villa suggests that General Francisco Franco, chief of general staff under Gil Robles, might have been who really encouraged the expansion of mountain units: Fernando Puell de la Villa, *Historia del ejército en España*, 2nd edition (Madrid: Alianza, 2005), p. 186.
34. The first official reference to the brigade in June 1935 mentioned a single artillery battalion, while the official announcement of creation mentioned an artillery regiment: *Diario Oficial del Ministerio de la Guerra* (hereafter *DOMG*), 30 June 1935, 27 September 1935.
35. *DOMG*, 27 September 1935.
36. Ramón Salas Larrazábal, *Historia del Ejército Popular de la República*, 4 vols. (Madrid: Editora Nacional, 1973), I. 556 note 36.
37. Actually, the Italian infantry division on the eve of the Second World War was reduced to two infantry regiments as well; however, this smaller size did not bring any advantage in practice: López Muñiz (ed.), *Diccionario*, *sv* 'Division'.
38. RGU 1925, Article 25.
39. See also Michael Alpert, *The Republican Army in the Spanish Civil War, 1936–1939* (Cambridge: Cambridge University Press, 2013), pp. 73–4.
40. The description of the organization of the Republican army in this section is based on Salas Larrazábal, *Historia del Ejército Popular*, I. 504–7, 697–8, 718 note 15; Ramón Salas Larrazábal, 'Los efectivos del Ejército Republicano', *Historia y Vida*, 66 (September 1973), pp. 15–19; Ramón Salas Larrazábal, 'El Ejército popular de la República', in Mario Hernández Sánchez-Barba and Miguel Alonso Baquer (eds), *Las fuerzas armadas españolas. Historia institucional y social*, 8 vols. (Madrid: Alhambra, 1986), VII. 81–155. For details of each mixed brigade's war record, see Carlos Engel, *Historia de las brigadas mixtas del Ejército Popular de la República, 1936–1939* (Madrid: Almena, 1999).
41. Salas Larrazábal, 'El Ejército popular', pp. 94, 97; Engel, *Historia de las brigadas mixtas*, p. 9.
42. Moreover, brigade batteries were often equipped with anti-tank or light infantry guns rather than field cannon.
43. See, for instance, 'La polémica Mortera-Salas contra Howson en la Revista Española de Historia Militar', *Sociedad Benéfica de Historiadores*

Aficionados y Creadores, http://www.sbhac.net/Republica/Fuerzas/Armas/Polemica/Polemica.htm (accessed 2 November 2016).
44. See, for instance, Salas Larrazábal, 'Los efectivos', pp. 19–27.
45. Salas Larrazábal, *Historia del Ejército Popular*, II. 1617, 2171–3.
46. Salas Larrazábal, *Historia del Ejército Popular*, I. 575–6; Alpert, *Republican Army*, p. 259.
47. Salas Larrazábal, *Historia del Ejército Popular*, I. 718 note 15, II. 1865.
48. Salas Larrazábal, *Historia del Ejército Popular*, I. 718 note 15, II. 1865.
49. *Enciclopedia universal ilustrada europeo-americana. Suplemento anual 1936–1939*, 2 vols. (Madrid: Espasa-Calpe, 1944), *sv* 'España' [II. 1462]. The information about the Republican army in the article is based on Republican official publications, military information bulletins, and periodicals, contrasted with witness testimonies.
50. See, for instance, Table 1 in Juan Miguel Campanario, 'Los proyectos fallidos del Ejército Popular de la República para dividir en dos la zona ocupada por el enemigo: El Plan P del general Vicente Rojo', working paper, http://www3.uah.es/jmc/planp.pdf (accessed 14 August 2016).
51. Alpert, *Republican Army*, pp. 82–3. For a firsthand account of the war experiences of a Republican non-professional officer who became a mixed brigade, division, and army corps commander, see Manuel Tagüeña Lacorte, *Testimonio de dos guerras* (Barcelona: Planeta, 1978), pp. 101–98. Tagüeña (1913–1971) was a reserve NCO, a graduate in physics and mathematics, and a Communist militant in 1936; when the Civil War broke out, he joined the militias and had risen to divisional commander one year later; his performance won him the command of an army corps from April 1938 to the end of the war; after a period in the Soviet army in the 1940s, Tagüeña broke with the Spanish Communist Party and left for Mexico in 1955, where he worked in scientific research jobs.
52. Alpert, *Republican Army*, p. 74. Republican Colonel Segismundo Casado, a harsh (post-war) critic of the mixed brigades, stated that they absorbed up to 10 per cent of the EPR strength in superfluous non-fighting assignments: Segismundo Casado, *Así cayó Madrid* (Madrid: Ediciones 99, 1977), p. 73. But on Casado's views, see Alpert, *Republican Army*, p. 73 footnote 39.
53. Archivo General Militar de Ávila, Fondos del Ejército Popular, Armario 55, Legajo 545, carpeta 8 (hereafter AGMA-EP No. Armario/No. Legajo/No. carpeta), Observaciones al empleo de la artillería en el ataque a Huesca del día 16 de Junio de 1937, 17 June 1937.
54. AGMA-EP 65/787/6, Las operaciones de Teruel, 25 February 1938. See also Alpert, *Republican Army*, pp. 259–60.
55. Ministerio de Defensa Nacional, Ejército de Tierra, *Instrucción número dos sobre la organización y el combate de una división* (n.p.: n.p., [1937?]), p. 1. The present autor used a copy filed in AGMA-EP 55/524/1.

56. Salas Larrazábal, *Historia del Ejército Popular*, II. 1861–2.
57. José Manuel Martínez Bande, *La ofensiva sobre Valencia* (Madrid: San Martín, 1977), p. 59. The pre-war regulations advised against too frequent reorganizations of major units, since they damaged the latter's *esprit de corps*: RGU 1925, Article 120. And in fact, a few Republicans divisions with elite status (for instance, 11th, 35th) had more stable orders of battle.
58. On the development of the Nationalist army's force structure, see 'El esfuerzo de guerra en ambas zonas durante la Cruzada', *Revista de Historia Militar* (hereafter *RHM*), 17 (1964), pp. 90–2, 96–101, 106–17.
59. Rafael García Valiño, 'La campaña del Norte', in *La Guerra de Liberación nacional* (Saragossa: Universidad de Zaragoza, 1961), pp. 282–3, 311; José Manuel Martínez Bande, *Vizcaya* (Madrid: San Martín, 1971), pp. 66–8; 'El esfuerzo de guerra', pp. 96–7; Jesús María Salas Larrazábal and Ramón Salas Larrazábal, 'La Guerra', in [Luis Suárez Fernández, Demetrio Ramos Pérez, José Luis Comellas, and José Andrés-Gallego (eds),] *Historia General de España y América*, 25 vols. (Madrid: Rialp, 1981–1992), XVII. 461, 509. On the Aragonese front by mid-1937, the 51st Division (22,000-strong) had 17 infantry battalions and two more attached, and a separate brigade (7700-strong) had nine battalions (plus some minor units attached): José Manuel Martínez Bande, *La gran ofensiva sobre Zaragoza* (Madrid: San Martín, 1973), p. 89. By June 1937, the VII Army Corps, on the northern Madrid front, had only two divisions but it was over 50,000-strong: Salas and Salas, 'La Guerra', XVII. 504.
60. García Valiño, 'La campaña del Norte', p. 283.
61. The non-territorial formations roughly resembled the pre-war division and mixed brigades in organization and strength, excepting the CTV divisions.
62. 'El esfuerzo de guerra', pp. 106–7, 109–12.
63. López Muñiz (ed.), *Diccionario*, *sv* 'División'; Salas Larrazábal, *Historia del Ejército Popular*, II. 1617; Salas and Salas, 'La Guerra', XVII. 630.
64. López Muñiz (ed.), *Diccionario*, *sv* 'División'. Indeed, the 4th and 5th Divisions went into action with this organization in January 1938: José Manuel Martínez Bande, *La batalla de Teruel*, revised edition (Madrid: San Martín, 1990), p. 126 footnote 158.
65. For instance, a report on the Huesca attacks of June 1937 stated that brigade commanders (who very often were not professional officers) had to be thoroughly taught to distinguish between the artillery's preparatory fire and support fire (the latter was delivered when the infantry moved forward): AGMA-EP 55/545/8, Observaciones al empleo de la artillería en el ataque a Huesca del día 16 de Junio de 1937, 17 June 1937.
66. Ministerio de Defensa Nacional, Ejército de Tierra, *Instrucción número dos sobre la organización y el combate de una división*, (n.p.: n.p., [1937?]), p. 8.

67. Salas Larrazábal, *Historia del Ejército Popular*, I. 1168 note 64.
68. *AME 1936*; *Situación del Cuerpo de Estado Mayor del Ejército en 25 de abril de 1937* (n.p: n.p., 1937). A school for ancillary staff officers was also created in July 1937; military experience was not necessary, but the candidates had to be 30 years old at least and university graduates. After a course of 30 or 40 days, they were commissioned as temporary first lieutenants and performed bureaucratic tasks at headquarters, thus freeing regular staff officers for operational matters. Nonetheless, only 417 out of 1170 course vacancies (35%) offered during the war were filled: Luis Grávalos González and Restituto Valero Ramos, 'La formación de oficiales de Estado Mayor en la Guerra de Liberación', *RHM*, 37 (1974), pp. 155–60, 167–9; José María Gárate Córdoba, *Alféreces provisionales. La improvisación de oficiales en la guerra del 36* (Madrid: San Martín, 1976), pp. 170–4.
69. Alpert, *Republican Army*, p. 91. An exception in the army corps commanders was Juan Yagüe, who was an infantry lieutenant colonel: *AME 1936*.
70. Michael Alpert, *El ejército republicano en la guerra civil*, 2nd edition (Madrid: Siglo XXI, 1989), pp. 100–2, 126. In an effort to overcome the handicap in the middle and higher levels of command Republican divisions, army corps, and armies ran courses on battalion and brigade command (for captains and majors), and divisional and higher command (for senior officers); in May 1937 the Republican army also set up a staff college, whose 45-day courses were passed by 198 students (out of 320 course vacancies offered): Grávalos and Valero, 'La formación de oficiales', pp. 160–6, 169–74; Alpert, *Republican Army*, pp. 148–9; José María Gárate Córdoba, *Tenientes en campaña. La improvisación de oficiales en la guerra del 36* (Madrid: San Martín, 1976), pp. 181–94, 248.
71. Kindelán to Franco, 6 January 1938, in Fundación Nacional Francisco Franco, *Documentos inéditos para la historia del Generalísimo Franco*, 4 vols. (Madrid: Fundación Nacional Francisco Franco, 1992–1994), I. 174–6.
72. Martínez Bande, *Valencia*, p. 82 footnote 70. On the Nationalist elite divisions, see Stanley G. Payne, *Politics and the Military in Modern Spain* (Stanford: Stanford University Press, 1967), pp. 389–90.
73. Archivo General Militar de Ávila, Fondos del Cuartel General del Generalísimo (N), Armario 7, Legajo 375, carpeta 103 (hereafter AGMA-CGG No. Armario/No. Legajo/No. carpeta), Observaciones entregadas en 'Términus' [codename for the Nationalist general headquarters] por el Comandante Beltrán, a su regreso de enlace con el Cuerpo de Ejército de Castilla, 21 May 1938; AGMA-CGG 7/375/103, Resumen de las observaciones hechas por el Jefe que suscribe en relación con las operaciones realizadas por el C. de E. de Castilla durante los días 22 a 28 de Mayo, ambos inclusive, 30 May 1938.

74. Quoted in Alberto Guerrero Martín, 'Análisis y trascendencia de la Colección Bibliográfica Militar (1928–1936)' (doctoral dissertation, UNED [Universidad Nacional de Educación a Distancia], 2015), p. 241. The aim of developing a genuine Spanish military doctrine has been seen in the series of professional literature (*Colección Bibliográfica Militar*) published by Vicente Rojo and Emilio Alamán before the Civil War: Guerrero, 'Análisis y trascendencia', pp. 241–5.
75. House, *Toward Combined Arms Warfare*, pp. 169–71.
76. Martínez de Merlo, 'De la Brigada de Cazadores', pp. 42–5.
77. *Wikipedia, The Free Encyclopedia*, *sv* 'Transformation of the United States Army', https://en.wikipedia.org/wiki/Transformation_of_the_United_States_Army (accessed 16 July 2016).

CHAPTER 9

The Tank, 1909–1936

This is the first of two chapters which explore the attitudes of the Spanish military towards the tank and mechanized warfare before and during the Civil War. While they do not aim to provide a revisionist history of the tank in Spain during those periods, they do argue that there is ground for a more nuanced view. Spain was far behind the main European states in terms of tank strength and innovative thinking by 1936, but she was not a hopeless military wilderness during the 1920s and 1930s either.[1] It is hoped that both chapters will fill a gap in the historical literature on the Spanish military and the Spanish Civil War.

Though there is no shortage of works (mostly in Spanish) on armour in Spain until 1939, their authors have shown a penchant for a 'trees rather than forest' approach, focusing on the history of the technical development of the tanks and their service record.[2] They have neglected a more integrated treatment, which should include a closer examination of the doctrine and the professional debate on armour before the Civil War.[3] This chapter, in particular, hopes to shed new light on how the attitudes towards armour were affected by factors such as the war experience in the Moroccan campaigns and the knowledge of foreign thinking. It will also survey arguments both for and against the establishment of a tank force and how some officers thought about ways to embrace mechanization suitable to Spain's conditions.

J.V. Herrero Pérez, *The Spanish Military and Warfare from 1899 to the Civil War*, DOI 10.1007/978-3-319-54747-3_9

The Spanish Army and Armour up to 1926

The Beginnings of Spanish Armour, 1909–1922

The first tanks went into action in the First World War, but the idea of a self-propelled armoured fighting device had not been alien to military writers and pundits before this time and Spain can display contributions to this pioneering thinking about armoured vehicles.[4] In 1909 Ramón Jiménez Bonilla, a veteran of the last Cuban War, presented a project of a wheeled armoured vehicle, armed with cannons and machine guns, for use against the Moroccan natives to the War Ministry. Moreover, the natives would be electrocuted if they came into contact with the vehicle because its armour plate was to be connected with an internal electric battery. Needless to say, such a proposal came to nothing.[5] Around 1912 a Lieutenant Colonel Emilio Gil Álvaro conceived of an armoured device moving on helicoidal rollers, which he called a 'war chariot'. His design similarly failed to arouse interest and only a passing reference in a professional journal a few years later rescued it from oblivion.[6] However, a year before his idea the Spanish army had recognized the usefulness of armoured vehicles for colonial warfare. The General Staff, in its conclusions about the campaign of 1909 around the northern African enclave of Melilla, recommended equipping motor vehicles with light armour and one machine gun for movement in hostile zones.[7] And a 'protected' lorry was indeed built in 1910 for use in Melilla.[8] This was the first manifestation of what would be henceforward a persistent interest in wheeled armoured or 'protected' vehicles.[9] The fact that these vehicles were easier to produce by the Spanish industry of the period was also related to such interest.

To date the present author's research has not found any evidence about the Spanish army's decision-making process to acquire tanks. It is certain that such a decision was made before the end of the First World War, since the General Staff communicated to the War Ministry undersecretary on 22 October 1918 that the military attaché in Paris, Colonel García Benítez, had reported that the French government would give permission to hand over a single Renault tank.[10] A few days later, the head of the Artillery Section, General Sanz, informed State (Foreign) Minister Eduardo Dato that the Spanish ambassador in Paris must ask for the permission; Sanz also pointed out that King Alfonso XIII had stressed the urgency of carrying out the operation.[11] The French government gave its permission in early January 1919.[12] Once the Spanish cabinet had approved its acquisition in March 1919, a Renault tank was delivered in May (there was a bureaucratic muddle about the payment, however, and the French Embassy in Madrid was still demanding it as late as May 1923).[13] This tank underwent trials on the training ground of Carabanchel, near Madrid, in late June 1919. Its performance was reckoned satisfactory by the Spanish army's ordnance experiment commission, which recommended starting negotiations to acquire ten tanks.[14] A decree dated 13 August 1919 authorized the purchase.[15]

The Spanish army was simultaneously looking for other suppliers of tanks. The military attaché in London, Colonel Rich, reported in early October 1919 that his inquiry (fulfilling an order of the General Staff) about the British authorities' reaction to a request to acquire a light tank had not yielded positive results. Nonetheless, Rich met with the British secretary of state for war, Winston Churchill, during a meal and they did discuss the subject of tanks during their conversation. After hearing of the negative answer to Rich's inquiry, Churchill said that he was ready to meet again with the attaché and to give his support if the Spanish government retained its interest in British tanks.[16] The Spanish efforts suffered a setback in November 1919 when Colonel García Benítez reported that the Allied powers had decided not to export some kinds of ordnance, including tanks.[17] This obstacle did not discourage the Spanish army and in late November 1919 the War Ministry ordered the military attaché in Washington, Lieutenant Colonel Pérez Vidal, to gather information in order to acquire tanks and artillery tractors made in the United States. Accordingly, Pérez Vidal contacted the American army and local manufacturers, but his efforts met with no success.[18]

The Allied ban on selling tanks did not last long and the Spanish army was eventually able to procure tanks in France. A contract for the supply of six Schneider CA 1 tanks was signed on 12 September 1921. This was followed two days later by another contract to purchase 11 Renault FT tanks and both acquisitions were approved by the Spanish cabinet on 16 September 1921.[19] Both models of tank had been developed for the French army during the First World War.[20]

One matter to be solved with the introduction of the tank was which branch of the army should take charge of the weapon. Each major operator of tanks in the Great War (Britain, France, and the United States) had grouped them into a specialized, ad hoc organization, manned by personnel from different branches of the army. In the aftermath of the conflict, the French and American tank forces were integrated into the infantry; only Britain retained an independent tank corps, albeit one that was much reduced in strength.[21] The Spanish experience differed from this, since in this case the tanks were divided between two army branches. Given the paucity of relevant official records, it is impossible to know why such a decision was made, but institutional rivalry seems the most logical explanation. The corps' contest for the extension of their technical responsibilities as a part of their competition for social and professional prominence had already divided military motor transport into heavy and light separate branches, run respectively by the artillery and the engineer corps.[22] It is unsurprising, therefore, that tanks became another area of professional responsibility to fight for. For instance, Major Pedro Jevenois (an artilleryman) stated in early 1921 that tanks were essentially accompanying artillery for the assault and that, consequently, the artillery corps should take charge of their organization—as had been reportedly the case in France.[23]

A General Staff paper of early 1922 assessed the idea of creating a separate tank branch with personnel seconded from the other corps of the army.[24] The artillery would provide gunners, machine gunners would come from the infantry, and the tanks would be driven by engineer corps personnel; tank units would be commanded by officers seconded from these branches and the staff corps.[25] The paper discarded such an arrangement in order to avoid the troubles of mixing personnel from different corps. Instead, it advocated assigning armour to different corps, depending on the characteristics and tactical missions of tanks. Light tanks, which were only armed with machine guns or a small calibre gun and provided close support to the infantry, should be assigned to this arm, since they would be much better commanded by its officers. Heavier 'breakthrough' tanks, armed with more powerful cannon, should be under the command of the artillery corps, whose officers were the most proficient for directing the fire of the main armament. Moreover, all of the corps had personnel qualified to become either drivers or machine gunners. The differentiation of tank roles is taken from the French army's 1920 field manual on tanks, but such differentiation did not prevent France from amalgamating its tank force into the infantry in the same year.[26] In any event the institutional partition put forward by the General Staff paper was actually the option chosen by the Spanish army. Tanks were divided between the infantry and the artillery, disregarding the advantages of a single authority for doctrine, training and service support.

First Blood: Morocco, 1922–1926

The Renault FT and Schneider CA 1 tanks were initially assigned to the Central Firing School (Madrid). In early 1922 the War Ministry decided to send the newly born armoured force to Morocco. On 6 March 1922, the War Ministry ordered the urgent deployment of five Renault tanks from the school to Melilla, in addition to the six Renaults sent the day before.[27] The 11 FTs were organized into an infantry tank company (Spain later purchased another six Renaults to replace lost tanks). The Schneiders had already reached Melilla on 6 March and been organized into a battery of artillery tanks. The Moroccan war provided the Spanish tanks their first combat experience.[28] The use of armour in Morocco had been strongly recommended in an article published by the infantry journal a couple of years earlier. Its author saw the tank as an impregnable mobile pillbox, capable of moving forward even through somewhat broken ground and overcoming points of concentrated resistance. The tanks would demoralize the Moorish warriors for sure.[29] However, the reality soon dispelled such hopes.

The first Spanish tank action is usually dated on 18 March 1922, but it actually took place four days before. The artillery tank battery was ordered to support the infantry in the seizure of Sbuch Sba on 14 March. The action was successful, but three Schneiders broke down during the approach march and a fourth one did it during the combat.[30] Much more dramatic and remarkable was the first combat of the Renault FTs on 18 March. This day the infantry tank company supported the assault of two Legion *banderas* (battalions) on the village of Ambar (or Anvar).[31] The FT's machine guns provided fire support to the infantry and the village was seized, but the tanks moved on and became separated from the legionaries. Under increasing pressure of the enemy, they withdrew to Ambar. Two FT tanks had to be abandoned because of mechanical breakdowns and were later blown up by the enemy. Major Francisco Franco, commanding officer of one of the *banderas* involved, provided some explanations for the FT tanks' poor performance. He asserted that they had been ordered into action too hurriedly. Their machine guns had been unpacked and installed the day before and the gunners did not seem trained enough.[32] Moreover, a single machine was an insufficient armament which left the tank defenceless in case of jamming, an incident too frequent in the Hotchkiss model installed due to its sensitiveness to the quality of ammunition. Tank crews had to be thoroughly trained, both technically and in tactical cooperation with other arms, in order to avoid the poor coordination displayed by the Spanish armour and infantry in Ambar.[33]

After such a disappointing experience, the Spanish army used its armour more carefully and effectively.[34] An attempt to increase the strength of the Spanish armoured force came to nothing the following year. Seven examples of the Chenillette Saint-Chamond model 1921 were acquired in 1923 in order to reinforce the infantry tank company. The Saint-Chamond was an armoured vehicle with alternative wheel and track propulsion (wheels being used for road mobility and tracks across country); it was armed with a machine gun. Clumsy and impractical like most vehicles of its type, the trials in Melilla were very unsatisfactory and the Saint-Chamond was immediately withdrawn from service.[35] The Spanish army eventually grew confident enough in the capacity of its armour as to think about a much bolder use of tanks in the amphibious landing in the bay of Alhucemas in September 1925. The Renault company would be part of the first wave. After hitting the beach, the tanks would disembark from the landing barges and precede the assault infantry troops. However, what it could have a historical event in armoured warfare turned into another anti-climax: the landing craft hit a shoal too far from the beach and the tanks could not land on the first day.

The experience of war in Morocco was one of the sources of Spanish thinking on tank warfare. Despite tactical improvement, the colonial experience of Spanish armour seems not to have been an outstanding success. After the disappointing early performance, it is unsurprising that an essayist concluded that the Moroccan conflict proved the need for a greater subordination of tanks to infantry, by using them as fire support for the lines of riflemen and under the cover of the latter.[36] Captain Vicente Guarner, who commanded the infantry tank unit during several months in 1925–1926, enumerated the mistakes made in the early period: using tanks in isolation as mobile blockhouses, too long periods in action (exceeding the nine hours' endurance of the Renault FT), inadequate reconnaissance, and losing contact with the infantry during the advance.[37] After the war, Goded noted that the rugged ground often found during the operations limited much the tactical mobility of tanks.[38] Another analyst of the war, Major José Díaz de Villegas, also pointed out that, besides experiencing tactical limitations, the tank saw reduced its strategic readiness because of the poor road network in Morocco. He asserted that armoured lorries armed with machine guns were more effective for road patrol missions than tanks.[39]

All the same, the tank still had advocates. Despite witnessing the Ambar fiasco and realizing that tanks were not a war-winning weapon, Franco held a positive opinion. He thought that tanks had a 'great usefulness' in the Moroccan war, which could be increased by developing a better-armed light tank. He also argued that their employment in larger numbers would help to reduce the Spanish manpower requirements for the African battlefields and 'it must not be forgotten that the most expensive thing in this war is not ordnance, but the men'.[40] Guarner for his part acknowledged that colonial warfare allowed the use of tanks in more imaginative ways: they could be used for assaults even in rough ground (if the enemy had no anti-tank weapons), long-range offensive reconnaissance (with supporting infantry), and line of communication protection (though it had to be used in numbers for this role).[41] And Goded concluded that, despite their limitations, it would be unwise dispense with tanks.[42]

The French army also used tanks in some of these ways sometimes during its Rif campaigns of 1925–1926.[43] The French army deployed two battalions of Renault FT tanks. Ground conditions and the absence of a serious anti-tank threat led to the piecemeal use of armour, attaching single platoons to the mobile columns. On the other hand, combat experience confirmed once again the importance of a good infantry–armour liaison, especially when tanks operated in penny packets.[44] For obvious reasons, amongst the colonial campaigns of the 1920s and early 1930s only the French experience in the Rif can be really compared with the Spanish one and it was not significantly different, so seeing the lacklustre tank tactics in Morocco as proof of the Spanish officer corps' professional ineffectiveness is unfair. The Italian army also used a company of light tanks against Libyan natives in the seizure of the oasis of Jarabub in 1926, but the operation took place in open desert, not in mountainous country like the Rif.[45] The British army, after failed trials with medium tanks in the aftermath of the Great War, tested light tanks in the mountains of India's North-West Frontier in 1930, albeit in peacetime conditions. Their performance was good enough to turn eventually a number of local armoured car companies into light tank units.[46] However, comparing this experience to the Spanish one is not fair either, since the British tanks were new models which profited from a decade of technological improvements.

The Influence of Foreign Thinking

French Thinking

Beside its own experience in Morocco, foreign military theories and experiences were the other source of thinking on tank warfare for the Spanish army. As mentioned in other chapters of this book, the Spanish army was mainly exposed to French military influence, so the greatest influence on their use of armour would plausibly be the French one. Above all, the Spanish army embraced the French official thinking stemming from the First World War as inspiration for its new battlefield doctrine developed in the 1920s, and this predictably affected its tank doctrine as well. The French had used tanks in large numbers in their successful offensives of 1918 and made some efforts to develop more innovative thinking on the use of armour in the early 1920s, but they failed ultimately to implement an effective doctrine of mechanized warfare, taking advantage of technological advances.[47] This may explain that the most thorough Spanish survey of mechanization (understood as the substitution of armoured fighting vehicles for foot and horse combat troops) published before 1936 did not mention any French theorist, in contrast to the British Ernest Swinton, J.F.C. Fuller, and Basil H. Liddell Hart, and the American K.B. Edmunds. Only Emile Alléhaut and Hubert Camon are referred to by name because of their proposals to improve the overall mobility of armies (through an increased use of motor vehicles), but these proposals had nothing to do with tank warfare proper.[48]

Actually the French experience was used to endorse conservative views on armour in the Spanish army in the 1920s. One year after the end of the Great War the infantry corps journal mentioned as an example to follow the French policy—inspired by Marshal Pétain—of keeping tanks confined to infantry support missions.[49] Majors Juan Seguí and José Martín Prat, who were students at the French *École Superieure de Guerre* just after the First World War, displayed an open dislike for armour. They reported in 1921 that the tank played a very secondary role in the French military exercises and concluded that it was a very limited weapon whose usefulness for the Spanish army was almost nil.[50] Several years later, Seguí, then a lieutenant colonel and military attaché in Paris, expressed his unfavourable view again after observing the French manoeuvres of July 1929, where he became reassured about the limitations suffered in using tanks. Moreover, he added that the improvement and increased employment of anti-tank weapons were 'to make precarious the achievements attained by this expensive equipment'.[51]

British Thinking

Britain was the other country whose experiences and theorists were assessed in the Spanish professional literature. An article in *La Guerra y su preparación* about the manoeuvres of the British army in 1925 acknowledged the technical improvements of the new Vickers tanks over those used in the First World War. Therefore, their main role could not be to support infantry, since they would lose the advantage of their greater mobility. The anonymous author or authors deduced that tanks were to be kept in reserve first, and later they would move forward quickly to attack points of resistance or to support mobile field forces.[52] This interpretation hints that the article's author or authors did not consider armour an integral part of the mobile field forces yet. Despite the new tanks' technical improvement and enhanced mobility, which made easier their integration in manoeuvre warfare, armoured units were still reckoned to be overly specialized forces and were not to intervene until a climax of battle was reached by the traditional arms.

Knowledge of the British mechanization pioneers was limited.[53] In general, their views were reckoned to be too radical. For instance, Captain Carlos Ruíz de Toledo, who designed the Spanish Trubia tank (see below), was particularly sceptical about some British theorists' visions of armoured forces fighting as 'land fleets'.[54] According to Ruíz de Toledo, they modified too radically the traditional features of battle validated during the First World War and underrated the value of ground, which channels movement and provides protection and vantage points to observe and attack the enemy. The seizure and defence of these positions was what made the difference between sea and land warfare. Armoured forces could conquer ground, but they were unable to hold it. Nonetheless, Ruíz de Toledo was no retrogressive thinker: on the contrary, he thought that infantry had to be as mobile as tanks to reach full effectiveness in modern warfare.

First Lieutenant Enrique García Albors, an infantry and tank officer who accepted a gradual increase of the mechanical factor in battle, criticized Fuller's and Liddell Hart's ideas on the role of infantry in mechanized warfare.[55] As described by García Albors, Fuller envisaged 'heavy' infantry on armoured carriers advancing with the tanks, and 'light' infantry for security and garrison duties and fighting in rough ground (in fact, Fuller pictured mechanized infantry in a limited and totally defensive role in open 'tank country' while light infantry would be the main arm in ground unsuitable for tanks). In his turn, Liddell Hart allegedly reduced the infantry to a defensive role or to offensive skirmishing missions (actually his ideas were broadly similar to those of Fuller, but he did envision an offensive role for mechanized infantry as 'land marines' landing from their armoured vehicles to clear enemy strongpoints). According to García Albors, both theorists' ideas led to an excessive specialization that deprived the infantry of their ability to fight whatever the ground and the situation. He accepted that the infantrymen had to be equipped with means that increased their mobility (for instance, light tracked armoured vehicles carrying infantry support weapons on the battlefield), but without introducing new infantry categories.[56]

García Albors also warned against too radical mechanization because of the dangers of overconfidence in machines, which harmed morale by making it too dependent on the quantity or quality of ordnance, or both.[57] In other words, the theorists of mechanization were subject to the charge of deliberately excluding the human, morale-related factors. This was actually one of the allegations wielded by the opponents of mechanization in general (see below) and illustrates the lingering strength of the paradigm of 'the psychological battlefield' in the Spanish military mindset.

Although the present author's research did not found solid evidence of it, it is very plausible that Fuller's ideas raised the interest of Vicente Rojo, who was a co-publisher of the Spanish edition of Fuller's *Lectures on F.S.R. III.* The memoirs of Manuel Tagüeña, a Republican army corps commander, recount an interesting anecdote. In late May 1938 the Republican army undertook an offensive on the Catalonian front. According to the plan, Tagüeña's corps were to exploit the initial breakthrough on motor transport, and in order to get the essentials of that kind of operation, Tagüeña and his senior officers were given copies of one of Fuller's works before the operation. Tagüeña does not mention the title, but since Rojo had planned the offensive, it seems very likely that this book was the Spanish edition of *Lectures on F.S.R. III.*[58] And it would be very strange that Rojo recommended such reading if he had not appreciated Fuller's ideas, at least in a partial way. Tagüeña's readings were in vain, however, for the Republican initial attack was a failure.

Tank Policy from Morocco to the Civil War

The Attempted Development of a Spanish Tank

After the defeat of Abd el-Krim in the spring of 1926 the infantry tank company returned to Spain. In late November of the same year the military authorities ordered the creation of an armoured group, attached to the Third Section of the Central Firing School. The organization of the unit was detailed on 12 January 1927. The group was formed with two companies. One of them would be equipped with the existing Renault FTs and would have a headquarters (one signals tank), two combat platoons (five tanks each), and a supply and service platoon (four tanks). The other would receive the first Trubia tanks (see below) and would have a similar organization (excepting the service platoon, with two tanks only). The total strength was 28 tanks (13 Trubias and 15 Renaults). However, the freeze of the development of the Trubia tank thwarted the project. The Renault company remained attached to the Third Section of the Central Firing School, where it was used as a demonstration and training unit in courses for officers. The artillery tank battery returned to the peninsula by 1930. It seems that it was also attached to the Central Firing School for some time before their tanks were eventually divided between the Heavy Motor Transport School in Segovia and the *Parque de Artillería* (Artillery Depot) in Madrid.[59]

While the small Spanish armoured force was still fighting in Morocco, the army thought of producing a tank of national manufacture.[60] An artillery officer and former commanding officer of the Schneider battery, Captain Carlos Ruíz de Toledo, was sent abroad to study foreign tanks in 1925. He designed a new model after his return and supervised the manufacture of the prototypes in the artillery factory at Trubia (six prototypes and pre-series examples were made from 1926 to 1934). The model was given the official name of 'Fast Tank', but it became generally known as the 'Trubia' tank. The first prototypes were finished in the summer of 1926 and sent to the Central Firing School (Third Section) for trials.[61] The military archives hold official records of the trials that the 4A prototype of the Trubia underwent at the Central Firing School in 1928. According to these records, the performance of the Trubia was highly satisfactory. It had good cross-country mobility and was fast (32 km/h on road) and well-protected (up to 15 mm armour) despite being a tank of moderate weight (9 tonnes); the armament it was supposed to carry (one 40 mm gun and two machine guns, or three machine guns) provided enough firepower. The conclusion was that the Trubia satisfied the requirements for a modern light tank and even outmatched other contemporary foreign models in many respects.[62]

The positive official judgement was not totally shared by Lieutenant Colonel Ladislao Ureña, of the engineer corps. The present author has been unable to find out if Ureña was well versed enough in the subject to be considered a reliable source, but this officer provides at least a contemporary contrasting assessment. According to Ureña, the Trubia had a relatively high speed, good vision capacity, and acceptable ability to surpass obstacles, but it was too bulky in relation to its performance, its armour was poor, and its tracks were too fragile.[63]

Whatever the Spanish military's assessment of the Trubia, the fact is that its series production never started. The confrontation between the artillery corps and Primo de Rivera after 1926 could not but have a negative effect on the project, whose production schedules were disrupted. And the subsequent republican regime had other priorities. The project seems to have simply withered rather than been officially cancelled.[64] In any case, financial reasons cannot be discarded as a major cause of this paralysis. The army budgets left too small a margin for ordnance procurement and since there were more immediate needs for basic equipment, it would have been rather odd to devote many resources to such a state-of-the-art (and still under question) weapon as the tank.[65]

Official Doctrine on Tanks

The early guidance for the Spanish tankmen was the Central Firing School's provisional instructions for the Renault tank company and the 1922 provisional regulations for 'assault artillery' (that is, the artillery tanks).[66] Later there were also some short references to armour in the 1924 *Doctrina* and the 1925 regulations for major units.[67] However, it is necessary to mention Captain Guarner's article published in *Memorial de Infantería* in February 1924, which advanced some of the principles for the employment of tanks that were sanctioned later by the official regulations. Guarner stressed the role of the tank as an offensive weapon. Tanks had to be used massively and simultaneously (in order to saturate the enemy's defences); if they could not be deployed all along the front, they had to be concentrated on the decisive point. The fighting order of tanks had to be organized in depth, thus allowing tanks in the rear ranks to suppress points of resistance which had been passed by the advanced ones and to relieve the latter in time.[68]

The 1924 *Doctrina*, in relation to which the 1925 regulations did not offer any new development, defined tanks as weapons which were essentially offensive.[69] Besides pointing out that armour must be used massively, it also defined the tactical roles of both the infantry and artillery tanks, by explaining that the breakthrough (artillery) tanks were those which first broke into the enemy lines, followed by the infantry tanks.[70] The *Doctrina* asserted that the employment of tanks could only be occasional. Armour had to be organized in separate units, which would detach their elements to support the infantry with the strength required for each mission; the number of armoured units recommended was one two-battalion regiment for two infantry divisions.[71]

The first detailed official doctrine on armour was finally issued in 1928 as a supplement to the 1926 Infantry Tactical Regulations. This doctrine confirmed the current light tank's subordinate status as an infantry weapon (it must be remembered that the regulations had to be compatible with the features of the Renault FT or similar models).[72] Therefore, all the light tank personnel had to belong to the infantry corps.[73] To dispel categorically any doubt on their role, the use of these tanks was banned in any mission other than those in direct relation to the infantry ones.[74]

The basic tank unit was the platoon. The so-called 'first echelon' (that is, fighting) platoon had five tanks (three armed with a 37 mm gun, two with a machine gun). The 'second echelon' platoon was a service and replacement unit with five tanks, whose role was to carry forward supplies to the tanks in the firing line and replace those lost because of enemy action or mechanical breakdown. Three first-echelon and one second-echelon platoons plus a headquarters (with one command and one signals tanks) and a tank-transporter platoon formed a first-echelon company. Two tank and one tank-transporter platoons formed a second-echelon company. The battalion was organized in three first-echelon and one second-echelon companies plus a headquarters (with two command and one signals tanks); the total battalion strength was 79 tanks. Two battalions formed a regiment, whose strength (after adding a regimental signals tank) was 159 tanks.[75]

The regulations recognized that the tank had features which were suitable for offensive action: tactical mobility, firepower and protection against light infantry weapons and shrapnel. On the other hand, they also enumerated its weak points: vulnerability to shelling, insufficient logistic (that is, non-tactical) mobility, limited observation means, quick wearing out of crews and mechanics, a noisy engine, mechanical unreliability, and a high fuel consumption.[76]

The tactical employment of tanks had to be ruled by the following principles:

- Use *en masse* on wide fronts, to reduce vulnerability to the enemy's artillery fire and to avoid exposed flanks
- Use in depth through successive echelons
- Preservation of a reserve
- Full cooperation with the infantry, whose commanders had to indicate targets to the tanks
- Surprise

This last point also stated that the intervention of tanks was momentary, so they had to be used in the last phase of combat (the assault); they were in great danger in case of premature deployment or if they were not withdrawn in time from the battle ground.[77] This prescription is worth noting because it signals that the tank was still branded as a 'weapon of opportunity'. Once again, new armament technology threatening to disrupt the prevailing tactical and operational orthodoxy is not turned down outright but it is denied a 'mainstream' condition and characterized as suitable for occasional use only, thus preserving the doctrinal status quo. In 1927 an article in the cavalry journal resorted to an overblown comparison in claiming that 'it is possible that [tanks] are useful sometimes, mainly in colonial wars, but battering rams and catapults can be useful as well and nobody thinks of unearthing these artefacts out of the pantheon in which they deservedly rest'.[78] But it is also remarkable to find tinges of the concept in Captain Ruíz de Toledo, an officer hardly hostile to armour, who wrote in 1929 that tanks tended to become a special arm for quick and violent interventions in the decisive moments of combat.[79] In fairness to the Spanish military, it must be said that such an attitude regarding the tank was not unique, since it can also be found in the British army of the 1920s at least.[80]

Despite the point about a massive use of tanks, there was no idea of large tank forces under the operational control of a single commander (as was proposed by the leading British and German theorists of armoured warfare). Fragmentation, instead of concentration, was the rule. Actually, as García Albors pointed out, there was a contradiction in the Spanish doctrine on tank operations, because it recommended attacks on wide fronts (that is, dispersal) and, at the same time, concentration of effort (that is, concentration of tanks) at the decisive point of the attack. Such a contradiction was confusing and dangerous for an army with limited tank strength (García Albors thought that, in case of conflicting needs, concentration of effort must prevail over attacks on wide fronts).[81]

The regulations did not conceive that a tank regiment or battalion could operate concentrated under its commanding officer.[82] The regiment was deemed an administrative unit.[83] The battalion commanders were reduced to an advisory and liaison role near the divisional or higher headquarters.[84] As regards the ratio of tanks to infantry in tactical operations, the 1926 infantry regulations reckoned that one tank platoon per infantry battalion was usually enough—though the armoured element could be augmented to a company if the enemy defence was very strong.[85]

The infantry tank regulations were virtually all the existing doctrine on armour, because, as far as the present author is aware, the artillery corps did not develop any regulations for its tanks, besides the 1922 provisional ones. Nevertheless, the 1924 *Doctrina* outlined the organization of the artillery tanks. The basic unit was the battery, consisting of the combat echelon (one command and four combat tanks) and the first (supply and liaison) and second (road transport) logistical echelons. Three batteries and a headquarters formed a *grupo* (battalion).[86]

Compared with the French regulations (which usually attached one tank company to an infantry battalion), the Spanish ones provided for smaller tank support (one tank platoon per infantry battalion as the standard ratio). On the other hand, the Spanish organization gave greater autonomy to tank companies and battalions, since they had more signals tanks and tank-transporters. But this advantage was limited by the fact that signals tanks were supposed to complete the divisional signals network, instead of being used for liaison between tank units and between these and the infantry (which seems the more logical role).[87] However, this did not seem very important because the official doctrine did not contemplate tank operations above the company level, for which coloured flags were reckoned enough at that time.

Developments Under the Second Republic, 1931–1936

The reorganization of the Spanish army after the advent of the Second Republic also had an impact on the armoured force. On 25 May 1931 the republican government created two separate infantry light tank regiments (two battalions each) and a cavalry armoured car two-squadron group (*Grupo de Autoametralladoras-Cañón*), organic to the Cavalry Division.[88] The tank regiments were based in Madrid (1st) and Saragossa (2nd) while the armoured car group was located after September 1931 in Aranjuez, near Madrid.[89] This expansion of the armoured force did not escape from criticism. Cebreiros thought that two tank regiments were too much armour, when the line infantry had been reduced to the 32 regiments of the standing divisions.[90]

The organization described in the 1928 regulations was not applied to the tank regiments created in 1931. These were formed according to tables of organization and equipment issued on 5 June 1931 (which contain no data on the armoured car group). Two first-echelon and one second-echelon platoons (five tanks each) formed a first-echelon company, two first-echelon companies formed a battalion, and two battalions formed a regiment. The regiment's total tank strength (after adding one signals and command tank for every company, battalion, and regimental headquarters) was 67 tanks.[91]

This establishment seems to contradict the decree of 25 May 1931 on the reorganization of the army, which stated that the organization of the basic tactical units would be in perfect agreement with the regulations.[92] There is no explanation of the new organization in the sources consulted, though this question becomes purely academic when talking about units which existed on paper only: in practice, there were just five Renault tanks available for each regiment.[93] On the other hand, since the platoon remained the basic tank unit, such a contradiction had no serious tactical consequences. Perhaps the answer is related to Azaña's personnel policy. Since the size of the officer corps had to be cut and there were no prospects in the near future to procure more tanks, a reduced establishment which saved a number of useless assignments might be acceptable.

García Albors thought that the 1931 organization could look after the equipment issues sufficiently well through the War Ministry's Ordnance Section and the Central Firing School (for trials of new equipment). But he pointed out the need for an inspectorate of tank units with responsibilities for doctrine and training. In practice, the tank regiments were subordinate in these matters to the army general inspectors. García Albors proposed as a temporary remedy that training rules be given to the general inspectors or the tank units directly by the General Staff.[94]

After 1930 the Spanish army seemed to develop a greater interest in motorization, but with no real consequence in practice.[95] The lack of efforts to pursue a clear-cut policy of motorization—even on a small scale—led to a dismal state of affairs. According to Lieutenant Colonel Ureña, the Spanish army used a motor pool of 2655 vehicles of 92 different models in 1933. Moreover, the civil motor resources would not be a great help in wartime. Ureña reckoned that only 30% of vehicles available for requisition would be at best of any use because of the diversity of models and spare parts.[96] Under the right-wing cabinets of the period 1933–1935 there were two attempts to create a fully motorized major unit, but they did not become more than projects. In late 1934 Prime Minister Alejandro Lerroux thought of organizing a new motorized division. It would be an all-volunteer and army general reserve formation, ready for deployment in home territory or Morocco. The second attempt was planned in June 1935, when it was decided to motorize one of the existing divisions as a first step towards a gradual motorization of the army as a whole.[97]

The interest in developing a tank made in Spain did not disappear in the 1930s either, although it bore almost no fruit. The production of Trubia prototypes proceeded at a snail's pace. Meanwhile, Major Víctor Landesa, of the artillery corps, had designed an armoured tractor in 1928. In the next years the Trubia factory produced a few prototypes and pre-series examples of the tractor, named after its designer. The factory also began to develop a fighting version of the Landesa, armed with a single machine gun. Two examples of this Landesa tank had been produced by 1934, with the exception of their engines, whose manufacture suffered troubles that could not be solved before the outbreak of the Civil War.[98]

Another issue related to armoured warfare and neglected in the Spanish army was anti-tank defence. There were not any regulations on the subject. The Spanish army was supposed to be equipped with an anti-tank gun designed in 1926, the Ramírez Arellano gun (named after its designer, an army officer), which was theoretically able to pierce the Renault FT's armour at 2000 metres. A series of about a dozen guns was made in the artillery factory of Trubia, but they remained untested by

1931. These weapons were still stored in the factory when the insurrectional forces took them during the leftist uprising of Asturias in October 1934 for use against army troops, which included—ironically—the gun's designer.[99] Incidentally, the Asturias uprising also saw the construction of a number of artisan protected lorries by the rebels. The government forces used protected lorries as well, including a few sent from Morocco and later reassigned to the Tank Regiment 1.[100]

The Debate on Armour Until 1936

Against Mechanization

The development of a Spanish tank force in the 1920s and 1930s faced objective obstacles which some contemporary military essayists did not fail to highlight. One of these was the underdevelopment of the Spanish industry. In the early 1920s a military textbook written by General José Villalba pointed out Spain's lack of an industrial base solid enough to produce and maintain this kind of ordnance.[101] Major Federico Beigbeder, of the engineer corps, wrote in 1928 that any attempt to implement mechanization would end in failure because Spain was not an industrial country capable of affording such a luxury, not to mention the problems of getting a supply of fuel.[102] If truth be told, the Spanish governments were not indifferent to this last problem and by the eve of the Civil War had started taking preliminary steps towards adopting a policy of oil self-sufficiency.[103]

Beigbdeder also mentioned the other major factor limiting the development of armoured forces in Spain: the ground. He stated that mechanization could certainly be discussed and supported on paper, but he also said of its supporters that they often disregarded the rugged Spanish terrain. Beigbeder was sure that if mechanized forces had to operate on the latter, their supporters' enthusiasm would turn into 'the bitterness brought by reality, which cannot be discussed'.[104] In fact, this contention was not unreasonable. An apostle of armoured warfare as J.F.C. Fuller himself also recognized terrain as a restraint for developing a mechanized army in Spain in his prologue to the Spanish edition of *Lectures on F.S.R. III*:

> If I had written this book for the Spanish army, I would have written it in a different way, because Spain's topography is different from that which we usually found in the British Empire. From a topographical point of view,

> Spain is a unique country, divided, as it is, by a series of almost parallel mountain chains from west to east. In my view, such a country is ideal to combine the methods of muscular and mechanized warfare; for the older weapons can occupy the mountains, while the newer ones can go through the plains.[105]

This awareness of Spain's ground as a restraint to the use of armour seems to have been forgotten by field commanders during the Civil War, however. The Nationalist army records include complaints about the misuse of tanks on unsuitable ground, submitted by armour commanders in different stages of the war. Fighting too frequently on too rugged ground limited the tactical effectiveness of tanks, wore them down excessively, and imposed long repair and maintenance periods.[106] As late as the battle of the Ebro (25 July–15 November 1938), Nationalist field commanders carried on using armour on unsuitable ground that forced tanks to move forward in single file along paths, with no possibility for deployment.[107]

The restraints of the ground were also used before the Civil War as an argument to keep any armoured force created in the Spanish army small. Seguí and Martín had already recommended, as early as 1921, the setting up of a unit to be used only for occasional operations and experimental purposes.[108] A decade later Cebreiros wrote that the two tank regiments created by the republican government were an oversized force, since their employment on the rugged Spanish territory had not yet been tried.[109] However, Cebreiros in making such a remark contradicted himself since several pages earlier he had alluded to Spain's large plains in which the cavalry still played a prime role.[110] If these plains were adequate for horse troops, they could also prove suitable for the deployment of a sizeable tank force.

Another source of criticism of mechanization lay in the latter's alleged unfavourable effects on morale. Seguí and Martín pointed out in 1921 that tank crews had to be hand-picked, but few soldiers had the highest military virtues. These soldiers had to extend these virtues to the mass of troops through their exemplary role on the battlefield. According to Seguí and Martín, a soldier's heroic example is most effective when it can be seen by his comrades. But how could this be achieved, Seguí and Martín asked, when the fighting men became confined within armoured shells with minimal external communication?[111] Major Beigbeder also wrote with regret in 1928 about the 'madness' (*locura*) which had taken hold of some foreign armies pursuing mechanization theories which, according to him, aimed to eliminate the human factor in warfare. These new armies would turn mechanics into their god.[112]

Cebreiros admitted the need not to lose track of a new military development such as tanks. But they must not be given too much importance, because these *cacharros* (pieces of junk)—as he called them—were not essential for the effective operation of the army. By contrast, the essentials of the army were qualities such as morale, discipline, training and so on. Without them, any material improvement was of no avail.[113] Thus Cebreiros scorned the military value of armour if it was not accompanied by the soldierly moral values and he hinted that the latter were likely to be threatened by the reforms introduced by the republican regime. It is likely that General Mola had a similar view of the likely impact of mechanization threatening traditional military virtues, since he regarded the persistent talk about mechanization and motorization as 'suspicious'.[114]

Backing Mechanization

Not all the attitudes towards armour and mechanization were hostile or retrogressive, however. Despite an unpropitious institutional environment, a few essayists foretold that mechanization was to bring a major transformation to warfare sooner than expected. Nonetheless, with the exception of García Albors, they merely expressed in vague terms an awareness of the changes brought by mechanization but did not display thorough thinking about it. For instance, Cándido Pardo, a retired general, wrote in 1934 about how mechanization was to transform warfare. In future wars the mobility of troops would be more important than their numerical strength alone. Armies would seek to carry out their strategic manoeuvres quickly through the use of motor transport. Although the increase of firepower on the battlefield had forced the adoption of more dispersed fighting orders, the offensive would not lose strength since it would use cross-country fighting vehicles invulnerable to small arms fire.[115]

Interestingly, two essayists argued in the same year (1930) that the future of the cavalry as a tactical arm lay in mechanization. Lieutenant Colonel José Monasterio, a cavalry officer, challenged both the idea that the use of cavalry was in decline and the idea that mechanization would never substitute for this arm. On the contrary, he felt it very likely that technology would provide a mechanical replacement for the horse, capable of carrying out the cavalry's missions in a more effective way. Monasterio thought that perhaps mechanization would offer a final solution to the problem of mobility in the offensive, thus preventing armies from engaging in trench warfare again.[116] Major Vicente Montojo (whose essay's bibliography included Fuller's *On Future Warfare* and Liddell Hart's *The Remaking of Modern Armies*) for his part argued that the stalwarts of traditional cavalry thought as horsemen, instead of thinking of the *mission* of cavalry as the mobile fighting arm. If cavalrymen did the latter, they would seek to implement mechanization at any cost. For tanks, according to Montojo, were the modern version of heavy cavalry, to be used in massed attacks against weak points and lines of communication of the enemy army as it was pinned down by artillery fire and infantry fighting.[117]

García Albors, the most thoughtful essayist on the issue, can also be included among those advocating for motorizing and mechanizing the Spanish army. García Albors thought that motorization was a commendable ideal which had to be pursued as much as capacities and needs allowed, but he did admit that traditional cavalry and horse-drawn transport still had to be preserved in the army. García Albors recognized that motorization, which also needed a supporting industrial base, was expensive in its initial stage, so it had to be implemented with moderation and completed in wartime through requisition. Despite these difficulties, he thought that most units of the army could be motorized to a great extent and their needs for armoured support on the battlefield could be satisfied with an adequate number of armoured fighting vehicles.[118]

He was rather less optimistic in relation to the feasibility of mechanization. This had to be implemented gradually and required sizeable national economic resources. Despite its promise, mechanization could not assure the quick end of a war. An army of armoured fighting vehicles could not decide the result of a campaign, since the infantry would always be necessary to protect the conquered ground. Only non-mechanized units could operate in all terrain and combat situations. Therefore, total mechanization had to be limited to units carrying out specialized missions, such as raids or attacks against the enemy's lines of communication and headquarters. The mechanized forces would be comparatively small, since it was not normal that the enemy army gave up fighting just because a major unit or headquarters had been surprised or surrounded. The clash with the enemy's main force had to be the basic aim for a field army and raids were only supporting efforts in this endeavour.[119]

An Intermediate Alternative

If horse, armour and other motorized troops were to live together at least for the near future, how would their coexistence be organized? Would they lead separate lives or could they be grouped into composite units? Lieutenant Colonel Juan Beigbeder (Federico's brother), who had seen manoeuvres of the German army with cavalry, concluded that the future lay in combining motor and horse, instead of opting for either motor or horse, since they could act in a complementary fashion. The motor vehicle eased the logistical support for the cavalry and therefore extended the range of this arm, whose fighting capacity was also increased by attached armoured cars and motorized infantry. The horse for its part provided greater tactical mobility.[120]

Major Epifanio Gascueña recognized that the motorized division had more strategic mobility than the cavalry division, but this superiority was only significant in long-distance marches and behind a screen of other units. When a motorized division had to provide its own protection, its speed would diminish and its mobility would also be restricted by its dependence on roads. Moreover, the logistic support for a motorized force would be more complex. Thus Gascueña concluded that composite formations combining motor and horse units were superior to the fully motorized ones. At the tactical level, he took the superiority of the mounted forces for granted. However, Gascueña inadvertently undermined his arguments by writing that the cavalry would have to demand fast tracked armoured vehicles for fighting and logistic support missions.[121] If it was accepted that these vehicles were able (or would be able in the near future) to carry out tactical (that is, fighting) missions, there was little point in advocating the survival of mounted units.

Moreover, Gascueña admitted that troops on horseback were relatively fragile (horses wear out quickly without a lot of proper care) and vulnerable to modern weapons.[122] So the cavalry had to adapt their doctrine to technical progress. However, he insisted that this evolution had to be prudent and the cavalry division had to remain a horse and motor formation due to the automotive industry's actual level of development. Gascueña proposed a series of improvements for the division: the motorization of the units' transport trains; the substitution of cross-country armoured cars for the current models; the building of a fast cavalry tank able to follow the movements of mounted units; and the increase of the artillery component.[123] Actually, all this implied an increased mechanization of the division.

These arguments also reflected a trend, which started receiving notice in the Spanish professional literature after the late 1920s, in support of motorizing the cavalry (at least partially) and equipping this arm with armoured vehicles (initially armoured cars). For instance, against the argument that the armoured car limited the cavalry's mobility due to its dependence on roads, it was alleged that armies tended to fight—for both economic and geographical reasons—in areas with extended road networks. On the other hand, mobility also meant being able to advance quickly under one's own steam against enemy resistance, which armoured cars could help to overcome. Moreover, armoured cars allowed a more economical use of horses.[124]

Major Gascueña stated that the cavalry division could not be engaged in prolonged attacks because of its limited fighting means. Its offensive operations had to be based on surprise and directed against the enemy flanks, looking for short, intense clashes.[125] Under these circumstances, Gascueña pointed out the usefulness of adding some tank units to the cavalry division; the tanks, armed with field guns, would be able to follow the mounted troops. Thus the cavalry attacks—though carried out on foot—would have enough power to be really sudden, strong and impetuous.[126]

Captain Joaquín Izquierdo, an infantryman who had passed the tank course, considered that the cavalry needed, at least occasionally, an increase in speed that horses could not give, but the armoured car could—with the advantage of a degree of protection against fire. The armoured car was therefore suitable for use on short missions to be carried out quickly, such as raids, screening, road reconnaissance, and harassing enemy withdrawals. Izquierdo also proposed the use of a modified version of the Trubia tank as a way to overcome the limitations of the armoured car (road dependence, limited vision and excessive size).[127] In another article, Izquierdo suggested that, if the enemy had not consolidated a resistance line yet, tanks could be temporarily attached to an attacking cavalry unit in order to act as close support artillery.[128]

If Gascueña thought that cavalry troops would actually fight on foot as a rule, other essayist were more sanguine about actions on horseback and proposed that mounted troops supported tanks in the attack. After all, if the infantry could advance under the cover of smoke screens, why not the faster cavalry?[129] This proposal was answered by Izquierdo, who stated that mounted troops would be destroyed in any attack against an organized front. The tank needed the support of foot troops to fight against the anti-tank defences. Of course, the cavalry could dismount to do it, but this action would turn the cavalry into second-class infantry, instead of remaining the 'fast arm'. Using as an example for his arguments the battle of Cambrai (November 1917), Izquierdo concluded that only infantry and tanks could attack organized defensive lines. Therefore the idea of using cavalry as a replacement for infantry had to be discarded. Once a breach had been opened, the cavalry would exploit the success (supported by tanks, if possible), but not before a complete breakthrough.[130]

This possible *modus vivendi* between the horse and the motor vehicle was not seen as a happy compromise by other essayists. Major Antonio Hernández Ballester rejected the mixture of horse and motor small units to carry out tactical missions. Mechanization provided fighting means and procedures which were very different to those of the forces relying on horses. Horse units were adequate as mobile tactical reserves and for liaison and rear security missions. Motorized troops, in their turn, could achieve strategic surprise and move quickly to threatened sectors of the front, but they would have to give up exploiting their speed in order to keep contact with the slower elements.[131] Hernández Ballester went further and proposed the creation of fully motorized major units in order to make an effective use of their capacity for strategic mobility, instead of spreading motor resources in small batches in every major unit. Moreover, in operating beside strong motorized units, generals would learn to assess the latter's real capacity and to combine these forces with traditional, non-motorized troops.[132] In a similar vein, García Albors, who believed that talking of the end of the cavalry was premature, also thought that mixing mounted and motorized troops in a cavalry division could damage its effectiveness, since its components tended to act in a dissociated way.[133] It is worth noting that in this period the French army organized composite divisions of armour, cavalry and motor troops, which in practice failed to operate as integrated fighting formations.[134]

It was no coincidence that this debate on cavalry, tanks and mechanization took place in the early 1930s, when other major armies were starting (or at least testing) the mechanization their cavalry arm.[135] It shows that a least a section of the Spanish officer corps was aware of developments abroad which deserved to be taken into consideration, even if their implementation in Spain was still remote. These officers certainly cannot be accused of lackadaisical professional attitude.

Writing in 1935, García Albors offered a view about how motorization and mechanization could be tackled by the Spanish army in the near future. He considered a total motorization unfeasible, since the Spanish territory was too varied and there was neither enough motor industry nor raw materials. Therefore, although it was possible to substitute motor transport for horse-drawn transport, the latter would remain important; this seemed to be confirmed by the trend of the General Staff to increase the proportion of mountain troops in the army force structure.[136] Instead of creating new kinds of divisions, completing and improving the existing ones was a better policy.[137] García Albors also urged the procurement of more infantry tanks (at least the minimum number needed for training purposes). Nonetheless, mechanization in Spain should be set aside for the moment, but without neglecting its theoretical study. At best, mechanization should be initially implemented on a small scale, following contemporary orthodox thinking and discarding bolder, unproven theories.[138]

Conclusion

This chapter has shown that the Spanish army had an interest in the tank before the end of the First World War, but its efforts to purchase a first batch were unsuccessful until 1921. However, this initial interest did not turn into sensible plans for building up a sizeable armoured force and the Spanish military remained satisfied by adopting a conventional doctrine in which tanks operated only to support infantry. All the same, the evidence discussed in this chapter shows that by the early 1930s motorization was a concept gaining ground within the Spanish military environment, although its implementation was slowed by financial constraints. More doubts were expressed about the impact of mechanization. The use of tanks was accepted only in the infantry support role. Although a small section of the contemporary essayists shared the ideas on future warfare envisaged by the mechanization pioneers, the rest openly disagreed or saw mechanized armies as unsuitable for a country like Spain. Industrial backwardness and rugged ground were certainly tangible obstacles to the development of a mechanized force. More questionable were the criticisms based on mechanization's alleged harmful influence on morale.

Nevertheless, the professional literature assessing the future of cavalry by the early 1930s perhaps offered a promising, unperceived way to make mechanization more acceptable. Such literature shows an incipient acceptance of mechanization to support cavalry missions; initially, by encouraging the use of armoured cars; then, by recommending the integration of some kind of fast tank in cavalry major units. This was probably a logical conclusion for an arm which had to fight like infantry and whose last claim to a distinctive status—mobility on horseback—was challenged with increasing success by motor transport.

However, it is also true that mechanization could not be implemented overnight. In periods of technological transition like the one surveyed in this chapter, innovation—even if recognized as desirable—must be pursued without damaging the morale of the existing forces which have to carry on operating for the immediate future (and functioning in this 'somewhat schizophrenic mode' is difficult for a military organization, admittedly).[139] The Spanish cavalry arm was indeed moving towards such a transitional stage in the 1930s and had to be reassured about the usefulness of the horse in contemporary warfare. Therefore, it is no surprise that the professional literature still displayed defences of the horse beside acknowledgements of the rise of the machine. In fact, the arguments for keeping strong mounted forces were still widespread in the Spanish military after the Civil War. However, fear of institutional demise if the cavalry relinquished horses outright cannot be discounted either.[140] In any case, the history of mechanization in the Spanish army must no longer be presented as a Manichean account of retrogressive defenders of the warhorse versus forward-thinking advocates of the tank.

Notes

1. There is much literature on the development of armour and mechanized warfare in the First World War and the interwar period which can be resorted to as background reading. The following list, besides other works mentioned in later notes, is only a small sample of those consulted by the present author: Charles Messenger, *The Blitzkrieg Story* (New York: Charles Scribner's Sons, 1976); Williamson Murray, 'Armored Warfare: The British, French, and German experiences', in Allan R. Millett and Williamson Murray (eds), *Military Innovation in the Interwar Period* (Cambridge: Cambridge University Press, 1996), pp. 6–49; Azar Gat, *British Armour Theory and the Rise of the Panzer Arm: Revising the Revisionists* (London: Macmillan, 2000); Richard W. Harrison, *Architect of Soviet Victory in World War II: The Life and Theories of G.S. Isserson* (Jefferson: McFarland, 2010), Chaps. 3–5.

2. A good deal of literature (with its fair share of regurgitation) has appeared in the last few decades on armour development and manufacture in Spain before and during the Civil War. For a sample of works, see F.C. Albert, *Carros de Combate y Vehículos Blindados de la Guerra 1936–1939* (Barcelona: Borras, 1980); Francisco Fernández Mateos, *Carros de combate y vehículos acorazados en la historia de España* (Madrid: Servicio de Publicaciones del Estado Mayor del Ejército, 1986); Francisco Marín, 'Blindados y Carros de Combate Españoles (1906–1939)', Special Issue 45, *Defensa* (1996); Javier de Mazarrasa, *Blindados en España. 1ª Parte: la Guerra Civil 1936–1939* (Valladolid: Quirón, 1991); Javier de Mazarrasa, *Los Carros de Combate en la Guerra de España 1936–1939* (Valladolid: Quirón, 1998); Artemio Mortera Pérez, *Los medios blindados en la Guerra Civil Española. Teatro de operaciones del Norte 36/37* (Valladolid: AF Editores, 2007); Josep María Mata Duaso and Francisco Marín Gutiérrez, *Blindados autóctonos en la Guerra Civil Española* (Valladolid: Galland Books, 2008). The subject has also been covered in Spanish military history periodicals such as *Serga: historia militar del siglo XX* and *Revista Española de Historia Militar*, some of whose articles will be mentioned in later notes.
3. An exception is Alberto Guerrero Martín, 'La Colección Bibliográfica Militar y el debate sobre la mecanización y la motorización (1928–1936)', *Revista Universitaria de Historia Militar*, 3, 6 (2014), pp. 174–88.
4. T.H.E. Travers, 'Future Warfare: H.G. Wells and British Military Theory, 1895–1916', in Brian Bond and Ian Roy (eds), *War and Society: A Yearbook of Military History* (London: Croom Helm, 1975), pp. 72–4; Charles E. Gannon, *Rumors of War and Infernal Machines: Technomilitary Agenda-setting in American and British Speculative Fiction* (Lanham: Rowman & Littlefield, 2005), pp. 63–74; Antulio J. Echevarria II, *Imagining Future War: The West's Technological Revolution and Visions of Wars to Come, 1880–1914* (Westport: Praeger, 2007), pp. 58–63; Richard Ogorkiewicz, *Tanks. 100 Years of Evolution* (Oxford: Osprey, 2015), pp. 11–24.
5. Francisco Marín and Josep María Mata, *Blindados. Los Medios Blindados de Ruedas en España. Un Siglo de Historia (Vol. 1)* (Valladolid: Quirón, 2002), p. 10.
6. Ricardo Caballé, 'El carro de asalto ó tanque', *Memorial de Infantería* (hereafter *MI*), 6, 2 (1918), p. 465.
7. Estado Mayor Central del Ejército: *Enseñanzas de la campaña del Rif en 1909* (Madrid: [Depósito de la Guerra], 1911), p. 196.
8. Carlos Medina Ávila, 'El armamento. Visión histórica del material de Artillería en España', in Aurelio Valdés Sánchez (ed.), *Al pie de los cañones: la Artillería española* ([Madrid:] Tabapress, 1994), p. 617.

9. For more information on wheeled armoured vehicles in Spain, see Albert, *Carros de Combate*; Fernández Mateos, *Carros de combate*; Marín, 'Blindados'; Marín and Mata, *Medios Blindados de Ruedas.*
10. Archivo General Militar de Segovia, 2ª Sección, División 1ª, Legajo 67 (hereafter AGMS No. Sección/No. División/No. Legajo), Estado Mayor Central del Ejército, Sexta Sección to Subsecretaría del M° de la Guerra, 22 October 1918.
11. AGMS 2/1/67, Ministerio de la Guerra, Sección de Artillería to Ministro de Estado, 28 October 1918.
12. AGMS 2/1/67, Estado Mayor Central del Ejército, Sexta Sección to Subsecretaría del M° de la Guerra, 15 January 1919.
13. AGMS 2/1/67, Sección de Artillería, 2° Negociado, Nota para el Señor General, n.d.; AGMS 2/1/67, [State Minister Santiago] Alba to [War Minister Niceto] Alcalá Zamora, 23 February 1923; AGMS 2/1/67, M. Barrail [French financial attaché in Madrid] to General [Luis] Hernando [head of the Artillery Section], 9 May 1923.
14. AGMS 2/1/67, Ministerio de la Guerra, Artillería, Comisión de experiencias, proyectos y comprobación del material de Guerra, Adquisición de 8 carros-ametralladora tipo Renault y dos carros-cañón de 37, 2 July 1919.
15. *Diario Oficial del Ministerio de la Guerra* (hereafter *DOMG*), 14 August 1919.
16. AGMS 2/1/67, Estado Mayor Central del Ejército, Sexta Sección to Subsecretaría, 7 November 1919.
17. AGMS 2/1/67, Ministerio de la Guerra, Subsecretaría, 5° Negociado to Sección de Artillería, 25 November 1919.
18. AGMS 2/1/39, Embajada de España en Washington, Agregado Militar to General Jefe de la Sección de Artillería, 29 December 1919; AGMS 2/1/39, Ministerio de Estado to Ministerio de la Guerra, 8 January 1920.
19. The documents about these purchases are filed in AGMS 2/1/67.
20. On the technical development of the Schneider CA 1 and Renault FT tanks, see Steven J. Zaloga, *The Renault FT Light Tank* (London: Osprey, 1988); Steven J. Zaloga, *French Tanks of World War I* (Oxford: Osprey, 2010). See Table C.1 in Appendix 3 for specifications of the examples acquired by Spain.
21. For a sample of works on the history of these tank forces during the war and its aftermath, see J.F.C. Fuller, *Tanks in the Great War. 1914–1918* (New York: Dutton, 1920); A.J. Smithers, *A New Excalibur: The Development of the Tank, 1909–1939* (London: Leo Cooper, 1986), Kindle edition; Léon Dutil, *Les chars d'assaut. Leur création et leur rôle pendant la guerre. 1915–1918* (Paris: Berger-Levrault, 1919);

R. Lafitte, *L'Artillerie d'Assaut de 1916 à 1918* (Paris: Charles Lavauzelle, 1921); Tim Gale, *The French Army's Tank Force and Armoured Warfare in the Great War: The Artillerie Spéciale* (Farnham: Ashgate, 2013), Kindle edition; David E. Johnson, *Fast Tanks and Heavy Bombers: Innovation in the US Army, 1917–1945* (Ithaca: Cornell University Press, 1998), Kindle edition, Chap. 2; Robert Stewart Cameron, *Mobility, Shock, and Firepower: the Emergence of the US Army's Armor Branch, 1917–1945* (Washington, DC: US Army Center of Military History, 2008), Chap. 1.

22. Gabriel Cardona, *El poder militar en la España contemporánea hasta la guerra civil* (Madrid: Siglo XXI, 1983), p. 13 footnote 22.
23. Pedro Jevenois, 'Artillería de acompañamiento. Nuevos materiales que la integran', *Memorial de Artillería* (hereafter *MA*), 78, 1 (1921), pp. 104–5, 245. Actually the French tank force (designated *Artillerie Spéciale* or *d'assaut*) seemed to operate as a virtual separate arm during the Great War; see Dutil, *Les chars d'assaut*; Lafitte, *L'Artillerie d'Assaut*; Gale, *The French Army's Tank Force.*
24. AGMS 2/1/67, Estado Mayor Central, 1ª Sección, Carros de asalto. Informe en expediente sobre el empleo de dichas armas de combate en nuestro ejército, 2 February 1922.
25. Such a way of officering apparently came near to being adopted for the British tank corps in the aftermath of the First World War: Smithers, *A New Excalibur*, Chap. 20.
26. Ministère de la Guerre, État-Major de l'Armée, 3e Bureau, *Instruction provisoire sur l'emploi des chars de combat comme engins d'infanterie* (Paris: Charles Lavauzelle, 1920), Article 4; Robert A. Doughty, *The Seeds of Disaster. The Development of French Army Doctrine, 1919–1939* (Mechanicsburg: Stackpole Books, 2014), Kindle edition, Chap. 7 [originally published: Hamden: Archon Books, 1986].
27. AGMS 2/1/67, Ministerio de la Guerra to Capitán General [of the First Military Region], 6 March 1922.
28. On the Spanish armour in Morocco, see Francisco Marín, 'Los carros de asalto españoles en la campaña de Marruecos', *Defensa*, 144 (April 1990), pp. 62–7; Francisco Marín, 'La batería de carros de asalto de artillería (España, 1922)', *Defensa*, 156 (April 1991), pp. 60–4; José E. Alvarez, 'Tank Warfare During the Rif Rebellion, 1921–1927', *Armor*, 106, 1 (1997), pp. 26–8; Oscar Bruña, *Vehículos acorazados en el Tercio. De Cáceres a Sarajevo pasando por el Aaiún (Volumen I)* (Valladolid: Quirón, 1998), pp. 24–35; Dionisio García, 'Renault FT 17 en España (1): La Guerra de Marruecos', *Serga: historia militar del siglo XX*, 30 (July 2004), pp. 2–24.
29. Luis Berenguer, 'Facetas de la campaña marroquí', *MI*, 9, 1 (1920), pp. 76–7.

30. Francisco Marín and Josep M[aría] Mata, *Vehículos blindados en España* (Madrid: Susaeta, [2010]), p. 31.
31. For accounts of this combat, see [Francisco] Franco, *Diario de una bandera* (Madrid: Afrodisio Aguado, 1956), pp. 179–80 [originally published in 1922]; [La Legión,] *La Legión Española. 75 años de Historia (1920–1995)*, 3 vols. (n.p.: [La Legión], 2001–2003), I. 137–40; Bruña, *Vehículos acorazados*, pp. 27–9; José E. Álvarez, *The Betrothed of Death: the Spanish Foreign Legion during the Rif Rebellion, 1920–1927* (Westport: Greenwood, 2001), p. 78.
32. Franco, *Diario de una bandera*, p. 179.
33. Franco, *Diario de una bandera*, pp. 181–2.
34. For an account of an action in which artillery tanks were involved in May 1926, see [Andrés] Sánchez Pérez, *La acción decisiva contra Abd el-Krim. Operaciones en el Rif central en colaboración con el Ejército francés* (Toledo: Colección Bibliográfica Militar, 1930), pp. 33–40.
35. Marín and Mata, *Vehículos blindados en España*, pp. 32–3.
36. Manuel de Pazos, 'El ejército colonial. Lo que dicta la experiencia', *MI*, 12, 2 (1923), p. 326.
37. Vicente Guarner, 'Carros de combate', *MI*, 13, 1 (1924), p. 101. Guarner (1893–1981) was commissioned in the infantry and a staff college graduate; he also taught at the Academy of Infantry and served in Morocco; Guarner accepted the post of the Catalonian regional government's head of public order in early July 1936; during the Civil War he held senior positions in the Republican army on the Aragonese front, was commandant of the staff college and directed intelligence operations in Morocco from Tanger; exiled to Mexico, took this country's nationality and served in the Mexican army.
38. [Manuel] Goded, *Marruecos. Las etapas de la pacificación* (Madrid: Compañía Ibero-Americana de Publicaciones, 1932), p. 68.
39. José Díaz de Villegas, *Lecciones de la experiencia. (Enseñanzas de las campañas de Marruecos)* (Toledo: Colección Bibliográfica Militar, 1930), pp. 72–4.
40. Franco, *Diario de una bandera*, pp. 181–2.
41. Guarner, 'Carros de combate', pp. 101–2.
42. Goded, *Marruecos*, p. 68.
43. Anonymous, 'Les chars de combat au Maroc en 1925', *Revue militaire française*, 95, 4 (1925), p. 416.
44. On French tanks in the Rif campaign, see Anonymous, 'Les chars de combat', pp. 398–416; René Goubernard, 'Chars au Maroc en 1925', *Revue d'Infanterie*, 35, 68 and 69 (1926), pp. 619–49, 749–76, 110–31.
45. F. Cappellano and P. P. Battistelli, *Italian Light Tanks 1919–45* (Oxford: Osprey, 2014), p. 8.

46. Smithers, *A New Excalibur*, Chap. 20; David Fletcher, *British Light Tanks 1927–45. Marks I-VI* (Oxford: Osprey, 2014), pp. 6–7, 11.
47. On the development of French armour after 1918, see Jean Delaunay, 'Chars de combat et cavalerie', *Revue historique des armées*, 155 (1984), pp. 2–24; Michel Pescqueur, 'L'évolution du concept français d'emploi des chars entre 1917 et 1924', *Revue internationale d'histoire militaire*, 81 (2001), pp. 283–312; Doughty, *Seeds of Disaster*, Chaps. 7 and 8; Brian Bond and Martin Alexander, 'Liddell Hart and De Gaulle: The Doctrines of Limited Liability and Mobile Defense', in Peter Paret (ed.), *Makers of Modern Strategy from Machiavelli to the Nuclear Age* (Oxford: Oxford University Press, 1986), pp. 602–4.
48. Enrique García Albors, *Motorización y mecanización del Ejército* (Toledo: Colección Bibliográfica Militar, 1935), pp. 7–8, 14–15. On the ideas of Alléhaut and Camon, see, for instance, Emile Alléhaut, *Motorisation et armées de demain* (Paris: Charles-Lavauzelle, 1929); [Hubert] Camon, *La Motorisation de l'armée et la manoeuvre stratégique* (Paris: Berger-Levrault, 1926).
49. 'La Infantería y los tanques', *MI*, 8, 2 (1919), pp. 463–6.
50. Juan Seguí and José Martín Prat, 'Evolución del armamento de la infantería francesa durante la gran guerra', *La Guerra y su preparación* (hereafter *LGP*), 6, 1 (1921), pp. 250, 253. They even used a paper on the employment of airpower in colonial warfare to show their negative views on tanks: Juan Seguí and José Martín Prat, 'Utilización de la aviación en las campañas coloniales', *LGP*, 6, 1 (1921), p. 36
51. [Juan] Seguí, 'Maniobras del Ejército francés en el Campo de Mailly', *LGP*, 15, 1 (1930), p. 53.
52. 'Las maniobras del Ejército británico', *LGP*, 10, 1 (1926), p. 366.
53. Only J.F.C. Fuller raised enough interest to prompt the translation of some of his works (*Training Soldiers for War* [*Educación del soldado para la guerra*], *On Future Warfare* [*La Guerra futura*], and *Lectures on F.S.R. III* [*Operaciones entre fuerzas mecanizadas. Comentarios al F.S.R. III*]) before 1936: Michael Alpert, *La reforma militar de Azaña (1931–1933)* (Madrid: Siglo XXI, 1982), pp. 71–2. This may have not been necessarily a handicap; in the Second World War ignorance of the ideas of British all-tank theorists allegedly helped the Italian mobile forces in North Africa to learn faster than the British ones the lesson that armour, infantry, and artillery must operate as a team on the battlefield: McGregor Knox, *Hitler's Italian Allies. Royal Armed Forces, Fascist Regime, and the War of 1940–43* (Cambridge: Cambridge University Press, 2000), p. 154.
54. Carlos Ruíz de Toledo, 'El problema del carro de combate', *MA*, 86, 2 (1929), pp. 182–4. Although Ruíz de Toledo does not mention

individually any theorist, the analogy of 'land fleets' of armour is recurrent in J.F.C. Fuller's writings; see, for instance: J.F.C. Fuller, *Tanks in the Great War*, pp. 306, 313–15; J.F.C. Fuller, *Lectures on F.S.R. III (Operations Between Mechanized Forces)* (London: Sifton Praed, 1932), pp. 29–30.

55. Enrique García Albors was born in 1906, commissioned in the infantry (1925), and took part in the last Moroccan campaign (1927); trained in the command of tank units (in 1926 and 1929), he served in the Infantry Tank Regiment 2 in 1931–1935; in July 1936 he was stationed in Valencia and allegedly willing to join the military uprising, but its failure compelled him to serve in the Republican army; after holding staff and teaching assignments (including the tank school) he defected in March 1938; he was court martialled by the Nationalists and, after a short spell in prison, dismissed from the service in 1944.
56. Enrique García Albors, *Carros de Combate*, 3 vols. (Toledo: Colección Bibliográfica Militar, 1932–1933), III. 560; García Albors, *Motorización*, pp. 185–6. For Fuller's ideas, see, for instance, J.F.C. Fuller, *The Reformation of War* (London: Hutchinson, 1923), pp. 161–4; J.F.C. Fuller, *On Future Warfare*, (London: Sifton Praed, 1928), pp. 233–7, 248–51, 368–70; J.F.C. Fuller, *Lectures*, pp. 14–16. For Liddell Hart's ideas, see, for instance, Basil Liddell Hart, *Paris, or the Future of War* (London: Kegan Paul, 1925), p. 88; Basil Liddell Hart, *The Remaking of Modern Armies* (London: John Murray, 1927), p. 9, 15.
57. García Albors, *Motorización*, pp. 169–71.
58. Manuel Tagüeña Lacorte, *Testimonio de dos guerras* (Barcelona: Planeta, 1978), p. 133; José Manuel Martínez Bande, *La batalla del Ebro* (Madrid: San Martín, 1978), p. 28.
59. García Albors, *Carros*, III. 527–8; Ministerio de la Guerra, *Colección Legislativa del Ejército 1926* (Madrid: n.p., 1926) (hereafter *CLE plus the year of publication*), Item No. 408; *CLE 1927*, Item No. 14; Marín, 'Carros de asalto españoles', p. 67; Marín, 'La batería', p. 64.
60. On the development of the Trubia tank, see Artemio Mortera Pérez, *Los carros de combate 'Trubia' (1925–1939)* (Valladolid: Quiron, 1993); Dionisio García, 'Trubia: El Primer Carro de Combate Español', *Serga: historia militar del siglo XX*, 52 (March 2008), pp. 54–64. See Table C.1 in Appendix 3 for specifications of the Trubia.
61. García Albors, *Carros*, III. 527–8; Fernández Mateos, *Carros de combate*, pp. 13, 21.
62. AGMS 2/1/39, Negociado de estudios, proyectos, experiencias y fabricación de material de guerra, Acta nº 10, 23 May 1928; AGMS 2/1/39, Ministerio de la Guerra, Sección de Industrias y Construcciones Militares, 1er Negociado, Expediente relativo a informe sobre pruebas realizadas con el carro ligero de combate Trubia de 75 HP serie 4A, 8 June 1928.

63. Ladislao Ureña, 'Motorización, mecanización', *Memorial de Ingenieros*, 88 (1933), pp. 523–4.
64. Mortera, *Los carros de combate 'Trubia'*, p. 32.
65. For instance, the Spanish soldiers were not provided with an issue steel helmet until 1935: Stanley G. Payne, *Politics and the Military in Modern Spain* (Stanford: Stanford University Press, 1967), p. 305.
66. García Albors, *Carros*, III. 530–1. The present author could not find any copy of these provisional instructions.
67. García Albors, *Carros*, III. 531.
68. Guarner, 'Carros de combate', pp. 92–6.
69. Estado Mayor Central del Ejército, *Doctrina para el empleo táctico de las Armas y los Servicios* (Madrid: n.p., 1924), Title I, Chap. II, Article 4.
70. Estado Mayor Central, *Doctrina*, Title I, Chap. II, Article 3.
71. Estado Mayor Central, *Doctrina*, Title II, Chap. II, Article 4.
72. Dirección General de Preparación de Campaña, *Instrucción y empleo táctico de los carros de combate ligeros o de acompañamiento. Anexo III al Reglamento Táctico de Infantería* (Madrid: n.p., 1928) (hereafter DGPC, *Instrucción*), Article 1.
73. DGPC, *Instrucción*, Article 2.
74. DGPC, *Instrucción*, Article 128.
75. DGPC, *Instrucción*, Articles 47, 66, 76, 77, 80. This organization had already been outlined (with a lesser attention to second echelon units) in 1924: Estado Mayor Central, *Doctrina*, Title I, Chap. II, Article 4.
76. DGPC, *Instrucción*, Article 82.
77. DGPC, *Instrucción*, Article 86.
78. A., 'La guerra en el porvenir', *Memorial de Caballería* (hereafter *MC*), 12 (1927), p. 226.
79. Ruíz de Toledo, 'El problema del carro de combate', p. 197.
80. Smithers, *A New Excalibur*, Chaps. 20 and 21.
81. Enrique García Albors, 'Ligero parangón y comentario de los Reglamentos de carros franceses y españoles', *MI*, 25, 1 (1936), p. 101.
82. DGPC, *Instrucción*, Article 122.
83. DGPC, *Instrucción*, Article 124.
84. DGPC, *Instrucción*, Article 79.
85. Dirección General de Preparación de Campaña, *Reglamento Táctico de Infantería*, 2 vols. (Madrid: n.p., 1926), II. Article 726.
86. Estado Mayor Central, *Doctrina*, Title II, Chap. IV, Article 3.
87. García Albors, *Carros*, III. 531–3, 537; García Albors, 'Ligero parangón y comentario', p. 102.
88. *DOMG*, 26 May 1931.
89. *DOMG*, 23 June 1931; *CLE 1931*, Item No. 729.

90. Nazario Cebreiros, *Las reformas militares. Estudio crítico* (Santander: J. Martínez, 1931), p. 61.
91. *CLE 1931*, Item No. 314.
92. *DOMG*, 26 May 1931.
93. Fernández Mateos, *Carros de combate*, p. 22. The cavalry group lacked armoured cars until 1935, when it started receiving the Bilbao model (made in Spain and armed with one machine gun), which had been developed for the paramilitary riot police force: Fernández Mateos, *Carros de combate*, p. 16; Steven J. Zaloga, *Spanish Civil War Tanks. The Proving Ground for Blitzkrieg* (Oxford: Osprey, 2010), pp. 11–12.
94. García Albors, *Carros*, III. 536.
95. García Albors, *Motorización*, p. 221.
96. Ureña, 'Motorización, mecanización', pp. 522–3. There were 306,062 registered motor vehicles in Spain in 1933 (about half were cars and buses, the rest were lorries, vans, and motorcycles); however, it was reckoned that only 200,000 were actually circulating; the ratio of inhabitants to cars was 116:1 (the ratio was 27:1 in France, 97:1 in Germany, 153:1 in Italy): José Díaz de Villegas, *Geografía militar de España, países y mares limítrofes* (Madrid: Servicio Geográfico y Cartográfico, [1939]), p. 141.
97. García Albors, *Motorización*, pp. 221–2; Mariano Aguilar Olivencia, *El ejército español durante la Segunda República (claves de su actuación posterior)* (Madrid: Econorte, 1986), p. 406.
98. Mazarrasa, *Carros de Combate*, 118–19.
99. García Albors, *Carros*, III. 541–3; Cardona, *El poder militar*, p. 168.
100. Mortera, *Los medios blindados en la Guerra Civil*, pp. 12–18. See also Lucas Molina Franco, 'Blindados en la Revolución de Asturias, 1934', *Revista Española de Historia Militar*, 52 (October 2004), pp. 207–10.
101. J[osé] Villalba, *Táctica de las tres Armas (Aviación y carros de asalto)*, 9th edition, 3 vols. (Toledo: n.p., 1923), I. 485 footnote 1.
102. Federico Beigbeder, 'Los transportes militares en la guerra', *LGP*, 13, 2 (1928), pp. 103, 116–18.
103. On the eve of the Civil War Spain imported annually slightly over 600,000 tonnes of oil and there were plans to produce up to 200,000 tonnes of synthetic oil per annum in a first stage: Díaz de Villegas, *Geografía militar de España*, pp. 104–6.
104. Federico Beigbeder, 'Los transportes militares', p. 103.
105. J.F.C. Fuller, *Operaciones entre fuerzas mecanizadas. Comentarios al F.S.R. III* (Toledo: Colección Bibliográfica Militar, 1933), p. 11. Present author's translation of the text in the Spanish edition.
106. For instance, see Archivo General Militar de Ávila, Fondos del Ejército Nacional, Armario 15, Legajo 21, carpeta 24 (hereafter AGMA-EN No.

Armario/No. Legajo/No. carpeta), Cuartel General del Generalísimo, Estado Mayor, Sección Tercera to General Jefe del Ejército del Norte, 26 May 1937; AGMA-EN 15/27/24, Coronel von Thoma [commander of the German tank training unit] to General Jefe de Movilización, Instrucción y Recuperación, 3 October 1938.

107. AGMA-EN 15/27/32, La Legión, Bandera de Carros de Combate, Mando to General Jefe del Ejército del Norte, 10 October 1938.
108. Seguí and Martín, 'Evolución del armamento de la infantería francesa', p. 253.
109. Cebreiros, *Las reformas militares*, p. 61.
110. Cebreiros, *Las reformas militares*, p. 39.
111. Seguí and Martín, 'Evolución del armamento de la infantería francesa', p. 253.
112. Federico Beigbeder, 'Los transportes militares', p. 103.
113. Cebreiros, *Las reformas militares*, pp. 55–6.
114. Emilio Mola, *Obras completas* (Valladolid: Librería Santaren, 1940), p. 1160.
115. Cándido Pardo González, *El problema militar de España; su resolución más racional, económica y nacional* (Madrid: n.p., 1934), p. 246. Pardo had been a staff corps officer and one of the planners of the Alhucemas landing.
116. J[osé] Monasterio, *El momento de la Caballería* (Toledo: Colección Bibliográfica Militar, 1930), pp. 93–4. Monasterio (1882–1952) fought in Morocco, taught in the Saragossa military academy, and commanded the Nationalist 1st Cavalry Division in the Civil War, achieving a notable success in the battle of the Alfambra; he held senior assignments after the war, despite supporting a prompt restoration of the monarchy.
117. Vicente Montojo, *Ejército moderno* (Madrid: Suc. de Rivadeneyra, 1930), pp. 148–50.
118. García Albors, *Motorización*, pp. 174, 218–19.
119. García Albors, *Motorización*, pp. 173–4, 219–20.
120. [Juan] Beigbeder, 'Las maniobras del Oder', *Revista de Estudios Militares*, 1, 2 (1932), pp. 427–8. For the evolution of the German cavalry doctrine and organization during this period, see David R. Dorondo, *Riders of the Apocalypse: German Cavalry and Modern Warfare, 1870–1945* (Annapolis: Naval Institute Press, 2012), Chaps. 4 and 5.
121. Epifanio Gascueña, *La moderna división de Caballería. Sus misiones; su organización; aplicación de ésta a nuestro ejército* (Toledo: Colección Bibliográfica Militar, 1931), pp. 23–6.
122. During the Civil War, the nationalist army's general headquarters reminded that more attention should be paid to the care of cavalry

mounts; many of them had been requisitioned and were of medium quality and poorly trained: AGMA-EN 14/24/13, Instrucciones para el empleo de la Caballería, 24 March 1938.

123. Gascueña, *La moderna división*, pp. 26–9.
124. Juan Camps, '¿Cuándo será un hecho la creación de las Unidades de Auto-ametralladoras en Caballería?', *MC*, 13 (1928), pp. 292–4.
125. Both sides in the Civil War issued instructions stressing these points: Archivo General Militar de Ávila, Fondos del Ejército Popular, Armario 55, Legajo 540, carpeta 2, Inspección General de Caballería, Directivas para la instrucción de las Brigadas de Caballería, Directiva nº 1, 10 July 1937; AGMA-EN 14/24/13, Instrucciones para el empleo de la Caballería, 24 March 1938.
126. Gascueña, *La moderna división*, p. 143.
127. Joaquín Izquierdo Jiménez, 'Necesidad de auto-ametralladoras-cañones y carros de combate en cooperación con la Caballería', *MC*, 16, 1 (1931), pp. 12–15.
128. Joaquín Izquierdo Jiménez, 'Carros de combate en cooperación con la Caballería', *MC*, 16, 3 (1931), pp. 22–5.
129. Julián Gómez Seco, 'La Caballería en las campañas modernas. Nuevas modalidades de su empleo eficaz', *MC*, 16, 5 (1931), pp. 19–23.
130. Joaquín Izquierdo Giménez, 'Colaboración de los carros de combate. Infantería con la Caballería', *MC*, 17, 1 (1932), pp. 15–23.
131. Antonio Hernández Ballester, 'La combinación de las Armas', *MI*, 22, 2 (1933), pp. 377–8.
132. Hernández Ballester, 'La combinación de las Armas', pp. 376–7.
133. García Albors, *Motorización*, pp. 96–102.
134. Bond and Alexander, 'Liddell Hart and De Gaulle', p. 608; John Wheldon, *Machine Age Armies* (London: Abelard Schuman, 1968), pp. 63–4.
135. On the modernization of the cavalry in, for instance, the French and American armies in this period, see Pierre-Henri Delorge, 'Pourquoi avoir gardé une cavalerie à cheval (1918–1939)?', *Guerres mondiales et conflits contemporains*, 225 (2007), pp. 21–36; Doughty, *Seeds of Disaster*, Chap. 8; Faris R. Kirkland, 'The Aristocratic Tradition and Adaptation to Change in the French Cavalry, 1920–1940', *European Studies Journal*, 4, 2 (1987), pp. 1–17; Johnson, *Fast Tanks*, Chap. 10; Cameron, *Mobility, Shock, and Firepower*, Chaps. 2 and 3.
136. García Albors, *Motorización*, pp. 223–5.
137. García Albors, *Motorización*, pp. 225–8.
138. García Albors, *Motorización*, pp. 228–30.

139. R. James Woolsey, 'The Sources of Technological Innovation', in Shai Feldman (ed.), *Technology and Strategy: Future Trends* (Boulder: Westview Press, 1990), p. 21.
140. Juan Carlos Losada Malvárez, *Ideología del ejército franquista (1939–1959)* (Madrid: Istmo, 1990), pp. 59–66. Losada states that these views were due to the stress on the morale-related factors in warfare in Franco's army: Losada, *Ideología*, pp. 48–59. But cf. Raúl Lión and Juan M. Silvela, *Del Caballo a los Medios Acorazados. Una historia reciente de la Caballería española, 1939/90* (Valladolid: Qirón, 1999), pp. 42–7.

CHAPTER 10

The Tank in the Civil War

This chapter completes the study of the subject addressed in Chap. 9 by surveying the reaction (or lack of it) on both sides to the actual operational experience with armour in the Civil War. It will look at the Spanish forces' doctrinal responses to battlefield experience and their occasional attempts to exploit the potential of mechanized forces in that conflict. As a matter of fact, tackling thoroughly the use of armour in the Spanish Civil War would require a monograph of its own. But it is still possible within the limits of this survey to provide some evidence showing that there was some evolution from the pre-war doctrine in the Republican army and that both sides launched offensives that tried to exploit the principles of mechanized warfare.

The Spanish Civil War became an opportunity for testing new military technologies and doctrines developed in the interwar period, whose implementation was watched intently by contemporary military analysts in search for 'lessons'. So it is unsurprising that the military historians' attention has been mostly focused on what the belligerent armies of the Second World War learnt (or did not). As regards armour, this has meant devoting a lot of attention to the experiences of Germany, Italy, and the Soviet Union, which sent tanks and specialist personnel to support their respective local partner.[1] Unfortunately, this focus induces to see the experience of the Spanish military, who were the main actors in the conflict, reduced to the passive role of following the foreign experts' guidance because Spain's backward army had an underdeveloped armoured arm in 1936. This chapter aims to amend such an underrating view

J.V. Herrero Pérez, *The Spanish Military and Warfare from 1899 to the Civil War*, DOI 10.1007/978-3-319-54747-3_10

and to prove that the Spanish military were something more than mere unthinking servants of foreign tank masters and could even see at times the potential of mechanized forces.

The Spanish army had around 20 tanks available in July 1936: five Renault FTs in each infantry tank regiment, at least four Schneider CA 1s in the Madrid Artillery Depot, three Trubia prototypes attached to the *Milán* Mountain Infantry Regiment 32 in Oviedo, and a few isolated examples in other places.[2] The government got hold of the Tank Regiment 1's FTs and also captured the Schneiders initially joining the military uprising.[3] The Tank Regiment 2 and the Oviedo garrison joined the rebellion.[4] The performance of this armour was unremarkable and gave little advantage to both sides in the early clashes. The attempts to produce tanks in Spain were not of much use either.[5] But once the war was in full swing, its ideological nature provided each side with the military assistance of sympathizing foreign powers. The Republican side got aid from the Soviet Union, while Germany and Italy assisted the Nationalists.[6] The effect of this foreign intervention in the evolution of mechanization in Spain was limited, however.

For a start, both sides were supplied too few tanks to attempt real, large-scale mechanized operations. The Soviet Union sent 281 T-26 infantry light tanks and 50 BT-5 fast tanks; a part of them were captured and used by the Nationalist army.[7] Germany supplied 122 Panzerkampfwagen I (PzKpfw I) light tanks.[8] Italy sent 157 CV 3/33 and CV 3/35 tankettes, but their crews were mostly Italian and fought under the tactical control of the *CorpoTruppe Volontarie* (CTV), so their operations are outside the scope of this survey.[9]

Nor did the presence of Soviet and German advisers and instructors seem to have been much of an opportunity for the Spanish military to learn forward-looking doctrines of mechanized warfare. It must be reminded that such doctrines still were untested in real battlefield conditions by 1936, so a cynic might assert that, rather than on-the-job learners, Spanish tankmen were guinea pigs in what has been described as a 'laboratory' where the Soviet Union, Germany, and Italy tried out their tanks and tactical methods on the eve of the Second World War.[10] In truth, the Soviet advisers in the Republican side could hardly teach a doctrine of mechanized 'deep battle' that they did not plan to implement fully in Spain.[11] On the other hand, Nationalist records in the Spanish military archives indicate that the German advisers (aware of their armour's limitations in quality and numbers) seem to have been principally worried about organizational matters and basic technical and tactical training.[12]

The Tank in the Republican Army

Organization and First Battles

The Republican armoured force was really born in mid-October 1936 after the arrival of the first Soviet tanks (50 T-26s) and armoured cars. The shortage of time and the Soviet tankmen's eagerness to get combat experience meant that only one company could be manned with Spanish crews when the Republican tank force (grouped into two battalions) went into action before the end of October. Though the number of Spanish tankmen increased in the following months, the tactical command of Republican armour was often held by Soviet officers well into 1937. This makes assessment of the pre-war Spanish doctrine's effectiveness difficult, since Soviet thinking and procedures weighed much in practice. Suffice to say, for this survey's purposes, that the Soviet-led Republican tanks did not achieve any offensive success in the early months, but they proved effective for delaying and defensive actions during the defence of Madrid and the battles around the capital (November 1936–March 1937).[13]

The *Ejército Popular de la República* (EPR) carried on expanding its armoured forces during 1937 and 1938. According to Salas Larrazábal, in the first half of 1937 the Republicans organized one tank brigade (at least three battalion-strong), one four-battalion tank regiment in the northern front, and one organic tank battalion in all the other field armies; according to Zaloga, they organized three new brigades (one tank and two armoured cars battalions each) in addition to the existing three-battalion tank brigade. By the following October the tank and armoured car forces were consolidated into a *División de Ingenios Blindados* (Armour Division) with one four-battalion tank and one armoured car brigades. After the Republican territory became split in April 1938, one armour division of two combined brigades of armoured cars and tanks was organized in the eastern army group, whereas the central army group had a division of three combined brigades. In any event brigades and divisions operated as administrative commands only.[14]

The EPR's first major strategic offensive which committed a sizeable tank force (three battalions, two of them supporting the main effort) led to the battle of Brunete (6–26 July 1937).[15] The Republicans failed to take advantage of its superiority in armour. It is worth noting that an order for the Tank Brigade to send tanks to carry out tactical exercises with the units spearheading the offensive was cancelled two days later.[16] An unexpected problem may have prevented the implementation of this training schedule, but this incident helps to understand their troops' performance during the battle.

On 19 July 1937 the Republican Army of Manoeuvre (whose headquarters conducted the battle) issued instructions on the use of armour after the experiences of the first stage of the battle.[17] The instructions first listed the mistakes made by the Republican forces: tank commanders were too independent, while the infantry failed to keep contact with armour when the latter gained temporary advantages during the fighting; there was no effort to achieve tactical surprise; and tank units were engaged in frontal assaults on defended villages which caused too many casualties. The instructions urged field commanders to seek a closer liaison between tanks and infantry, more secret preparation of attacks, and the use of enveloping manoeuvres whenever villages were attacked. The instructions' spirit generally followed the pre-war doctrine and still saw tanks as a support weapon for infantry units, whose orders should be obeyed by armour. However, the instructions also emphasized the effectiveness of using tanks in comparatively huge numbers and recommended not to use them below company strength. This last point was a change from the pre-war doctrine, which reckoned the platoon as the basic fighting unit.[18] And it was also a necessary reminder to correct the tendency shown during the battle by the Republican commanders to retain tanks temporarily attached to their troops.[19] Such conduct led to the fragmentation of armour, which rendered it less effective.

The Offensive on Saragossa

The Republican army made a bolder attempt to use mechanized forces for an in-depth advance during the offensive on Saragossa (24 August–7 September 1937).[20] The operation was planned by Vicente Rojo, at the time a colonel and the EPR's chief of general staff, whose knowledge of Fuller's ideas might well have encouraged him to achieve an operational-level success through the speed of mechanized forces.[21] Saragossa was one of the largest cities under Nationalist control, enticingly near the front, and defended by an overextended enemy. Seizing it would help to disrupt the successful Nationalist advance in the northern front.

Rojo planned for a three-pronged converging offensive. In the northern sector, once a conventional infantry attack in reinforced division strength had broken through the Nationalist front and taken the town of Zuera, a motorized task force (two tank companies, ten armoured cars, and one mixed brigade, one field artillery battery, and one sapper company on lorries) was to advance southwards to seize the northern quarter of Saragossa and the bridges on the Ebro river (about 24 kilometres from Zuera). Another division in the central sector, after breaking through the front, would organize a motorized group of undefined size with the lorries and armour available, which would rush westwards to link with the northern task force in the outskirts of Saragossa. In the southern sector (on the south bank of the Ebro), while other units cut off and attacked the main Nationalist positions, a third motorized task force (30 tanks, ten armoured cars, and two lorried mixed brigades) would dash through a wide gap in the thinly held enemy front and cover a distance of 36 kilometres from the starting line to Saragossa.[22] The motorized task forces were expected to enter Saragossa by the end of the day.

The operation has been called 'the most ambitious plan the [EPR] knew during its history'.[23] The ground (especially on the southern sector) was suitable for mechanized forces and the Nationalist front was a screen of separate strongpoints and fortified villages (in fact, there was a gap of 20 kilometres without permanent positions between Quinto de Ebro and Codo), although sooner or later any advance to Saragossa would meet a solid position that could not be simply bypassed.[24] As it was, things went badly for the Republicans from the start. The infantry attacks in the northern and central sectors were unsuccessful, so there was no exploitation. The motor transport for the southern task force arrived after a fateful six-hour delay and then the Republican troops wasted too much time in subduing all the Nationalist resistances they met, instead of advancing past them. And the second day (when a fast advance to Saragossa was still possible, according to Rojo) the southern task force engaged in the fighting for Fuentes de Ebro on its right flank. The delay allowed the Nationalists to bring reserves and close the gap. The Republican offensive made no further progress and was eventually reduced to crush the Nationalist resistance in Belchite, a small town which was surrounded in the first day and no longer had military value.

The operations plan for the Saragossa offensive probably was too ambitious for the EPR's real capabilities and credit must also be given to Nationalist garrisons standing fast against the odds. But a more resolute advance of the southern task force, bypassing the smaller enemy positions, might have turned the battle for the Nationalists into a nearer-run thing than was actually the case. Anyway the offensive is noteworthy as it marks the Republican army's most significant attempt to use combined forces of armour and motorized infantry, and it shows that Vicente Rojo, the mastermind of the operation, saw the potential of mechanized forces.[25]

The offensive had an epilogue in the disastrous attack on Fuentes de Ebro next October, the best-known tank action of the war and the first appearance of the Soviet BT-5 tank.[26] It is not clear who ordered and planned the operation, but there are grounds to think that it was a Soviet experimental endeavour. With Stalin's Great Purges underway in Moscow at this time, the Soviet advisers certainly thought it wise to explain away their responsibility and to blame the Spanish commanders for the fiasco.[27] On the other hand, it appears that there is no Spanish documentation about the planning, which suggests that the latter was made by some organization separate from the Republican army's usual chain of command (and armour was one of the areas over which the Soviet advisers exercised autonomous control).[28] Moreover, the BT-5 tank was intended for the deep manoeuvre operations which the Soviet army was testing at the time.[29] And the technique of carrying infantry on the tanks during the attack tallied with those operations better than with the contemporary Spanish tactical doctrine. This set of circumstantial evidence points strongly to an overriding Soviet role in the conception, planning, and implementation of the attack.

Assessments

The EPR assessed the performance of its armour during the summer battles of 1937. In the instructions on offensive operations written by Colonel Rojo in late September 1937 he stressed that armour had to avoid frontal attacks against strongpoints and use envelopment instead. Tanks must operate through speed, bypass points of resistance whenever possible, and advance deeply into the enemy's rear. Moreover, they must be closely supported by anti-tank guns.[30] In these instructions, Rojo's thinking looks more open-minded about the capabilities of armour than the 1928 regulations, and perhaps his knowledge of Fuller's ideas was not unconnected with his stress on speed and deep penetration. The mention of close cooperation between armour and anti-tank guns (a feature of the German doctrine of which the Nationalists were also aware) suggests that Rojo knew (possibly through captured papers) the other side's ideas on armour tactics and thought it wise to follow some of them.[31]

Later on, the Republican command continued to recommend the use of armour in larger groupings. On the eve of the offensive on Teruel in December 1937 the forces of the Army of the Levant were ordered always to use tanks *en masse*, usually in battalion strength.[32] However, this kind of instruction was in contrast with the teaching given about the same time in the Republican staff college, where the students still learnt that the tank company did not fight as a single tactical unit under its commander, who had no responsibility for the actions of its platoons. The students of the Republican staff college were also taught that, once the objective was reached, tanks had to carry out a mopping-up operation and refrain from pursuing the enemy.[33] This contradiction reveals problems of organization in the EPR, whose field commanders might well feel confused by the lack of a clear, unified doctrine on the employment of armour.

A Nationalist assessment (undated, but surely written sometime after the summer of 1937) put the Republican armour's poor performance down to limited technical and tactical training and a dearth of offensive spirit. Together with the Soviet armour's technical limitations, these factors meant that the tanks' role depended heavily on their firepower, so they ended up being used as mobile guns.[34] However, this mobile artillery role was far from ineffective. During the battle of Brunete, Republican tanks used on the defensive as mobile artillery screens were a serious obstacle for the Nationalist troops. Claiming that the range of their anti-tank guns was too short, the Nationalist units demanded Italian light guns, which were reckoned more effective against enemy tanks.[35] And the Nationalist command was not unaware of the importance of tank firepower. The assessment of Republican armour mentioned above also complained of the lightness and weak firepower of the German and Italian tanks, which had to rely too much on the close support of infantry to fight against enemy strongpoints.[36]

The summer of 1937 seems to have been the heyday of Republican tanks. They no longer played so important a role for the rest of the war and their tactical performance seems not to have differed in a significant way from that displayed in 1937. According to General Kindelán, the Republican tank force had even lost its offensive spirit by the time of the battle of Teruel. Republican tanks did not operate as aggressively as before and limited themselves to the accompanying artillery role (though their firepower still intimidated the Nationalist troops).[37]

The Tank in the Nationalist Army

Organization and Early Times

The Nationalist army's tank force was truly born on 1 October 1936, when the brand new Light Tank Battalion, under the command of Major José Pujales, was created with the first batch of machine gun-armed Panzerkampfwagen Is (PzKpfw Is) and initially affiliated to the *Argel* Infantry Regiment 27 for administrative purposes. Its training was the responsibility of the German advisory *Panzergruppe Drohne* (it appears that German tankmen occasionally took part in operations as well). The Tank Regiment 2 provided cadres to the unit, which initially had two tank (16 PzKpfw Is each) and one anti-tank gun companies, and which became a two-battalion group by the end of the war.[38] This unit provided the armoured support for the Nationalist armies in all the main campaigns and battles from the autumn of 1936 to the end of the war.

Although their tank strength was very small in the early weeks of war, the Nationalists realized that even a small number of tanks could mean a significant advantage. One month after the outbreak of the conflict, Lieutenant Colonel Juan Yagüe, commander of the group of Nationalist columns advancing towards Madrid from the south, wrote to Lieutenant Colonel Franco Salgado-Araujo (then General Franco's aide) that half a dozen tanks would be very useful for his troops. They would save casualties in the occupation of villages and small towns and provide more speed in the envelopment of enemy positions.[39] At this stage of the war, the bulk of the Republican forces was poorly trained and led militiamen, who often panicked under the threat of being cut off by the Nationalist troops.

The Nationalist high command seems to have reached somewhat contradictory conclusions after its troops' early experiences in fighting with or against tanks. The first operations of the Republican armour in the autumn of 1936 had been unsuccessful due to the lack of coordination with the infantry. As a result of these combats, on the eve of the assault on Madrid (7–20 November 1936) the Nationalist Army of the North headquarters concluded that tanks required the close support of solid, aggressive infantry.[40] At the same time, however, the Nationalists probably became too sanguine about the capacities of their own armour. Forest and built-up areas strengthened the fortifications in the outskirts of the capital and the assault troops were to face better-organized Republican forces. The assault on Madrid revealed the limitations of the Nationalist armour and prompted the Nationalist general headquarters—*Cuartel General del Generalísimo* (CGG)—to send instructions to the Army of the North about the tactical use of tanks. Tanks were said to be unsuitable for street-fighting: they were too exposed to enemy weapons while their own field of fire became more limited, and they could not give mutual support. It was better to keep them outside towns as a reaction force against enemy counterattacks. Tanks should not be scattered among columns either, because their effectiveness lay in concentrated use.[41] But even concentration could not suffice for tanks that were not a match for the Republicans' Soviet armour. On 29 November the Light Tank Battalion's two existing companies attacked together the Republican front in Aravaca, northwest of Madrid, and were severely mauled by T-26s and gun-armed BA-6 armoured cars, as well as antitank guns; they reportedly lost about 15 tanks. The battalion (which would be reinforced by its new third company of PzKpfw Is) lost 36 tanks overall during the battles for Madrid -the worst losses of the whole war.[42]

Tactical Evolution and Organizational Teething Problems

The Light Tank Battalion was deployed to the Biscay front in the early spring of 1937. The experience of previous operations was summed up and disseminated through information sheets reminding the basic principles for the use of tanks. Experience had proved that, once the surprise effect had disappeared, using armour in the same place for several days in a row was disadvantageous, since the defence could be strengthened quickly. The sheets also stressed the need for cooperation with motorized anti-tank units. Moving in successive, concealed bounds, the anti-tank guns provided a defensive screen against enemy armour. Both tanks and anti-tank guns were integral parts of the armoured units.[43] These guidelines certainly distilled some teachings of the early months of war. The Nationalist army lacked enough capacity to exploit thoroughly initial successes and the Republican forces had proved in the battles around Madrid their resilience in the defence. Therefore, it made no sense to insist on using tanks in sectors where the enemy could be reinforced.

In addition, the tanks supplied by Germany were no match for the Soviet-built armour in firepower and protection, so the cooperation with anti-tank units had become essential.

The Nationalist field commanders were provided with these sheets, but they seem not to have paid enough attention to them. Lieutenant Colonel Wilhelm von Thoma (commanding officer of the German tank training unit, who became commander of the *Afrikakorps* in 1942) wrote to General Emilio Mola (commander of the Army of the North) that the Spanish commanders' determination to issue orders on their own to the tank units had prevented the Nationalist infantry from being much more successful.[44] Spanish commanders on the Biscay front also had a tendency to overuse armour. Von Thoma complained to the CGG of the misuse of tanks, which fought on too rugged ground and were forced to cover long distances from a point of intervention to another. This wrong use wore tanks down and forced long repair and maintenance periods. Therefore, the CGG reminded Mola that tanks had to be preserved to operate on ground more suitable for their tactics and mechanical performance.[45] If the Nationalist Spanish commanders were not well-versed in basic tank tactics, they could hardly be innovative in the use of armour.

It was pointed out above that close cooperation with anti-tank guns was stressed in order to improve the chances of the undergunned PzKpfw I facing Republican gun-armed armour or targets requiring a heavier punch than machine gun fire. However, anti-tank guns could not always move fast enough to provide support when needed. One attempted answer was the substitution of an Italian 20 mm automatic gun for the standard machine guns. Very few PzKpfw Is were modified in Spanish workshops in the summer of 1937 before the plan was discontinued.[46] Fortunately for the Nationalists, using captured enemy armour provided a handier solution. An early example was the use of a recently captured BA-6 armoured car by the single PzKpfw I company which fought in the campaigns of Santander and Asturias. By early autumn of 1937 the Nationalist army had enough T-26s to substitute one of them for one PzKpfw I in every platoon. And a full platoon in every PzKpfw I company was re-equipped with T-26 s in early summer of 1938.[47] It is worth noting that the British army resorted to a similar arrangement of substituting a Sherman Firefly for an undergunned standard tank at a rate of at least one per troop (platoon) in 1944–1945, in order to strengthen the panzer-killing capability of many of its armoured units.[48]

The Nationalist Spanish commanders' neglect was not limited to doctrinal issues alone. Little attention was also paid to maintenance and logistical problems, to von Thoma's chagrin. In January 1938 the CGG relayed a message from von Thoma to the Army of the North to remind the Nationalist command of the need to withdraw tanks from the firing line by nightfall. If this was not done, mechanical maintenance was impossible and fuel supply became very difficult because the lorries had to stop at a distance of several kilometres from the front.[49]

The administrative management of the Nationalist tank force was a motive for criticism as well. Colonel von Thoma affirmed, in a report dated 29 April 1938, that the state of affairs in personnel and training issues was very deficient.[50] The training period was too short due to the continuous need for replacements in the front line. This affected especially the training of tank drivers, who needed several months' training. Moreover, once the tank force became affiliated to the Spanish Legion, the pool of trainees allegedly decreased in numbers and quality; personnel unsuitable for the Legion infantry units were often sent compulsorily from the depots, with no regard for their educational qualifications. Assignment to tank units seemed like a punishment. This must have been shocking for the German officer, because in his country tank crews were hand-picked from the best manpower available.[51] Von Thoma thought that the standards of the officers were not better than the enlisted ranks', either in technical or tactical issues. Moreover, since trained officers were transferred from time to time from the tank units to new assignments, the basic courses for officers had to be repeated again and again, with almost no chance for advanced training.

As regards the tactical use of armour, von Thoma stated that battalion and company commanders should stay with their units all the time. When the companies fought separately, the battalion commander should join the one whose tactical mission was most difficult. Von Thoma's advice on this matter is in clear contrast with the 1928 Spanish regulations, which assigned the battalion commander to higher headquarters with no tactical control over his units. On the other hand, the separation of the tank companies was undesirable because it complicated their logistical support. The knowledge of the crews about mechanics was so poor that breakdowns which could have been solved at the front often necessitated moving the tank long distances for its repair, since there was just one mobile workshop. As a consequence of the personnel problems and the lack of tanks for training, the report concluded, the tank force had become more and more deficient, both tactically and technically.

Von Thoma kept on stressing the poor maintenance and training of the Nationalist tank force by the time of the battle of the Ebro. He sent a new memorandum, dated 13 September 1938, to Franco and General Luis Orgaz (head of the manpower, training and ordnance recovery command), a copy of which was later sent to General Fidel Dávila by Lieutenant Colonel Pujales (commanding officer of the Legion Tank Battalion). Von Thoma stated that no decision had been made about the proposals he put forward in his April report, because the tank units had been continuously in action. As a result, now it was almost impossible to supply enough spares and the replacement personnel was untrained because nobody had been sent to the tank school in half a year. Von Thoma proposed to withdraw one tank company at a time from the front line, in rotation, in order to follow a short course. Over the course of 12 days, officers (von Thoma emphasized this point) and enlisted men would be taught the essentials of tanks. Furthermore, the tank school needed to train replacements continuously in one month's courses. Von Thoma proposed that the school had 25 drivers and 25 gunners (including officers) in training at any time, plus 15 drivers and 15 gunners under reassessment in order to replace those who had lost proficiency.[52]

On 3 October 1938 von Thoma sent a report on the tank strength of the Legion Tank Battalion to Orgaz, which provides further evidence regarding the German officer's statements. 11 out of 64 PzKpfw Is were being repaired and another nine could not be used because there were no drivers for them, meaning that nearly a third of the unit's German tanks were out of service. The situation of the captured Soviet tanks was better, though one-sixth of them (five out of 32) were under repair. The reasons for this situation were poor maintenance (because it received insufficient time and the crews lacked training) and the use of armour on unsuitable ground (due to the incompetence of the senior commanders).[53]

The tactical misuse of armour had been pointed out by von Thoma in his September memorandum. He openly disapproved of the use of tanks on the Ebro battlefield. Despite the losses suffered for no gains, the Nationalist forces carried on resorting to armour on unsuitable ground. Von Thoma attributed it to the Nationalist commanders' limited knowledge of tanks. The presence of enemy tanks was no argument, because the Republican army used them as artillery pieces in fixed positions, rather than for attacks on rugged ground. Lieutenant Colonel

Pujales, in his introductory note to von Thoma's memorandum, also complained about the unsuitability of the ground, which forced tanks to move forward in single file along paths with no opportunity for deployment. Pujales concluded his note by asking Dávila for the withdrawal of his battalion from the Ebro sector.[54] Later operations proved that the Nationalist infantry still fell short of being trained sufficiently in cooperation with armour. During defensive operations in the Serós sector (November 1938), for example, the Nationalist tanks carried out support missions. However, after the day's fighting was over the infantry withdrew in such a way that the tanks remained isolated on the battlefield. Nine tanks had broken down and the recovery of five of them during the night involved a number of casualties. The CGG urged the Army of the North command to draw the field commanders' attention to the misuse of armour by punishing those responsible for leaving behind nine tanks on the battlefield.[55]

The March 1938 Offensive in Aragon, High Point of the Nationalist Armour

But despite such an indifferent record, the Nationalist armour still achieved a remarkable success in the same ground where the Republicans had failed miserably six months before and using quite similar methods of mechanized warfare. This accomplishment took place in the first phase (9–17 March 1938) of the Nationalist offensive in the Aragonese front, which moved the front line a hundred kilometres forward in a few days.[56] The Nationalist high command had concluded that the Republican forces were in a weak condition after their defeat in the battle of Teruel (15 December 1937–22 February 1938) and ripe for an offensive that would exploit the initial success with fast, deep advances into the enemy rear.[57] The March offensive's remarkable feature was that, after a conventional breakthrough of the front in roughly the same area on the south bank of the Ebro where the Republican had attacked the previous summer, three Nationalist infantry divisions (1st, 5th, and 150th) sent motorized task forces to plunge deeply along roads into the enemy territory. Each task force had tanks, three or four infantry battalions, one artillery battery, and the essential engineer and logistical elements.

Whenever a task force met resistance, the infantry troops got off the lorries and, supported by the accompanying artillery, they and the tanks manoeuvred on the enemy's flanks and rear and overrun the opposition.[58] The rapid seizure of the main crossroads by the Nationalist advance left straggling Republican troops isolated in pockets which were later mopped up by follow-on forces advancing across country.[59] Nonetheless, these pockets were small and a consequence—by default—of the fast Nationalist advances. There was no attempt to carry out major envelopments of Republican forces through the concerted manoeuvre of Nationalist major units.[60] When the Nationalists finally halted, they had advanced over 100 kilometres, smashed six Republican divisions and several additional mixed brigades to pieces, and captured over 10,000 prisoners (50,000 Republican casualties overall seem a plausible figure) at an almost negligible cost (3000 Nationalist casualties).[61]

Apart from the material weakness and poor morale of the defenders, the Nationalist success was made easier by the quick collapse of the enemy command and control structure (one of the effects sought by the theorists of mechanized warfare).[62] A Republican report after the offensive stated that frontline commanders became more concerned by the security of their headquarters than the conduct of operations. The report acknowledged that the Nationalist tactics (deep advances along several axes with little regard to flank liaison among the columns) prevented the Republican forces from organizing new defensive lines.[63]

Although it would be exaggerated to consider this Nationalist operation as fully-fledged blitzkrieg, the outcomes were outstanding by the standards of the Spanish Civil War. For instance, the Moroccan Army Corps (to which the 5th and 150th Divisions belonged) advanced 38 kilometres on 12 March.[64] This success seems to vindicate von Thoma, who claimed several years later that he improved the effectiveness of the Nationalist tank force by pressing the Spanish commanders to use armour in a concentrated way.[65] Yet even if the Nationalist army did not develop a doctrine for mechanized forces of its own, the March offensive in Aragon proved that it could be more open-minded than it is usually thought and give its tank force (otherwise rather attached to the pre-war doctrine) an opportunity to show its mettle in mechanized warfare (albeit in a limited scale).[66]

Nonetheless, this type of operation was probably the most the Nationalists could achieve with the means available. It seems doubtful that they could have attempted it a few more times and in consequence won the war sooner. The ground is one factor which must be taken into account. The Nationalist army carried out successively its main offensive efforts from the spring of 1937 to the summer of 1938 in three large geographical areas: the Basque-Cantabrian region (provinces of Biscay, Santander and Asturias), middle and lower Ebro valley, and southeastern Aragon and northern Valencia region. Only the Ebro valley was suitable for the employment of sizeable mechanized forces. The other two areas had too much rugged ground constraining the mobility of armour, which too often could not operate off-road.[67] It must be also remembered that Spanish battlefield doctrine after the mid-1920s was oriented towards methodical operations (see Chap. 7), and a departure from it was hardly to be expected from an officer corps which had not been very interested in mechanized warfare theories before 1936.

Moreover, the Spanish official doctrine matched Franco's penchant for playing safe and avoiding unnecessary risks, which was not incompatible with the favourable view of armour that he had displayed before (see Chap. 9). Such a penchant seems to be a legacy of his experience in the Moroccan campaigns, where he learned to appreciate the effectiveness of the deliberate methods of Dámaso Berenguer and Oswaldo Fernando Capaz (two able Spanish colonial commanders) in contrast with the hasty advances of General Manuel Fernández Silvestre, which eventually led to the Annual disaster in 1921.[68] The Italian fiasco in Guadalajara was also a cautionary tale about undertaking large-scale mechanized operations with poor man's means: despite the contemporary hype (reproduced in later accounts), the CTV was pitifully lacking in assets such as radio communications and off-road motor transport for its infantry and artillery.[69] The Nationalist army had improved by 1938, but not as much. For instance, command was exercised mostly over the telephone during the whole war, due to the primitive level of radio communications. This obstacle placed a considerable limit on the scale of mechanized operations.[70] So to argue that Franco deliberately protracted the war for his self-seeking, political benefit because he allegedly refused to implement blitzkrieg is groundless. The Nationalist *generalísimo* simply lacked the means to conduct a war-winning, large-scale campaign based on mechanized warfare methods.

Conclusion

This chapter has not intended to disprove entirely the assertion that during the Civil War 'tank operations in Spain were… led by officers with little or no tactical training, on missions conceived by senior commanders who had no grasp of the tactical potential or shortcomings of tank units in combat'.[71] Spanish armour's operations were certainly flawed on many occasions, a consequence of the military's limited interest in the tank before the war. But there also were other factors that prevented the Spanish Civil War from becoming a successful proving ground for mechanized warfare. Much of the terrain where the main campaigns and battles were fought was unsuitable for using armour *en masse*. Nor was contemporary tank technology still able to produce models suitable for conducting the sort of operations envisaged by the mechanization theorists of the 1920s and 1930s (and both sides never had enough tanks to attempt such kind of operations anyway). Therefore, it must come as no surprise that the Spanish commanders usually did not think of any other use for tanks beyond the infantry support role prescribed in the pre-war regulations. The only exceptions were the Republican plan to seize Saragossa in the summer of 1937 and the initial Nationalist offensive on the Aragonese front in March 1938, and these were operations limited in time and space. Nonetheless, they show that Spanish commanders could sometimes grasp the potential of mechanized forces.

On the other hand, there was a difference between the Republicans and the Nationalists. The Nationalists remained attached to the Spanish pre-war doctrine and their German advisers (surely aware of their tanks' limitations) seem to have been satisfied with introducing minor tactical innovations (using slightly larger tactical units, employing anti-tank guns in support of armour). In fact, the evidence shows that the Germans were mainly worried about organizational matters and the Spanish commanders' poor understanding of elementary tank tactics. But above all, there was a single, coherent doctrine.

By contrast, the evidence shows less coherence on the Republican side. Officers were trained following the Spanish regulations in force before the conflict, whereas the Republican command simultaneously issued instructions (based on battlefield experience) which differed significantly in some points from the pre-war doctrine. How did an officer reconcile the teachings of the staff college (where he learnt that tanks must not pursue the enemy) with Rojo's instructions about advancing deep into the enemy rear? This problem was worsened by the nature of the Republican officer corps. If even fully regular officer corps abroad were hard put to assimilate the procedures of armoured warfare, it is easy to understand why the improvised officers of the Republican army so often failed to use armour effectively.

Notes

1. An overview from this perspective is Steven J. Zaloga, *Spanish Civil War Tanks. The Proving Ground for Blitzkrieg* (Oxford: Osprey, 2010), which focuses more on tank design than on doctrine. A Spanish contribution is José Luis Infiesta Pérez, 'El empleo de los carros de combate en la guerra de España', *Revista de Historia Militar* (hereafter *RHM*), 78 (1995), pp. 149–208. For a sample of more limited treatments in English language, see George F. Hofmann, 'The Tactical and Strategic Use of Attaché Intelligence: The Spanish Civil War and the US Army's Misguided Quest for a Modern Tank Doctrine', *Journal of Military History*, 62 (1998), pp. 101–33; John L.S. Daley, 'The Theory and Practice of Armored Warfare in Spain, October 1936–February 1937', *Armor*, 108, 2 (1999), pp. 30, 39–43; John L.S. Daley, 'Soviet and German Advisors Put Doctrine to the Test: Tanks in the Siege of Madrid', *Armor*, 108, 3 (1999), pp. 33–7; Antonio Candil, 'Soviet Armor in Spain: Aid Mission to Republicans Testing Doctrine and Equipment', *Armor*, 108, 2 (1999), pp. 31–8; Daniel Kowalsky, *Stalin and the Spanish Civil War* (New York: Columbia University Press, 2004), Chap. 15; Steven J. Zaloga, 'Soviet Tank Operations in the Spanish Civil War', http://bobrowen.com/nymas/soviet_tank_operations_in_the_sp.htm (accessed 7 January 2014) [originally published in *Journal of Slavic Military Studies*, 12, 3 (1999), pp. 134–62]; James J. Sadkovich, 'Some Considerations Regarding Italian Armored Doctrine Prior to June 1940', *Global War Studies*, 9, 1 (2012), pp. 40–74. Opposing viewpoints on what was learned during the war are briefly presented in Steven J. Zaloga and Kenneth W. Estes, 'Armor: Was the Spanish Civil War a testing ground for the military use of tanks?', in Kenneth W. Estes and Daniel Kowalsky (eds), *History in Dispute. Volume 18: The Spanish Civil War* (Detroit: St. James Press, 2005), pp. 33–40.
2. For an overview of the tanks used in the war, see José María Manrique and Lucas Molina Franco, *Las armas de la guerra civil* (Madrid: La Esfera de los Libros, 2006), pp. 291–337.
3. Servicio Histórico Militar, *Heráldica e historiales del Ejército. Tomo VI: Infantería* (Madrid: Servicio Histórico Militar, 1984), p. 22; Adolfo Meléndez Jiménez, 'Apuntes para la Historia de las Unidades de Infantería de Carros de Combate del Ejército español', *Ejército*, 312 (January 1966), p. 27; Ramón Salas Larrazábal, *Historia del Ejército Popular de la República*, 4 vols. (Madrid: Editora Nacional, 1973), I. 133–4; Francisco Fernández Mateos, *Carros de combate y vehículos acorazados en la historia de España* (Madrid: Servicio de Publicaciones del Estado Mayor del Ejército, 1986), p. 13. On the operations of Republican armour in the initial stage of the war, see Infiesta, 'El empleo

de los carros', pp. 149–51; Salas Larrazábal, *Historia del Ejército Popular*, I. 532–3, 574; Fernández Mateos, *Carros de combate*, p. 13.

4. Servicio Histórico Militar, *Heráldica e historiales*, p. 25; Fernández Mateos, *Carros de combate*, p. 22. On the operations of these units see Meléndez, 'Apuntes', p. 27; Fernández Mateos, *Carros de combate*, p. 14. Until the end of 1937, a Renault platoon (enlarged with captured FTs) was the only armoured element of the Tank Regiment 2, turned into a training depot for new infantry battalions which retained the regimental name: Servicio Histórico Militar, *Heráldica e historiales*, p. 27.
5. Lucas Molina Franco and José Manrique García, 'Carros de combate autóctonos en el Ejército Español. 1936–39', *Revista Española de Historia Militar*, 31–2 (2003), pp. 51–60; Lucas Molina Franco and José Manrique García, *Blindados españoles en el Ejército de Franco* (Valladolid: Galland Books, 2009). See also the works listed in Chap. 9, note 2.
6. The Republican government also acquired some FTs from France and another 32 from Poland in the early months of the war: Zaloga, *Spanish Civil War Tanks*, pp. 29–30. See Table C.1 in Appendix 3 for specifications of Italian, German, and Soviet tanks supplied during the war.
7. Zaloga, *Spanish Civil War Tanks*, p. 27. See also Lucas Molina Franco and José Manrique García, *Blindados soviéticos en el Ejército de Franco* (Valladolid: Galland Books, 2007); Steven J. Zaloga, *T-26 Light Tank: Backbone of the Red Army* (Oxford: Osprey, 2014), pp. 4–17; Steven J. Zaloga, *BT Fast Tank: The Red Army's Cavalry Tank 1931–45* (Oxford: Osprey, 2016), pp. 33–6.
8. Zaloga, *Spanish Civil War Tanks*, p. 22. See also Lucas Molina Franco and José Manrique García, *Blindados alemanes en el Ejército de Franco* (Valladolid: Galland Books, 2008); Bryan Perrett, *German Light Panzers 1932–42* (Oxford: Osprey, 1998), pp. 5–8, 35.
9. Zaloga, *Spanish Civil War Tanks*, p. 40. See also Lucas Molina Franco and José Manrique García, *Blindados italianos en el Ejército de Franco* (Valladolid: Galland Books, 2007); F. Cappellano and P.P. Battistelli, *Italian Light Tanks 1919–45* (Oxford: Osprey, 2014), pp. 12–19, 28–9.
10. Zaloga, *Spanish Civil War Tanks*, p. 4.
11. Zaloga, 'Soviet Tank Operations'.
12. See, for instance, Archivo General Militar de Ávila, Fondos del Cuartel General del Generalísimo (N), Armario 8, Legajo 388, carpeta 10 (hereafter AGMA-CGG No. Armario/No. Legajo/No. carpeta), Teniente coronel von Thoma, Informe del arma de Carros de Combate, 29 April 1938; von Thoma's memorandum dated 13 September 1938 and attached to Archivo General Militar de Ávila, Fondos del Ejército Nacional, Armario 15, Legajo 27, carpeta 32 (hereafter AGMA-EN No. Armario/No. Legajo/No. carpeta), La Legión, Bandera de Carros de

Combate, Mando to General Jefe del Ejército del Norte, 10 October 1938; AGMA-EN 15/27/24, Coronel von Thoma to General Jefe de Movilización, Instrucción y Recuperación, 3 October 1938. German records also substantiate that Spain was not a real testing ground for the Panzer forces: see, for instance, Thomas L. Jentz, *Panzerkampfwagen I. Kl.Pz.Bef.Wg. to VK 18.01* (Boyds: Panzer Tracts, 2002), p. 176.

13. See, for instance, Salas Larrazábal, *Historia del Ejército Popular*, I. 533–4; José Luis Alcofar Nassaes [pseud. of José Luis Infiesta Pérez], 'Los carros sovieticos durante la guerra de España', *Defensa*, 96 (April 1986), pp. 55–60.
14. Salas Larrazábal, *Historia del Ejército Popular*, I. 1103, 1164 note 30, II. 1867–8; Zaloga, 'Soviet Tank Operations'. See also Infiesta, 'El empleo de los carros', pp. 160–1, 172–5.
15. Salas Larrazábal, *Historia del Ejército Popular*, II. 255–6; Zaloga, *Spanish Civil War Tanks*, pp. 27–8.
16. Archivo General Militar de Ávila, Fondos del Ejército Popular, Armario 59, Legajo 664, carpeta 9 (hereafter AGMA-EP No. Armario/No. Legajo/No. carpeta), Ejército del Centro, Estado Mayor, Sección de Operaciones, General Ejército Centro to Jefe Brigada de Carros, 28 June 1937; AGMA-EP 59/664/9, Ejército del Centro, Estado Mayor, Sección de Operaciones, General Ejército Centro to Jefe Brigada de Carros, 30 June 1937.
17. AGMA-EP 64/778/22, Ejército de Maniobra, Estado Mayor, 3ª Sección, Instrucción reservada número 28, 19 July 1937.
18. See, for instance, Dirección General de Preparación de Campaña, *Reglamento Táctico de Infantería*, 2 vols. (Madrid: n.p., 1926), II. Article 726.
19. Salas Larrazábal, *Historia del Ejército Popular*, II. 1264.
20. For accounts of the offensive on Saragossa, see José Manuel Martínez Bande, *La gran ofensiva sobre Zaragoza* (Madrid: San Martín, 1973), pp. 77–126; Ángel Aparicio Cámara, 'La gran ofensiva republicana sobre Zaragoza', in Fernando Martínez de Baños Carrillo (ed.), *Guerra Civil. Tomo I. Aragón* (Cuarte de Huerva: DELSAN, 2004), pp. 272–85; Fernando Martínez de Baños Carrillo, 'Ofensiva republicana sobre Zaragoza y desenlace (1937 y 1938)', in Fernando Martínez de Baños Carrillo (ed.), *Guerra Civil. Aragón. Tomo VII. Zaragoza* (Cuarte de Huerva: DELSAN, 2010), pp. 181–251.
21. Martínez Bande, *Zaragoza*, p. 97; Salas Larrazábal, *Historia del Ejército Popular*, II. 1309–10. Antonio Cordón (then chief of staff of the Republican Army of the East) later claimed unconvincingly the operational concept of the offensive: Antonio Cordón, *Trayectoria (Recuerdos de un artillero)* (París: Librairie du Globe, 1971), pp. 352–4.

22. This sector's total tank strength was 40 tanks, but ten of them were used in the attack on Quinto de Ebro: Gonzalo Lorén Garay, 'La batalla de Quinto de Ebro en la ofensiva republicana sobre Zaragoza', *RHM*, 115 (2014), p. 105.
23. Martínez Bande, *Zaragoza*, p. 99.
24. Francisco Escribano Bernal, Ángel Aparicio Cámara, Juan Carlos Cortés Hidalgo, Fernando Martínez de Baños, Valentín Solano Sanmiguel, and Miguel Flores Pintado, 'Las ofensivas republicanas del 37', in Francisco Escribano Bernal (ed.), *Guerra Civil. Tomo II. Aragón. Imágenes* (Cuarte de Huerva: DELSAN, 2005), p. 164; Lorén, 'La batalla de Quinto de Ebro', p. 131.
25. Rojo's later plan for a projected Republican offensive in Extremadura to split the Nationalist territory (as well as a previous plan by Colonel Aureliano Álvarez Coque) also relied on fast advances of large task forces of lorried infantry and armour: see Juan Miguel Campanario, 'Los proyectos fallidos del Ejército Popular de la República para dividir en dos la zona ocupada por el enemigo: El Plan P del general Vicente Rojo', working paper, http://www3.uah.es/jmc/planp.pdf (accessed 14 August 2016); Juan Miguel Campanario Larguero, 'Los proyectos fallidos del ejército popular de la república para dividir en dos la zona ocupada por el enemigo: El Plan P del general Vicente Rojo', *RHM*, 106 (2009), pp. 35–62.
26. For accounts of this attack, see Zaloga, 'Soviet Tank Operations'; Infiesta, pp. 178–82; Martínez de Baños, 'Ofensiva republicana', pp. 251–8. See also Escribano Bernal et al., 'Las ofensivas republicanas del 37', pp. 199–209.
27. Zaloga, 'Soviet Tank Operations'.
28. Infiesta, 'El empleo de los carros', p. 180; Stanley G. Payne, *The Spanish Civil War, the Soviet Union, and Communism* (New Haven: Yale University Press, 2004), pp. 170–1.
29. Zaloga, *Spanish Civil War Tanks*, p. 34.
30. Ministerio de Defensa Nacional, *Instrucción número cuatro. Normas generales para el desarrollo de las maniobras ofensivas* (Madrid: n.p., 1937), p. 15. A previous typescript of these instructions is filed in AGMA-EP 55/472/8, Instrucciones generales para el desarrollo de la maniobra ofensiva de conjunto, 21 September 1937.
31. On German doctrine, see Jonathan M. House, *Toward Combined Arms Warfare: A Survey of 20th Century Tactics, Doctrine, and Organization* (Fort Leavenworth: United States Army Command and General Staff College, Combat Studies Institute, 1984), p. 54. For an example of how German doctrine was spread to the Nationalist troops, see AGMA-CGG 7/358/49, Hoja recordatoria para carros blindados, n.d. The Republican

staff college actually included references to Nationalist instructions on armoured forces in its teaching, according to evidence that can probably be dated from late 1937 onwards: AGMA-EP 55/520/1, Escuela Popular de Estado Mayor, Táctica de Infantería, Resumen de las instrucciones del enemigo a las Unidades de Tanques, n.d.

32. AGMA-EP 64/778/16, Ejército de Maniobra, Estado Mayor, Sc. III y IV, Directivas al Ejército de Levante para el desarrollo de la operación dispuesta en su frente (extracto), 8 December 1937.
33. AGMA-EP 55/520/1, Escuela Popular de Estado Mayor, Carros de combate, n.d.; AGMA-EP 55/520/1, Escuela Popular de Estado Mayor, Carros blindados, Segunda promoción, Conferencias 3ª y 4ª, n.d.
34. AGMA-CGG 8/387/57, C.T.V., Centro Complementi e Adiestramento, Carros empleados en España, n.d.
35. AGMA-CGG 7/369/3, unidentified handwritten note for the chief of the Italian mission at the Nationalist general headquarters (*Cuartel General del Generalísimo*), 19 July 1937. A report of the German training unit, dated 6 December 1936, admitted that the German-supplied anti-tank gun had a very poor effectiveness against long-range targets because of its low accuracy (its sight ended at 900 metres) and its round's reduced penetration power; by contrast, the gun in Soviet tanks could engage targets at up to 3000 metres: Jentz, *Panzerkampfwagen I*, p. 175.
36. AGMA-CGG 8/387/57, C.T.V., Centro Complementi e Adiestramento, Carros empleados en España, n.d.
37. Alfredo Kindelán y Duany, *Mis cuadernos de guerra* (Madrid: Plus Ultra, c. 1947), p. 123.
38. October 1937: the unit was affiliated to the Tank Regiment 2 and renamed as its first battalion; by the end of the year, it had six companies (five equipped with PzKpfw Is and the sixth one with pre-war and captured Renault FTs) divided into two groups, and a company of anti-tank guns; there also was a non-organic company of captured T-26 s, whose tanks were attached to PzKpfw I sub-units on an temporary basis. February 1938: affiliation to the Spanish Legion. March 1938: the unit was renamed the *Bandera de Carros de Combate del Tercio de Marruecos* (the Legion Tank Battalion); its FT company was later withdrawn from first-line service and replaced by the T-26 company; meanwhile, a new non-organic T-26 company was organized. October 1938: the unit is reorganized and upgraded as the *Agrupación de Carros de Combate* (Tank Regimental Group), with both tank groups turned into battalions (two mixed PzKpfw I/T-26 and one T-26 companies each), and affiliated back to the Tank Regiment 2. Meanwhile, a battalion-sized separate mixed unit of tanks and armoured cars operated with the Army of the South. This summary is based on Oscar Bruña, *Vehículos acorazados en el Tercio.*

De Cáceres a Sarajevo pasando por el Aaiún (Volumen I) (Valladolid: Quirón, 1998), Chap. 2. See also Luis Eugenio Togores, *Historia de La Legión. La infantería legendaria. De África a Afganistán* (Madrid: La Esfera de los Libros, 2016), Chap. 7.

39. The letter is reproduced in Francisco Franco Salgado-Araujo, *Mi vida junto a Franco* (Barcelona: Planeta, 1977), p. 351.
40. AGMA-EN 15/18/73, Ejército del Norte, Estado Mayor, Defensa contra tanques, 5 November 1936.
41. AGMA-EN 15/18/72, Cuartel General del Generalísimo, Estado Mayor, Tercera Sección, Instrucciones sobre el empleo de los carros de combate, 19 November 1936.
42. Jesús [María] Salas Larrazábal, *Intervención extranjera en la guerra de España* (Madrid: Editora Nacional, 1974), p. 549; Fernández Mateos, *Carros de combate*, pp. 33–4; Infiesta, 'El empleo de los carros', pp. 161–3; Bruña, *Vehículos acorazados*, pp. 69–70.
43. AGMA-CGG 7/358/49, Hoja recordatoria para carros blindados, n.d.
44. AGMA-EN 15/21/24, von Thoma to Mola, 9 May 1937.
45. AGMA-EN 15/21/24, Cuartel General del Generalísimo, Estado Mayor, Sección Tercera to General Jefe del Ejército del Norte, 26 May 1937.
46. Bruña, *Vehículos acorazados*, pp. 74–5; Zaloga, *Spanish Civil War Tanks*, pp. 30–1.
47. Artemio Mortera Pérez, *Los medios blindados en la Guerra Civil Española. Teatro de operaciones del Norte 36/37* (Valladolid: AF Editores, 2007), pp. 122, 137; Bruña, *Vehículos acorazados*, pp. 99, 103, 108–9.
48. The Sherman Firefly was armed with a harder-hitting 17-pdr gun and introduced into regiments equipped with standard Sherman and Cromwell tanks: John Buckley, *British Armour in the Normandy Campaign, 1944* (London: Frank Cass, 2004), pp. 16–17; Stephen A. Hart, *Sherman Firefly vs Tiger: Normandy 1944* (Oxford: Osprey, 2007), p. 44; David Fletcher, *Sherman Firefly* (Oxford: Osprey, 2008), pp. 37, 40.
49. AGMA-EN 15/23/49, Cuartel General del Generalísimo, Estado Mayor, Sección Tercera to General Jefe del Ejército del Norte, 7 January 1938.
50. AGMA-CGG 8/388/10, Teniente coronel von Thoma, Informe del arma de Carros de Combate, 29 April 1938.
51. Kenneth Macksey, *División Panzer. El puño acorazado* (Madrid: San Martín, 1974), p. 13 [original title: *Panzer Division: The Mailed Fist*]. A German General Staff's official report on the lessons of the Spanish conflict, which naturally had to be based on information from von Thoma's unit, stated that in the latter stages of the war the Nationalist army's captured Soviet tanks were manned by a mixture of 'inspired tank crews' and 'pardoned criminals or Spaniards who have been given a choice between a prison sentence or one attack trip in a tank': quoted in Jentz, *Panzerkampfwagen I*, p. 176.

52. AGMA-EN 15/27/32, La Legión, Bandera de Carros de Combate, Mando to General Jefe del Ejército del Norte, 10 October 1938.
53. AGMA-EN 15/27/24, von Thoma to General Jefe de Movilización, Instrucción y Recuperación, 3 October 1938.
54. AGMA-EN 15/27/32, La Legión, Bandera de Carros de Combate, Mando to General Jefe del Ejército del Norte, 10 October 1938.
55. AGMA-EN 15/28/17, Cuartel General del Generalísimo, Estado Mayor, Sección Tercera to General Jefe del Ejército del Norte, 20 November 1938.
56. For accounts of this operation, see José Manuel Martínez Bande, *La llegada al mar* (Madrid: San Martín, 1975), pp. 25–64; Valentín Solano Sanmiguel, *Guerra Civil Aragón. Tomo III. Teruel* (Cuarte de Huerva: DELSAN, 2006), Chap. 13.
57. Martínez Bande, *La llegada al mar*, pp. 18–20.
58. Martínez Bande, *La llegada al mar*, p. 61. On the advance of the Moroccan Army Corps, which the 5th and 150th Divisions belonged to, see Luis E[ugenio] Togores, *Yagüe. El general falangista de Franco* (Madrid: La Esfera de los Libros, 2010), pp. 429–41.
59. Luis María de Lojendio, *Operaciones militares de la guerra de España. 1936–1939* (Barcelona: Montaner y Simón, 1940), pp. 456–9.
60. Antonio Aranda Mata, 'La guerra en Asturias y en el frente de Aragón y Levante (El Cuerpo de Ejército de Galicia en la Guerra de Liberación)', in *La Guerra de Liberación nacional* (Saragossa: Universidad de Zaragoza, 1961), p. 342.
61. Republican casualties are taken from Martínez Bande, *La llegada al mar*, p. 60. Nationalist casualties are taken from Solano, *Guerra Civil Aragón. Tomo III*, p. 225.
62. The commander of the Republican 72nd Division, José María Enciso, uninformed as to the developments in the frontline, was captured when entering a village seized by the Nationalists a few hours before; Enciso, a leftist regular officer, was court martialled and shot four days later: Valentín Solano Sanmiguel, 'La guerra civil en Teruel', in Martínez de Baños (ed.), *Guerra Civil. Tomo I*, p. 329 note 19.
63. AGMA-EP 64/796/1, De cómo ganó Yagüe la batalla del Sur Ebro, para ganar después los franquistas la del frente del Este, n.d.
64. Martínez Bande, *La llegada al mar*, p. 61.
65. Basil H[enry] Liddell Hart, *The Other Side of the Hill* (London: Cassell, 1948), p. 99.
66. Motorized task forces (with a cavalry brigade covering their left flank) also spearheaded the Moroccan Army Corps' swift advance on the northern bank of the Ebro towards Lerida (26 March-3 April): Martínez Bande, *La llegada al mar*, pp. 103–4, 114–24; Togores, *Yagüe*, pp. 445–8.

67. José Díaz de Villegas, *Geografía militar de España, países y mares limítrofes* (Madrid: Servicio Geográfico y Cartográfico, [1939]), pp. 268–71, 336, 357–9.
68. Miguel Alonso Baquer, 'Las ideas estratégicas en la guerra de España', in Mario Hernández Sánchez-Barba and Miguel Alonso Baquer (eds), *Las fuerzas armadas españolas. Historia institucional y social*, 8 vols. (Madrid: Alhambra, 1986), VII. 29, 50.
69. Artemio Mortera Pérez, *Las tres Batallas de Guadalajara* (Valladolid: AF Editores, 2007), pp. 6–13.
70. Williamson Murray, 'The Luftwaffe Experience, 1939–1941', in Benjamin Franklin Cooling (ed.), *Case Studies in the Development of Close Air Support* (Washington, DC: Office of Air Force History, 1990), p. 76.
71. Zaloga, *Spanish Civil War Tanks*, p. 42.

CHAPTER 11

Conclusion

It was certainly a grim irony of history that, after almost four decades preparing for regular warfare, the Spanish officers' professional ability was tested by a large-scale civil war in which they had to fight each other. Nevertheless, this does not prevent the military historian from asking about the level of professional preparation of the Spanish officer corps. Preparing for the next war in peacetime is a major source of headache for modern professional officers. It is a journey through the realm of uncertainty. And the stakes are high, because there is no guarantee of a second chance if they get it wrong. If an army goes to war with a flawed battlefield doctrine, the result may be irreversibly fatal. When considering, deciding and articulating in peacetime how their armies must fight, makers of military doctrine must steer between the Scylla of conformist immobilism and the Charybdis of reckless change. They must weigh up tangible factors such as technology and the resources at their disposal, and intangible factors such as the training and moral fibre of the troops which will do the fight. But rarely, if ever, is this assessment a fully rational, unprejudiced exercise, because more often than not it is done through the filter of the worldview prevalent in the officer corps. And institutional preferences based on corporate self-interest and bias also are present.

This book has explored some of these factors and their influence on the Spanish military's professionalism and battlefield doctrine. If truth be told, the subject is extensive and there is still ample room for further research. Nevertheless, the present author thinks that the evidence

J.V. Herrero Pérez, *The Spanish Military and Warfare from 1899 to the Civil War*, DOI 10.1007/978-3-319-54747-3_11

examined makes it possible to reach a number of conclusions, some of which debunk several clichés about the Spanish army of the first four decades of the twentieth century and offer new perspectives for the understanding of the Civil War. It is also hoped that they will encourage new lines of research about the professional development of the Spanish military and the operational history of the Moroccan campaigns and the Civil War.

The Spanish army's limited preparedness for war during the early decades of the twentieth century was basically the result of insufficient funding for ordnance and training. And the officer corps was not free of responsibility for this condition. On the contrary, its conduct hindered the efforts to improve military preparedness. The politicians could be unaware of the professional requirements of the Spanish army, but when the latter asserted itself over the civil power, a prime mover was institutional self-interest, rather than professionalism. From the late 1870s onwards, the military enjoyed a large degree of autonomy within the state to deal with professional affairs in exchange for not intervening in politics. But the officer corps was unwilling to make the sacrifices required by effective policies of modernization. Even worse, the officers' internal struggles and their determination to preserve corporate interests prevented one cabinet after another from undertaking any effective military reform in depth until 1931. The course of Spain's politics after 1899 served the military's purposes by turning the army in the last safeguard of the Restoration regime. Thus the military's autonomy within the state was paradoxically a hindrance to the professional modernization of the army. Only the Second Republic achieved the implementation of a rather thorough reformist policy, but the outbreak of the Civil War brought a traumatic end to this effort.

Another point to be highlighted is the failure of the Restoration's military policy to centralize institutional power within the army. As stated earlier, the Restoration politicians gave the army autonomy to manage its professional affairs in exchange for its acquiescence to the political regime. But the branches of the service had developed a prickly corporate spirit throughout the nineteenth century and, accordingly, institutional power remained fragmented in a faction-ridden army. This corporate factionalism and a deeply rooted belief in the military's right to intervene in politics whenever the fatherland and the social order were allegedly under threat were a dangerous mixture because they could easily turn professional frictions within the army into major political issues, which could determine the survival of governments. Thus disagreements within the military about the promotion system unleashed the crisis of the *juntas de defensa* and the conflict of Primo de Rivera's government with the artillery corps.

The General Staff, as the highest professional body of the army, should have brought institutional power back to the centre, imposing its authority on the rest of the army. However, following its creation in 1904 it had many difficulties in asserting its institutional position, which produced friction with the War Ministry. Its disbandment in both 1912–1916 and 1925–1931 is a sign of the distrust it generated among politicians and generals who had become used to the Restoration's delicate balance of political-military power. It did not help either that the officers who had to man the General Staff were at each other's throats. The old staff corps survived at the price of sharing staff duties with the new *diplomados* after the compromise of 1893. This could not but begin a contest for the best assignments and open a new front of corporate rivalry, which did not disappear until the staff corps was consigned to natural wastage in 1932.

Another point in need of review and further research is the evolution of military education. At the start of the period under consideration, Spanish officers were trained in separate branch academies. This fragmented arrangement made it easier to implement syllabi which suited the parochial interests of each corps. Thus the learned corps could indulge in syllabi of heavy scientific and technical content to the detriment of military subjects. In addition, all the Spanish military education were relatively bookish in nature. Traces of these features can be found in the Republican army's officer training organization during the Civil War and may help to explain why the Republican side was less successful than the Nationalist one in producing wartime junior officers.

Such deficiencies were not ignored within the military. The period 1911–1913 saw a partial reform which attempted to implement more rigorous selection procedures and revised the academies' syllabi to introduce more practical training. However, this reform was derailed by the January 1918 decree and the abridged courses of the early 1920s. On the one hand, the decree, by suppressing the separate marking of practical content within syllabus subjects, allowed the inertia prevalent among many teachers to bring back the bookish methods. On the other hand, the increased need of junior officers in the early 1920s, which was caused by the escalation of war in Morocco, was met by facilitating the entry of military family applicants, which in practice lowered educational standards. Nonetheless, the reform of the early 1910s seems to have breathed some new life into the less conformist section of the military faculties.

This section's reformist spirit was inherited by the General Military Academy at Saragossa, whose teaching methodology was not as original as it is usually thought to be. For instance, it has been barely noticed that Vicente Rojo, one of the military educational reformists of the 1920s and often portrayed as the epitome of the 'enlightened officer', was a member of the working group which prepared the restoration of a general academy in the period 1924–1925. He advocated the substitution of guidelines written by the teaching staff for textbooks, one of the Saragossa academy's distinctive features and often associated with its head of studies, Miguel Campins. The latter's vindicating essay, which says nothing about the 1924–1925 groundwork in the pages on the academy's background, has contributed to cloud these antecedents after being almost consecrated by later historians as *the* primary source about the academy. In any case, the Saragossa academy actually turned out to be a step in a process of evolution of Spanish initial officer training towards the German model, which reached its destination through the reform of military education in 1932. This reform did not come out of the blue either. The main features of the German model (higher educational standards for selection, preliminary training in the enlisted ranks, academy training limited to junior officer duties) had been advocated in the professional literature during the three previous decades, and other elements (prevalence of practical learning, removal of textbooks, development of moral fibre) had actually been implemented in the Saragossa academy.

Regarding attitudes towards the evolution of warfare, the situation of the Spanish military was no different from that found in the main European armies at the beginning of the period under consideration. None of them had fought a regular war after the 1870s, but they realized that in the meantime an exponential increase in the firepower available on the battlefield had taken place after the introduction of a new generation of weapons (magazine rifles, quick-firing artillery and machine guns). The lack of recent combat experience generated uncertainty about how to cope with the new tactical conditions. Many officers would undoubtedly have shared the view of German General Helmuth von Moltke the Younger: 'It is in any case very difficult, if not impossible, to picture now what form a modern war in Europe would take.'[1] The clues provided by other conflicts before 1914 were either disregarded to some degree because the opponents were not proper European armies (the Boer War, the Russo-Japanese War, or the Melilla campaign of 1909) or came too late to be properly analyzed before the outbreak of the Great War (the Balkan Wars of 1912–1913).

The Spanish infantry tactical regulations after the 1880s tended to suit tactics to the increasing effectiveness of firepower. However, one section of the military, attached to what has been called the paradigm of 'the psychological battlefield', argued that a resolute advance in defiance of firepower was the best way to overcome the tactical difficulties arising from new firearm technology. Such a contention was not sheer retrogressive obstinacy and found intellectual support in contemporary ideas of anti-positivist irrationalism and social Darwinism. In fact, from 1905 on the doctrine of the offensive became more fashionable and took root in the Spanish infantry with the 1908 provisional tactical regulations and the even more offensive-minded 1913 regulations.

Another feature of the paradigm of 'the psychological battlefield' was the tendency to underrate technology. Such a tendency is shown clearly in the case of the machine gun. This weapon was a piece of firearm technology which seemed a symbol of the major contemporary tactical problem: moving troops forward in the firing line. Since technology was the cause of the problem, technology-based procedures could also provide a solution, at least in a partial way (in fact, the campaign of Melilla had shown the positive effects of using machine guns frequently on the battlefield). But the contemporary military mindset was unwilling to follow a path which overturned its intellectual foundations and its stress on psychological factors. The machine gun's tactical role was based on the delivery of a huge volume of firepower, but it was underrated by the military because it did not provide the moral virtues that armies needed to conquer on the psychological battlefield.

Corporate rivalry also influenced the way the Spanish military dealt with the machine gun, which became entangled in a controversy about who should be its operator. Such a controversy, on the other hand, was not unique to the Spanish army and may be inevitable whenever a new piece of ordnance displays features which do not fit the current modus operandi of the branches of the service. By 1900 it was not clear what kind of weapon the machine gun was. Was it an infantry support weapon or a special artillery piece? Or was it a weapon which required a brand new, separate branch? The infantry and the artillery were naturally rivals in the competition for institutional control of the machine gun. In the

Spanish case, however, the institutional arrangements for the testing and the procurement of new weapons meant that the artillery corps was in charge of the process to choose a model of machine gun. This corps, although claiming the right to be the legitimate operator, was unenthusiastic about the machine gun. The reason could not be technical, since there were already machine gun models which were reasonably reliable. Distrust of a weapon which might lessen the artillery's role on the battlefield and a small-minded self-interest in protecting corporate responsibilities are better explanations which can be concluded from the evidence available. They led to corporate procrastination which delayed the adoption of the machine gun, since the infantry could not intervene in technical responsibilities which the artillery corps kept zealously as its private domain.

By 1918 armies seemed to have learnt to fight a modern regular war. However, military professionals did not reach similar conclusions about their past experience. Some concluded that the methods which won the war for the Allied armies—massive firepower applied through carefully planned, methodical operations—were the key of future success on the battlefield. Other military theorists, by contrast, refused to accept the prolonged attritional contest these methods implied and argued that armies could return to decisive manoeuvre warfare through the implementation of fighting methods which exploited the new technology of the tank.

In contrast to the machine gun, the tank achieved a somewhat faster acceptance in the Spanish army because it certainly proved useful on the battlefields of the First World War. Nevertheless, as happened with the machine gun, the infantry and the artillery also contended for the tank—even though both corps displayed limited enthusiasm towards the weapon itself. There is evidence to show that the Spanish military considered the compromise of creating a functional service which grouped all tanks under a single command (creating a fully-fledged armoured corps was doubtless too radical an idea). But corporate factionalism finally prevailed and the Spanish armour was divided into two separate forces of infantry and artillery tanks. The actual consequences were not really that serious because of the Spanish tank arm's small size during the 1920s and early 1930s. And the artillery corps actually seemed to have lost interest in having its own armour when the Civil War broke out in 1936.

Although the Spanish army was certainly interested in the tank before the end of the First World War, this interest did not result in the adoption of a policy of mechanization. Financial constraints certainly ruled out the creation of any reasonably large armoured force. But neither did tanks—and mechanization in general—arouse a remarkable enthusiasm in the Spanish military. The official doctrine—articulated in the 1928 regulations—merely reckoned the tank as a support weapon for the infantry. Certainly, the Spanish military did not produce pioneers of mechanized warfare like J.F.C. Fuller or Heinz Guderian, but the latter were rather exceptional cases, even in their own armies. Not that the pioneers of mechanization and their ideas were unknown; they were not, but they were judged too radical even by the most conspicuous Spanish advocates of armour. Nonetheless, there were signs in the early 1930s (such as the cavalry's increasing interest in the use of armoured vehicles) which showed that the Spanish army was developing gradually a more positive attitude to mechanization.

During the Civil War, Spanish field commanders continued to be attached to the pre-war doctrine on armour and did not introduce any significant changes. In fairness, this is unsurprising. Armoured warfare could not be properly tested in the Spanish Civil War because the tanks used by both sides were inadequate (both in quality and numbers) for the operations of mechanized armies, nor was the terrain where many of the main battles and campaigns were fought suitable for armoured forces (a factor the pre-war essayists had warned about). Nonetheless, the Republicans also suffered from a problem in their policy on the use of tanks. The pre-war regulations were the official doctrine, but at the same time the Republican high command issued advice—based on operational experience—which was sometimes in conflict with the regulations. This perhaps would have been just a minor problem for an army whose officers were versed in the essentials of tank operations. But this was not the case in the Republican army, many of whose commanders were not even professional officers. Thus the underperformance of the Republican armour is hardly surprising.

If some officers saw the new technology of the tank as the means to restore manoeuvre and swift decision on the battlefield, others thought that armies would manoeuvre better by reducing their reliance on materiel and technology. It is relatively frequent to find in the Spanish professional literature of the late 1910s and early 1920s the view that the trenches of the Western Front were an aberration which was not to be repeated. Therefore, it is unsurprising that a number of military essayists voiced doubts about the French-inspired, firepower-based doctrine that was officially adopted in 1925. Moreover, they were aware of Spain's constraints to effectively implement a doctrine requiring many material resources, as well as the very different geographical context of mountains and rugged ground which cover much of the Spanish territory. These essayists envisaged that future military conflicts on Spanish soil would be fought through manoeuvre warfare since it would be impossible to develop long fronts with in-depth fortifications. This kind of warfare required a type of major unit combining self-reliance and mobility, but the contemporary Spanish 'organic division' was found wanting. Spanish professional essays advocated the mixed brigade as a better option in the 1920s and 1930s, and in fact the latter was partially introduced in the Spanish army force structure during the 1930s. Thus the Republican army did not need much, if any, foreign advice or inspiration to base its force structure on the mixed brigade in the Civil War.

Unfortunately for the Republican side, developing a well-thought-out force structure was not enough to win the war. Turning the rump regular army and the motley assortment of militias available in the late summer of 1936 into a working military organization was no mean achievement, and credit must go to the Republican professional officers. Building an effective fighting force was a different matter. One of the key requirements was an officer corps able to implement the 1925 field service regulations. They required major unit commanders and staff officers able to coordinate the different fighting arms and to make an effective use of artillery. But the Republican army did not have enough of these officers when the conflict turned into a full-scale regular war by late 1936 and they could not be made overnight. So it should come as no surprise to anyone that the Republican forces underperformed so often in offensive operations. Although an overall shortage of heavy artillery cannot be overlooked, there is archival evidence which illustrates that, independently of how much cannon the Republicans possessed, their employment of artillery would too often leave much to be desired due to the incompetence of field commanders and their staffs. And sneaking even full divisions through a thinly held Nationalist front was bound to be eventually unsuccessful if they were unable to overcome the resistances that sooner or later they would have to deal with.

The Nationalist army, due to its larger cadre of regular officers, was in a better situation to apply the 1925 regulations and take advantage of the potential of air power in support of ground operations. Its artillery's performance, however, was not always flawless and whenever this happened the Nationalist attacking troops were not much more successful than their Republican counterparts. Warfare according the 1925 regulations tended toward methodical—sometimes even plodding—offensives, and this defined much of the operational nature of the Spanish Civil War. But it must not be overlooked that the Nationalist offensives on Santander (1937) and in Aragon and the lower Ebro (March–April 1938) were fine exercises in operational level manoeuvre, and the plans for some of the Republican offensives (Brunete, Saragossa) were no less ambitious.

As regards the performance of both sides' field commanders, it has been suggested that its larger number of senior officers who acquired war experience in Morocco allowed the Nationalist army to implement a dynamic, offensive-minded style of warfare, in contrast with the Republican army, whose commanders allegedly conformed to a French-inspired, less aggressive doctrine.[2] Contemporary evidence does not support this argument. In fact, the conclusion of military essayists was that the concepts that the Spanish army adopted as its official doctrine for regular warfare in the mid-1920s brought definitive success in Morocco. And these concepts were inspired in the contemporary French doctrine. In this sense, a professional divide between Africanist and peninsular officers seems unfounded. The Morocco old hands fought the Civil War drawing on the same battlefield doctrine as their opponents. A different matter is that, as a general rule, officers with previous war experience were better prepared psychologically to cope with the demands of command on a modern battlefield than those used to undemanding peacetime soldiering.

One final point that is also worth considering, at least tentatively, is the military significance of the brutalization process which the Spanish colonial troops experienced during the Moroccan campaigns. Warfare in Morocco eventually led to extreme violence against the civilian population to reduce it into subservience. Nationalist commanders who also were veterans of those campaigns and used to such a style of warfare reproduced the latter in the early stages of the Civil War.[3] But it must be noted that such conduct was not unique to the troops of the Army of Africa. At the same time as the latter were advancing towards Madrid, Republican columns advancing into Aragon were taking similar action.[4] In fact, violence against actual or potential non-military opponents to secure one's own rear seems almost inherent in a situation of high uncertainty and undefined frontlines, such as the initial stage of a civil war.[5] And such a brutality was not unique to the Spanish army (or at least their colonial units) nor civil wars. For instance, it also can be found in the German and Austro-Hungarian armies before, and in the earliest stages of, the First World War.[6] Basing the moral censure of the Army of Africa's brutality in the Civil War on an alleged 'unique' character of such a brutality is unfounded.

To sum up, the Spanish army's limited preparedness for war in 1936 was basically the outcome of insufficient funding for ordnance and training during the previous decades (to which the oversized officer corps' refusal to accept cuts in its strength had also contributed). But it cannot be said that the Spanish army of the first four decades of the twentieth century was a retrogressive institution in terms of its attitudes towards new military technology and the changes it brought to warfare. These attitudes were not innovative, but they were on a par with those of the major Western armies. Such was the case of the debates about tactics and new firepower technology of the period 1899–1914. And it was not unsound to follow the French doctrines of firepower-based operations and armour employed as support for the infantry after the First World War. Failure to realize the potential of armour and mechanized warfare in the 1920s and early 1930s was a common feature among contemporary armies. And this does not mean that the Spanish official doctrine, used to fight the Civil War, was essentially outdated or flawed. In fact, the side that was able to implement it most effectively won the war. Therefore, the sheer lack of professional interest in the preparation for war was not the reason behind the military weaknesses of the Spanish army.

Notes

1. Quoted in David T. Zabecki, *The German 1918 Offensives: a Case Study in the Operational Level of War* (Abingdon and New York: Routledge, 2006), p. 41.
2. See, for instance, Michael Alpert, 'The Clash of Spanish Armies: Contrasting Ways of War in Spain, 1936–1939', *War in History*, 6 (1999), pp. 337–8.

3. Sebastian Balfour, *Abrazo mortal. De la guerra colonial a la guerra civil en España y Marruecos (1909–1939)* (Barcelona: Península, 2002) [original title: *Deadly Embrace: Morocco and the Road to the Spanish Civil War*].
4. Bartolomé Bennassar, *El infierno fuimos nosotros. La guerra civil española (1936–1942…)* (Barcelona: Taurus, 2005), pp. 87–93, 95–8, 102–21.
5. Stathis N. Kalyvas, 'Warfare in Civil Wars', in Isabelle Duyvesteyn and Jan Armstrong (eds), *Rethinking the Nature of War* (Abingdon: Frank Cass, 2005), pp. 94–5.
6. Ben Shepherd, *Terror in the Balkans: German Armies and Partisan Warfare* (Cambridge, MA, and London: Harvard University Press, 2012), Chaps. 1 and 2. See also Hew Strachan, *The First World War. Volume I: To Arms* (Oxford: Oxford University Press, 2001), pp. 1117–18.

Appendix A: Army Units

The translations do not follow the usage of any particular English-speaking army. Cavalry terms are in brackets (Table A.1).

The cavalry regiment had the administrative standing of a *cuerpo* (self-administering unit), but was actually a battalion-sized unit. In the artillery, the company is called battery and an artillery battalion is a *grupo*.

The infantry regiment was a *cuerpo* and a tactical unit (usually with three battalions in wartime). Two regiments formed a brigade. However, since the nineteenth century, the Spanish regiments rarely fought as tactical units. Instead, their battalions were distributed among other field commands, although battalions of the same regiment could be grouped together. This was also the case in the Civil War. In fact, the Nationalist infantry's organization resembled the British regimental system, since the pre-war regiments (and even separate battalions) left in the rear cadres which operated as depot units responsible for the organization and training of many of the new wartime battalions, which remained affiliated to their parent units. In the Republican army, each mixed brigade was the parent unit of its organic infantry battalions.

The infantry had other regiment-sized tactical units. The light infantry (*cazadores*) and mountain infantry separate battalions were grouped in half-brigades (two or three battalions each). And during the Civil War, the Nationalist army grouped their divisions' infantry in tactical units of three or four battalions, which were indistinctly named regiment, half-brigade or *agrupación*.

J.V. Herrero Pérez, *The Spanish Military and Warfare from 1899 to the Civil War*, DOI 10.1007/978-3-319-54747-3

Table A.1 Spanish army units and their translation

Spanish army	*Translation*
Regimiento	Regiment
Batallón (Regimiento/Grupo)	Battalion (Regiment/Group)
Compañía (Escuadrón)	Company (Squadron)
Sección	Platoon (Troop)
Pelotón	Section
Escuadra	Squad

A special case was the Moroccan troops. In the Spanish army's *regulares*, both the infantry battalion and the group of cavalry squadrons were called *tabor*, and infantry and cavalry *tabores* were grouped into mixed regimental groups (*grupos de fuerzas regulares indígenas*) for administrative purposes until the end of the Civil War. The *mehal-las* were also regimental groups of *tabores*, although their establishments were lower than those of the *regulares* units.

Appendix B: Army Officer Ranks

The translations do not follow the usage of any particular English-speaking army. The officer rank code of the North Atlantic Treaty Organization (NATO) is used as comparative reference for foreign equivalents (Table B.1).

The rank of *capitán general* is the highest one in the Spanish army, but became more of an extraordinary reward for a very distinguished career after 1900, in the style of *marechal de France* after 1870 (although this actually is a state dignity, rather than a military rank). This rank must not be confused with the position of territorial captain general, who was the general officer commanding one of the major military regions or districts (a new OF-9 rank of *general de ejército*—'army general'—was introduced in 1999 for the officer appointed chief of the army general staff or chief of the defence staff).

Before 1918 the ranks of *alférez* (which can be literally translated as 'ensign') and *teniente* were called respectively *segundo teniente* (second lieutenant) and *primer teniente* (first lieutenant).

During the Civil War, the Republican army adopted a single general officer rank, simply called 'general', from early 1937 to late 1938, and then reintroduced the ranks of brigadier general, major general and lieutenant general. It also substituted the term *mayor* for that of *comandante* (in the Spanish army, a *mayor* was the officer in charge of the administrative and financial affairs of a *cuerpo*; in a regiment, he usually had the rank of *comandante*).

J.V. Herrero Pérez, *The Spanish Military and Warfare from 1899 to the Civil War*, DOI 10.1007/978-3-319-54747-3

Table B.1 Spanish army ranks and their translation

Spanish army	*NATO rank code*	*Translation*
Capitán General	OF-10	Captain General
	OF-9	
Teniente General	OF-8	Lieutenant General
General de División	OF-7	Major General
General de Brigada	OF-6	Brigadier General
Coronel	OF-5	Colonel
Teniente Coronel	OF-4	Lieutenant Colonel
Comandante	OF-3	Major
Capitán	OF-2	Captain
Primer Teniente/Teniente	OF-1	First Lieutenant
Segundo Teniente/Alférez		Second Lieutenant

Appendix C: Basic Specifications of Tanks, 1921–1939

See Table C.1

J.V. Herrero Pérez, *The Spanish Military and Warfare from 1899 to the Civil War*, DOI 10.1007/978-3-319-54747-3

Table C.1 Basic specifications of tanks, 1921–1939

Model	*Weight (tonnes)*	*Max. speed (km/h)*	*Armament*	*Max. armour (mm)*	*Crew*
Schneider CA 1	13.5	6.7	1 x 75 mm gun 2 x machine gun	11.4	7
Renault FT	6.7	7.7	1 x machine gun	16	2
Trubia, Serie A	8.1	30	3 x machine gun or 1 x 40 mm gun 2 x machine gun	20	3
CV 3/35	3.2	42	2 x machine gun	15	2
PzKpfw IB	5.8	40	2 x machine gun	13	2
T-26 Model 1933	9.4	31	1 x 45 mm gun 1 x machine gun	15	3
BT-5	11.9	52[a]	1 x 45 mm gun 1 x machine gun	15	3

[a]On tracks; its maximum speed on wheels was 72 km/h

Sources Javier de Mazarrasa, ***Blindados en España. 1ª Parte: la Guerra Civil 1936–1939*** (Valladolid: Quirón, 1991), pp. 17, 22; Artemio Mortera Pérez, ***Los carros de combate 'Trubia' (1925–1939)*** (Valladolid: Quiron, 1993), p. 29; F. Cappellano and P.P. Battistelli: ***Italian Light Tanks 1919–45*** (Oxford: Osprey, 2014), p. 46; Bryan Perrett, ***German Light Panzers 1932–42*** (Oxford: Osprey, 1998), p. 48; Steven J. Zaloga, ***T-26Light Tank. Backbone of the Red Army*** (Oxford: Osprey, 2014), p. 20; Steven J. Zaloga, ***BT Fast Tank: The Red Army's Cavalry Tank 1931–45*** (Oxford: Osprey, 2016), p. 26

Glossary

Africanista in the Spanish military environment of the 1909–1939 period, an officer who volunteered for frontline service in Morocco, especially in forces like the *Regulares* (q.v.) and the *Legión* (q.v.), with the purpose of speeding up his career through battlefield achievements.

Agrupación (i) a task force or temporary grouping whose strength could go from a battalion-sized detachment to a field army-sized command; (ii) the name of some regiment or brigade-sized units.

Bandera (i) an alternative term for a company in a Spanish *tercio* (q.v.) of the sixteenth and seventeenth centuries; (ii) an infantry or tank battalion of the Spanish Legion; (iii) an infantry battalion recruited from volunteers from *Falange* (a Spanish fascist political party) in the Civil War.

Cazador (literally, 'hunter') a light infantryman or a light cavalry trooper.

Cuerpo (i) a self-administering unit (it was usually the regiment, but smaller units down to company size could also be *cuerpos*, as well as some artillery and engineer corps territorial commands); (ii) a term for the branches of the army which had been originally set up as single regiments (e.g., the artillery corps) or were not a tactical arm.

Estado Mayor the body of officers performing staff duties at the headquarters of a major unit or an equivalent military command; the army's General Staff was called *Estado Mayor Central,* which must not be confused with the *Estado Mayor General* (the general officer list).

J.V. Herrero Pérez, *The Spanish Military and Warfare from 1899 to the Civil War*, DOI 10.1007/978-3-319-54747-3

Grupo (i) an artillery battalion; (ii) in the cavalry, either a tactical subunit (stronger than a squadron) of a regiment or a self-administering grouping of squadrons (usually smaller than a regiment); (iii) a self-administering grouping of companies in the engineer and service support corps.

Legión (i) a volunteer fighting force of the Spanish army, recruited from Spaniards and foreigners, and founded as *Tercio* (q.v.) *de Extranjeros* in 1920; (ii) a regimental group of *banderas* (q.v.) of *El Tercio* (q.v.) from 1934 to 1937.

Regulares regular infantry and cavalry troops recruited mostly from Moroccan native volunteers (specialist sub-units were manned by Spaniards) and led by Spanish officers.

Tercio (i) the term for the Spanish infantry regiment of the sixteenth and seventeenth centuries; (ii) the first name of the Spanish *Legión* (q.v.), founded as *Tercio de Extranjeros* and later retitled *El Tercio*, until 1937; (iii) the name of some regiment or brigade-sized units; (iv) an infantry battalion recruited from *requetés* (ultraconservative monarchist Carlist militiamen) in the Civil War.

Index

J.V. Herrero Pérez, *The Spanish Military and Warfare from 1899 to the Civil War*, DOI 10.1007/978-3-319-54747-3

The manufacturer's authorised representative in the EU is Springer Nature Customer Service Centre GmbH, Europaplatz 3, 69115 Heidelberg, Germany. If you have any concerns regarding our products, please contact ProductSafety@springernature.com

Printed and bound by CPI Group (UK) Ltd, Croydon, CR0 4YY
30/09/2026
02225656-0017